THE SCIENCE OF
WINE

To Fiona

University of California Press
Berkeley and Los Angeles, California

Published in association with Mitchell Beazley

ISBN: 978-0-520-27689-5

Library of Congress control number: 2013948637

Second edition published in Great Britain in 2014
by Mitchell Beazley an imprint of Octopus Publishing
Group Ltd, Endeavour House, 189 Shaftesbury Avenue,
London WC2H 8JY

Visit **www.ucpress.edu** for more information
about this book.

Commissioning Editor: Hilary Lumsden
Executive Art Editor: Yasia Williams-Leedham
Design: John Round Design
Senior Editor: Julie Sheppard
Editors: Jamie Ambrose & Constance Novis
Indexer: Ingrid Lock
Production: Caroline Alberti

Printed and bound in China

JAMIE GOODE

THE SCIENCE OF
WINE

FROM VINE TO GLASS

SECOND EDITION

UNIVERSITY OF CALIFORNIA PRESS

Contents

Note on the second edition

When I wrote the first edition of *The Science of Wine*, back in 2004 and 2005, I was still had a full-time job as a science editor. As a result, much of the writing work took place in the evenings, or on my commute, or on weekends. I was amazed by the success of the book. While I thought it was pretty good, I suspected its appeal would be quite niche, and I couldn't have hoped for it to be so well received by wine-lovers worldwide. It seems that science interests a lot of people—even those who come from a nonscience background. It just needs to be made accessible and interesting: something scientists aren't always brilliant at doing.

The book worked so well because it wasn't written as a textbook, covering the whole of wine science in a methodical manner. This could have made for a fairly dull read. Instead, I set out to tell the interesting wine science stories in a way that would engage people who weren't overly scientifically literate. I also tried to make it current, addressing science-related issues that hadn't yet been written about widely for a broader audience.

In the eight years since the first edition, I have left the day job and now work full-time as a wine writer and communicator. Since I made the transition in 2008, I have traveled extensively around the world of wine, and learned a great deal in the process. In revising *The Science of Wine* I wanted to add a lot of new material, but without expanding the book too much. (The 100,000-word limit imposed by the publisher is a sensible one—there are too many overlong wine books out there.) I also wanted to preserve the accessibility and readability of the first edition.

So I have added new chapters, and taken some out. I have rewritten the remaining chapters, some quite extensively. The major change has been an entirely new—and lengthy—chapter on the role of soils in shaping wine, which was a notable omission from the previous book. There's also a new chapter on oxygen and wine, and an expansion of the section on our perception of wine, as well as the role of language in our attempts to describe what we are experiencing.

First and foremost, I am a lover of wine. I have been well and truly bitten by the wine bug, and I hope that through the book, the fascination I have for this most remarkable of drinks is the dominant theme that places the scientific content in context.

Finally, here is a necessary note on conflicts of interest. Be assured that the opinions expressed in this book are my own genuinely held ones, and the topics covered here are those I consider to be the most interesting and relevant. In weighing up the science I have reported, I have tried to be as even-handed as possible. But I do need to mention some potential conflicts of interest, which readers deserve to know about. As well as writing about wine, over the last few years I have done some paid work for a range of wine-related businesses. This work includes running tastings, giving talks, taking part in brainstorming sessions, chairing debates, and acting as a discussion panelist. However, I am not retained by any company. The companies I have worked for that are relevant here are DIAM (closure manufacturer), Nomacorc (closure manufacturer), Conetech (wine technology company), Lallemand (yeast and oenological product company) and New Zealand Wine Growers (generic body). It is important that readers know this so they can make up their own minds.

Finally, since I wrote the first edition, social media has become part of the world we live in. This means I'm very happy to interact with readers via facebook and twitter (I'm @jamiegoode), as well as good old-fashioned e-mail (jamie@wineanorak.com).

JAMIE GOODE, MAY 2013

Introduction: why wine science?

Wine is remarkable. Consider the following questions and statements. How can this drink of fermented grape juice have assumed such an important place at the center of many cultures, and maintained this place through millennia? How can it have spread from its origins in Eurasia some 5,000 years ago to become a frequent fixture on dinner tables across the world? People collect it, read books about it, spend large chunks of their disposable incomes on it, and some even give up their well-paid day jobs to go and make it. It has even survived (so far) the technological advances of the twentieth century and the shift from largely rural-based economies to city living. Despite their best efforts, the branders and marketing wizards of modern retailing haven't been able to kill it. In countries that do not produce it, wine has begun to shed its predominantly elitist image and shows signs of becoming the drink of the masses.

From just one species of vine, *Vitis vinifera*, thousands of different varieties have emerged, each with its own characteristics. The grapevine even has the capacity of transmitting some of the character of the site on which it is grown into the wine that it produces. As well as making drinks with myriad flavors, textures, and degrees of sweetness and astringency—many of which make perfect foils for different foods—the vine gives us a naturally alcoholic product with pleasant mind-altering and mood-mellowing characteristics. To top it all, it seems that wine also has health benefits when used in moderation.

While this book is about wine, its focus is to explore this remarkable substance through a particular lens: that of science. As an ancient drink, wine has been produced through the ages without the help of a modern scientific worldview. And many will argue that what science has brought to wine hasn't really helped it at all—some will go so far to suggest that the so-called "advances" promoted by scientists, such as the use of pesticides, herbicides, and mechanical harvesting to help in the vineyard, and filtration, cultured yeasts, enzymes, and reverse osmosis machines in the winery, have actually been detrimental to wine quality. Certainly, there is little doubt that the potential interventions that science has made possible have been abused.

But science has a lot to offer wine at all levels, from industrial production of mega-brands to artisanal, handcrafted boutique wines. In this introduction, I'm going to outline why I think science is a fantastically useful tool for winegrowers. Like all tools, though, it can be used correctly or abused. Indeed, one of the goals of this book will be to show how it is possible to integrate many of the most interesting and absorbing topics in wine with a scientific understanding of these issues, and that such an integration will assist in the production of more interesting, compelling wines at all levels. Even if your goal is to produce manipulation-free "natural" wine, a good grounding in wine science will help you reach this target with fewer disasters along the way. As an example, people pay a lot of money to buy wines coming from a particular patch of ground, or "terroir." Wine science will help us understand what is special about that vineyard site, and may thus facilitate the identification of similarly endowed sites, or help in the production of better wines from vineyards less blessed by nature.

SCIENCE IS USEFUL

The scientific method is an incredibly useful tool. It helps us overcome our biases and prejudices, and allows us to answer difficult questions. It helps us to be objective. It presents a coherent model of the world around us that assists our understanding of this environment, and enables us to develop new technologies that actually work.

It needs to be emphasized that objectivity is one of the keys to the successful practice of science. By nature we are not objective. We are pulled and pushed in various directions by our built-in preconceptions, predilections, and prejudices, often subtly, sometimes not so subtly. We frequently display confirmation bias, recruiting pieces of evidence that fit with our narrative of the world around us. Good scientists will step aside and try, as much as is possible, to be ruthlessly objective about the phenomena they are studying—the two arms of scientific enquiry being observation and experiment. Scientists look at what's there, formulate hypotheses, and then test those hypotheses by experiment, trying their hardest to disprove them. This is the only way they can be sure that they are correct.

Let's make this practical. Imagine you had a novel chemical treatment that you thought would protect your vines from mildew. How would you test it? Well, you could try treating all your vines with it and then see how they do. There's a problem with this approach, though. If you get positive results, how do you know they are attributable to your treatment, and not, for example, to the benevolent conditions of this particular vintage? The answer is, you don't. This is where the scientific method helps.

A more rigorous and useful approach is to compare the treated population of vines with what scientists term a "control": in this case, a group of vines that have not been treated, or more precisely a group of vines that has been sprayed with an inert substance according to the same schedule as the test group, to rule out the possibility that it is the act of spraying that is having the effect, rather than anything specific to the chemical. So you split your vineyard into two and treat just one half. Still, there is a problem with this experiment. In any vineyard there will be natural variation, and any significant results might be because one part of the vineyard enjoys better conditions than the other (it might be slightly warmer, or have different drainage properties). The answer? Subdivide the vineyard further into dozens of different plots, and randomize the treatment such that plots that are treated are interspersed with those that aren't in a way that evens out the environmental variation.

So do we go ahead? Not yet. Once we get our results we will need to know whether any beneficial effect is significant. That is, what is the likelihood that such a benefit could have been obtained by chance, through natural variation in the measured populations? This is where statistical analysis steps in. Statistics is intrinsic to any good experimental design. Good experiments should be designed from the start with statistical analysis in mind; how many replications (repeated observations) will be necessary to produce a significant result? This can be worked out in advance. Whenever you see a graph or table presenting experimental results, your first question should be: how significant are the differences between the control and experimental treatments?

The number of experimental replications needed depends on the variation in the populations being studied. The variation in a set of results is defined by a statistical term, standard deviation. It's not necessary to go into details about how this is calculated because all we need to know for our purposes here is that measures such as this allow scientists to work out whether their results are meaningful or not.

Let's take a slightly different example that will throw some more light on how scientists work and think. You suspect that wine drinking may be beneficial for health by protecting against heart disease. But how do you study this? For ethical and practical reasons it is rarely possible to do direct experiments. You can't easily isolate a group of people and vary just one parameter in their environment, such as whether or not they drink wine, especially when you are looking at a disease process that takes many years to develop.

You might want to start by doing animal experiments, looking at the cardioprotective effects of wine consumption on rats or rabbits kept under controlled conditions. The advantage here is that you can study the physiological effects of your treatment in depth; the disadvantage

How science works

The scientific community is a remarkable global enterprise. Researchers across the world are united by a common currency: data published in peer-reviewed scientific journals. It's an inclusive club, open to anyone, as long as they have good data and are prepared to play by the rules. How does it work? Universities, government institutions, or private companies employ scientists. The latter will typically be paid a salary, but will need to fund their work by means of grants, usually awarded by government-supported funding bodies or industry. To gain credibility and status, researchers need to publish their work in reputable peer-reviewed journals, and their publication record is how they are assessed. There are many thousands of these journals, and they vary in their scope from broad to very narrow.

Not all journals are created equal; some have much higher reputations than others. Typically, a scientist (or more commonly, a group of researchers) will write up his or her results and then choose the most appropriate journal to send them to. He or she will want to have them published in the highest-ranking journal possible (journals are ranked, for example, by the average number of times a paper published there is cited, with the status of the citing article borne in mind—it is called an "impact factor"), but won't want to send their paper to a journal where it will be rejected, because of the delay in publication that will ensue.

How do journals decide which papers to accept? This is where "peer review" kicks in, a process vital to the integrity of the scientific literature. Each journal has a board of editors made up of leading researchers in the field covered by the journal, and also a larger pool of scientists willing to act as referees for papers in their chosen subject areas. A paper coming in will be assessed by one of the editors: if it is clearly unsuitable it will instantly be rejected, but if it is potentially good enough, it will be sent out to two or more scientists for review. They will prepare a report on the paper, checking that it is correct, is suitable for the journal it has been submitted to (if it is a high-ranking journal, are the results exciting, novel, and significant?), and that the science is good. If they recommend it to be accepted, they might also suggest possible revisions or further experiments. Then the paper and the referees' reports are sent back to the editor, who makes a final decision whether to accept it, accept it with revision, or reject it. Journals with good reputations are fussier than others. Getting your paper into one of the elite band of leading journals can make your career.

It should be pointed out that peer review is a slightly controversial process because (1) it involves scientists reviewing the work of their peers, who may well be their competitors, (2) it can take a long time, and (3) some consider it not to be as rigorous as it should be because good papers are sometimes rejected while less good ones get through. Worth noting is that the scientific publishing process is a highly collaborative venture. Neither the scientists publishing work nor those reviewing it get paid, although some publishers make a lot of money from their journals.

Science is highly competitive. The entry ticket into the scientific community is a doctorate (a PhD), which is awarded by a university for the successful completion of an acceptable thesis (a written account of research undertaken on the subject of choice). This typically takes from three to five years to achieve. But getting a PhD doesn't guarantee a research job. After you complete your PhD you need to do what's known as a postdoc (postdoctoral research position), a short-term (usually three years) contract working as a researcher in someone else's lab. After two such positions (preferably with one abroad), if you've been reasonably successful and have published several papers in good journals, then it's time to try to land a proper research job. These are few and far between, and competition for them is fierce. The pay isn't great, either. This is a stage at which many people look for an alternative career or leave academia, opting to work in industry instead. For those who succeed, though, running a successful research group is a highly rewarding career, albeit one that requires grueling hours and absolute commitment. Despite the competition and the race to be first with each new discovery, it's refreshing to find that scientists are generally quite an open bunch, sharing their results at conferences and distributing reagents freely to other laboratories.

is that while animal models are helpful, mice, rats, and rabbits are different from people, a factor that significantly limits the utility of any knowledge obtained in this way. Another avenue of investigation might be to identify a specific physiological process involved in the development of human heart disease, and then study the effects of wine consumption on this "surrogate" process over a limited period in human volunteers—perhaps over a couple of days. Of course, identifying a reliable surrogate process or marker is the key here, and this is a nontrivial challenge.

Instead, you could study large populations over time and try to correlate behaviors, such as wine drinking, with changes in health status, such as the progression of heart disease. This is the science of epidemiology, and it was precisely such a study conducted in the 1950s by Sir Richard Doll that showed conclusively for the first time what many people had suspected: that smoking is harmful to health. The key issues here are recruiting large enough populations to produce statistically significant results, controlling for confounding (more on this in a moment), and having a relevant, easily measurable endpoint (for example, in the case of heart disease, whether or not a heart attack occurs). So let's say you have decided to look at the influence of wine drinking on the incidence of heart disease in a population of 1,000 randomly selected adults, using the incidence of heart attack as your endpoint. You'd need to get the population to fill in a drinking questionnaire (and here's a source of potential error: most people will underreport the amount they drink), and then follow up the incidence of heart disease in the different groups (i.e. nondrinkers versus light drinkers versus heavy drinkers) over a period of time.

What if you find that wine drinkers have reduced levels of heart attacks? Then you'll need to show that the effect is a significant one by using statistics. But we're not finished there; it gets more complicated. Even if there is a significant association between wine drinking and the risk of heart attack, this doesn't prove that wine drinking protects against heart attacks. It might be that the population who choose to drink wine is associated with another trait that is linked to a reduced risk of heart attack. For example, on

average, wine drinkers might also eat a more balanced diet, or have higher levels of gym membership, or smoke less. It's also well-known that low income correlates with poor health status, for a variety of unspecified reasons, and people on low incomes might be underrepresented among the population of wine drinkers. These effects are known as confounding, and they need to be controlled for. One way might be to balance the different study groups by socioeconomic status, or do a study solely within one profession, to iron out any major discrepancies. It's complicated, but unless you take these sorts of precautions you'll end up with an unreliable conclusion.

If you want to know about the health effects of wine, you might also try a clinical trial where the effects of wine are tested on a group of patients or healthy volunteers. The key to success is using a placebo treatment and blinding the study: not letting the subjects know whether they are receiving the actual treatment or the placebo. Variations on this theme include crossover trials, where groups are switched from the treatment to the placebo halfway through. Studies can also be separated according to whether they are prospective (looking at the effects of interventions over a period following the beginning of the trial) or retrospective (using already gathered data to look back in time from a known endpoint).

Then there's the issue of mechanism. Epidemiology can tell you that a certain intervention or environmental factor has a particular effect on a population, but then you'll want to know why. In the case of wine, if it is clear that moderate drinking protects against heart disease, then what is the biological mechanism? Is it the effect of alcohol, or the effect of another chemical component of the wine? To answer these types of questions scientists frequently turn to animal experiments, simply because doing the equivalent tests on people wouldn't be ethical. The goal is that by understanding the mechanism, drug development or other targeted medical treatments might be possible.

THE RISE OF ANTISCIENCE
But despite the evident utility of science, we live in a culture that is now marked by a strong antiscience sentiment. Back in the 1960s and

1970s, scientists were largely revered. Now they are treated with suspicion. Part of the public disenchantment with science lies in the fact that people feel let down: science promised too much and couldn't deliver. The application of science has led to breathtaking technological advances that show no signs of losing pace. Moore's law—the idea that computer-processing power doubles every couple of years—is still holding very nicely. When I wrote the first version of this book in 2005, my cellphone could make calls and receive texts. As I write now, in 2013, I have a shiny smartphone that is a powerful computer and very able camera. My digital camera of 2005 is now looking very outdated when compared with my current digital SLR.

But despite this, scientific progress hasn't led to the nirvana of a happy, disease- and crime-free society. Medical advances against the chief killers in the West—cancer and heart disease—are slow and have included a large number of false dawns. Malaria is still the world's largest killer, and our treatments have advanced little. Bacteria are increasingly resistant in the face of our armamentarium of antibiotics, to the extent that we are facing a very real crisis where people are dying from infections which, a decade ago, would have been easily treatable. Bringing a new drug to market is hideously expensive, with myriad legislative hurdles, and the pipeline of new drugs in development is looking a little short. Consumers, disenchanted by the medical profession's perceived limitations, have turned increasingly to largely unproven alternative therapies. Even where science offers solutions for problems of the present and the future, such as genetically modified (GM) crop plants, consumers aren't sure they want them.

Perhaps we have expected too much of science—or maybe scientists themselves have been guilty of promising what they can't deliver. Science is a tool, and an incredibly useful one, but it is no more than that. Science can't address issues that belong in the realm of ethics, morality, religion, politics, or law. That scientists have sought to impose their ideas in these realms is not the fault of the scientific method, nor does it mean that science as a tool or process has failed. Instead, society has been wrong to look to scientists to provide enlightenment where it simply cannot.

Accessing the scientific literature

Science is published in peer-reviewed journals, of which there are many thousands. How can the nonscientist approach this scientific literature and glean something useful from it? The first port of call for most will be an online database. These contain the abstracts (a short summary of methods, results, and conclusions) of each published paper, along with other indexing data. Perhaps the most useful of these is PubMed, which can be found at www.ncbi.nlm.nih.gov/entrez/.

The skill in researching the literature comes from searching effectively, and then assigning the appropriate level of confidence to the results. Are they significant? Are they reliable? It is hard for nonspecialists to do this at all well, but there are a number of clues, the main one being how often is the paper cited by others, and the importance of the journal in which it is published. Not all journals are equal. Some are much harder to publish in than others, and will only take the very best studies that represent real breakthroughs.

Other journals may publish papers that come from less thorough or extensive studies, or where the interpretation of the data is flawed or subtly biased.

So while you may end up with a long list of articles on the subject you are researching, don't be surprised to find that there may be differing conclusions to the same questions. In any field, while there is a degree of consensus among scientists on some issues, there are usually just as many points of disagreement and intense debate. It's something of a minefield for the unwary, so tread carefully and ask the right questions before accepting something as fact.

Finally, here is a word of warning about pseudoscience. It's everywhere, and is especially common on the Internet and where treatments for human diseases are being discussed. If "results" have not been published in independent, peer-reviewed scientific journals, you should ask why. A useful resource in this regard is the quackwatch website, www.quackwatch.org.

To use a rather far-fetched analogy, if we are going on a journey, science is the engine that helps get us there, but it shouldn't be driving the car.

Scientists have often been guilty of undervaluing or ignoring things that cannot be measured, so let's be philosophical for a moment. Metaphorically speaking, many people would say that wine has a "soul." It's common in the production of wine to find people involved who possess a strong sense that there is a "spiritual" element to what they are doing. They believe that they need to operate with integrity and produce honest wines that reflect a faithful expression of the sites they are working with. Scientists typically find this sort of attitude hard to understand, because ideas like this can't be framed in scientific language. But isn't it best if we can establish some sort of dialogue between scientifically literate wine people and those who choose to describe their activities in other terms, such as proponents of biodynamics?

How does all this relate to wine? In this book I am going to be looking at wine through the particular lens of science. I'll be exploring how science is a useful (even vital) tool in the fields of viticulture, winemaking, and also in terms of helping us understand the human interaction with wine. But I am not suggesting for one minute that wine—this engrossing, culturally rich, life-enhancing and enjoyable liquid—should be stripped of everything that makes it interesting and turned into an industrially produced, technically perfect, manufactured beverage. Science is a tool that can help wine, but this doesn't mean that wine should belong to the scientists. For this reason, I'll be leaving the familiar, safe ground that you might expect a book titled *The Science of Wine* to cover, and venture into some of the more absorbing issues that get wine-lovers talking, such as terroir, biodynamics, and the production of "natural," manipulation-free wines.

Science has a lot to offer wine. My goal is that, by writing this book which is designed to be accessible to nonscientists yet still with enough meat to keep the scientists engrossed, I'll have helped some to an enhanced understanding of wine that will assist them in their pursuit of this culturally rich and fascinating beverage.

Section 1
In the Vineyard

1 The biology of the grapevine

ONE SPECIES, THOUSANDS OF VARIETIES

Agiorgitiko and Albariño, Baga and Bourboulenc, Cabernet Sauvignon and Chardonnay, Dolcetto and Durif; there are many thousands of different grape varieties, capable of making a bewildering array of different wines, but they are all cultivars of just one species: *Vitis vinifera*. Estimates are that across the globe there are some 14,000–24,000 different cultivars (the scientific term for variety), but because many of these are synonyms, these represent perhaps 5,000–8,000 varieties. A recent major book on the subject, *Wine Grapes*[1], identified 1,368 varieties used commercially to make wine. This single species is the source of almost all the wine consumed today. *Vitis vinifera* is commonly referred to as the Eurasian grapevine, because of its origin in the Near East, at the meeting point of Europe and Asia. This is where *Vitis vinifera* can still be found growing wild today.

The genus *Vitis* actually contains around 70 different species, many of which are found growing in the USA, such as *Vitis labrusca, V. riparia,* and *V. berlandieri*. These are sometimes used to make wine, but it has an unusual flavor. Their significance lies in that they have evolved in conjunction with the aphid phylloxera, and so can coexist with it. As a result, American vines are used for rootstock onto which almost all *vinifera* vines are grafted. Without this, viticulture in Europe and much of the rest of the world would have been finished by the phylloxera epidemic that occurred in the late nineteenth century, as the root-munching aphid found its way over to Europe from the USA. More on that later.

VINES IN THE WILD

When most people think about grapevines, they envisage pretty vineyards, with neat rows of vines arranged on a trellis system, or grown as bushes. Lovely. But this isn't how grapevines grow in the wild. Their natural growth form is as woodland climbers, using trees for support. Where the vine breaks through the canopy into sunlight, it flowers and produces grapes. These are eaten by birds, which then disseminate the seeds. Because of this growth form, vines need extensive root systems to enable them to compete for water and nutrients with the already established trees and bushes they are hitching a ride on; the ability to make the most of limited resources is a prerequisite to this sort of lifestyle. Vines also need shoots that are capable of rapid elongation to grow toward the outside of the host canopy to find sunlight. Then, when they are in the light, this is the right time for the shoots to produce flowers and thus fruit, an effort wasted if it takes place in the shade of the canopy. This makes it clear that the vine is designed to be a highly competitive plant with a flexible growth form; vines have to adjust to the shape of whatever host plant they are growing on. It's helpful to bear in mind this native state of the vine when considering viticultural issues, because knowing what the vine is programmed to do can help in uncovering the scientific basis of effective viticulture.

Above A wild vine growing on an olive tree in the Douro region of Portugal. This appears to be an American rootstock variety, and it is bearing fruit.

1 Robinson J, Harding J, Vouillamoz J. *Wine Grapes*. Allen Lane, London 2012

DOMESTICATING WILD GRAPEVINES

Ancient humans living in the right places no doubt would have been familiar with the wild grapevine and its attractive fruit. Mystery surrounds how the grapevine was first domesticated, though there have been plenty of guesses. One speculation, known as the Paleolithic hypothesis, seems plausible, although impossible to prove. Imagine some early humans foraging for food. They discover some brightly colored berries growing on vines suspended from the trees, so they pick them and eat them. They taste good, so these foragers collect as many as they can in whatever container they have to hand. On the journey home, the weight of the mass of grapes crushes a few, which then start fermenting. The result is a rather rough-and-ready wine that collects at the bottom of the container after a few days. If you found this sort of liquid mass at the bottom of a pot, you'd give it a try, wouldn't you? It's hard to imagine any wine produced in this fashion tasting terribly wonderful, but then these folks probably weren't all that fussy. When they experienced to a small degree the mind-altering effects of this liquid, you can imagine it catching on fairly quickly. Deliberate planting of grapevines would have likely soon followed these rudimentary first attempts at winemaking. Someone would likely have planted a few of the seeds, and with a little trial and error, have worked out how to make a vineyard. It is hard to be precise, but it is estimated that this grapevine "domestication" first occurred at least 7,000 years ago, and possibly as long as 10,000 years ago.

Before we examine the specific biology of the grapevine, it's worth taking a brief look at the principles of plant morphology and physiology; these will help put the biology of the vine in context.

VINE STRUCTURE AND DEVELOPMENT

There are six main challenges facing land-growing plants. First, find enough water, and then hang onto it while at the same time being able to exchange gases with the atmosphere. Second, defend against being eaten by herbivores or destroyed by pathogens. Third, find enough light for photosynthesis. Fourth, reproduce and disperse. Fifth, adapt to seasonal rhythms and the variability in the environment. Sixth, deal with competition.

It's the way that plants have met these challenges that has shaped and constrained their growth, form, and physiology. As well as these, the vine has further constraints and specializations resulting from its lifestyle as a woodland climber.

Let's take a look at how the grapevine works, beginning at the bottom with the roots. Roots serve two functions: anchorage and uptake. What exactly do vine roots take up from the soil? Like other plants, vines don't need much—they make everything themselves. But they do need an adequate supply of water and dissolved mineral ions, termed as macro- and micronutrients. These are inorganic (they don't contain any carbon). Root growth is determined by interplay between the developmental program of the plant and the distribution of mineral nutrients in the soil. The roots seek out the water and nutrients in the soil, sensing where they are and then preferentially sending out lateral shoots into these areas. Low nutrient levels in the upper layers of the soil cause the roots to grow down deeper. This is likely to improve the regularity of water supply to the vine, and such roots can reach depths of ten feet or more. The root system of one vine is capable of supporting an enormous mass of aerial plant structure.

Above ground, the vine has a growth form well-suited to life as a climber. Its shoot system is simple, adaptable, and capable of fast growth. The vine never intends supporting itself, so it doesn't waste resources on developing girth. Thin, long shoots are the order, which in turn can produce lateral shoots that eventually become woody. The formation of woody tissue is not for structural reasons but to provide protection, particularly during the dormant period. At regular intervals buds are formed. These buds are complicated structures, containing the potential for leaves, flowers, and tendrils, and develop over two seasons, with a rest over the dormant period.

SHOOT MORPHOLOGY

The stem is separated into sections by structures known as nodes. At each node a leaf is formed on one side, with either a tendril or a flower bud on the other. Thus both vegetative (leaf) and reproductive (flower) meristems (the growth region where cells are actively dividing) are formed

simultaneously on the same shoot. Light is the key to vine growth. In the absence of light, shoots show negative gravitropism (meaning they grow away from the ground). But light is the overriding growth cue: shoots are positively phototropic, growing toward light. Light is also the chief cue for flowering induction. The tendrils are important structures for the vine's climbing habit. They are modified stems that coil around supporting structures. Do tendrils signal to the shoot when they have coiled around something, in effect telling the shoot to keep on growing because it has adequate support? It's an interesting idea.

BUD DEVELOPMENT AND FLOWERING

The flowering process in grapevines is unusual, because it extends for two consecutive growing seasons. Flowering is first induced in latent buds during the summer, but initiation and floral development occurs the following spring. On flowering induction the shoot apical meristem (SAM; a meristem is a region in a plant where cells are actively dividing, producing new cells that haven't yet developed into a specific cell type, known as undifferentiated cells) produces lateral meristems; these will either give rise to flowers or tendrils. A research group led by Martínez-Zapater have proposed that *VFL*, a specific grapevine gene of the same type as a gene first identified in other plant species as *Floricaula/Leafy*, has a role in maintaining indeterminacy (keeping bud-fate options open) in meristems as well as specifying flower meristems, its normal role in annual plants. In this work they described grapevine

flower development in detail and correlated this with an expression of this *VFL* gene using in a technique known as *in situ* hybridization, which shows under the microscope precisely where genes are being expressed in the plant.

In grapevines, buds are formed and are first detectable in early spring in the axils (the inside of the join between the stem and leaf petiole) of the current year's leaves. These buds consist of several SAMs protected by bracts (scaly like structures). The earliest-formed SAM usually develops as a lateral shoot, while the rest remain dormant. In first months of development the SAM produces leaf primordia (cellular structures that will later give rise to leaves). Then, around May/ June in the northern hemisphere, it produces lateral meristems (as opposed to apical meristems, meristems not at the tip or apex, but further down the plant) opposite the leaf primordia; the first two or three of these will produce flowers, the rest tendrils. More meristems are produced, such that by the end of the growing season the bud contains a shoot that has inflorescence (flower structure) meristems, tendrils, and leaves. These are all protected by bracts, scales, and hairs and become dormant in the fall.

During the following spring the bud is reactivated, and more meristems are produced. Crucially, the lateral meristems giving rise to inflorescences or tendrils are indistinguishable at the time they form. This keeps the options of the vine wide open, allowing environmental cues to dictate at this late stage whether these lateral meristems should become flowers or

Above Flower clusters on a Muscat vine in Alsace, before the flowers have opened.

Left Tendrils, specialized structures that help the vine grow up other plants (in its native envrionment) or trellising systems in a vineyard.

Above Grapevine flowering. Almost all *Vitis vinfera* varieties are hermaphrodites, bearing both male (anthers, stamens) and female (ovule, carpel, pistil) flower parts. These flowers are now mostly open. A few "hats" can be seen sticking to the flowers that are just opening. These can end up as debris stuck within the cluster, and this can cause problems with botrytis infections later in the season.

tendrils. The decision affecting this choice between tendrils and inflorescences has been the subject of a separate series of studies by researchers in Australia led by Mark Thomas.

Thomas and colleagues have shown that tendrils and inflorescences are structures that are similar in origin, and can be converted either way by plant hormone application. Indeed, intermediate structures are commonly observed on vines. Their work has demonstrated that gibberellins (a type of plant hormone) are major inhibitors of grapevine floral induction. In the natural habitat of the grapevine, a woodland climber, gibberellin would have two roles. It would promote the elongation of the stem and the production of tendrils, while at the same time suppressing fruit production. This suggests that there may well be a connection between light sensing and gibberellin production (or responsiveness to gibberellin), because this is the sort of strategy that vines would use if they were in the shade of the host plant canopy. A vine in the shade would want to delay fruit production and maximize its energy on growing upward as fast as possible until it breaks through into the light.

When flowers are formed, their development and pollination occurs best during a period of warm, settled weather. Domesticated vines are hermaphroditic(meaning, each flower of each individual has both male and female structures) and can self-pollinate. Poor weather during this process can result in reduced or uneven fruit set, so flowering is one of the critical phases in the vineyard calendar.

GRAPE DEVELOPMENT

Grapes are for the birds. The wild vine "designed" grapes to enlist the help of birds for spreading seeds. With their flying habit, birds are pretty mobile, and the kinds of places birds go are promising locations for dispersing seeds. All those sugars in ripe grapes are a reward for potential seed-carrying birds. But the grapevine doesn't want the birds to be carrying the grapes off too soon, before the seeds are ready for dispersal, or before the onset of fall, with its rains and favorable conditions for seed germination. The maturation of grapes is therefore cleverly timed, and separated into three phases. The first phase involves the development of the grape structure; throughout this process the grapes accumulate acids and experience rapid cell division.

Above *Veraison* in red grapes. The berries begin to accumulate anthocyanins in their skins and change color from green to red/black. Accompanying this the skins begin to soften and the berries swell, and acid decreases while sugar accumulates. White grapes go through veraison, but less visibly because they lack anthocyanins.

This growth then slows for the second phase, *veraison*, which is when red grapes change color from green to red and the skins of white grape change from a hard green to a soft, translucent green. The third phase follows this, when cell growth begins again and grapes accumulate sugar and phenolic compounds, and acids decrease. It's all for the benefit of the birds: unripe grapes are camouflaged (green color) and unappetizing (highly acidic, containing high levels of leafy-tasting pyrazines, and with harsh tannins), while ripe grapes stand out with their attractive red or golden color and taste delicious (the sweetest of all fruits, with lower acidity, riper tannins, and the herbaceous pyrazine flavors degraded). The grapes are telling the birds that dinner is served. Seeds are the first part of the grape to reach physiological maturity, around veraison, and this makes sense.

GRAPE MATURITY

What is grape maturity in winemaking terms? It depends on the objective. Two types of maturity are talked about in wine circles: sugar accumulation and phenolic (or physiological) maturity, the latter also being referred to as flavor maturity by some. Match the right grape variety to your vineyard climate and do your viticulture well, and you'll reach the goal of perfect maturation: flavor maturity coinciding with a sugar level

that will yield a wine of some 12–13% alcohol. There is, however, a disconnection between the physiological processes that govern the rate of sugar accumulation and loss of malic acid, which is dependent on climatic factors, and the color, aroma, and tannin development (phenolic maturity) that is nearly independent of climate. The result is that in warmer climates, grapes only reach physiological maturity at sugar levels that are considerably higher than in cooler regions.

Typically, in the cooler, classic, Old World regions, grape harvest coincides with a shortening of day length and temperature, and thus sugar accumulation is more gradual. In these regions the measurement of sugar levels works well as a guide for when to harvest. It is also a simple measurement to make in the field. In these conditions, it is likely that by the time the grapes have accumulated 12 degrees of potential alcohol or so, they will have achieved satisfactory phenolic maturity. Pick at the same sugar levels in many New World regions and you'll end up with unripe flavors in your wines.

Light is crucial in the ripening process; grapes that are shaded contain less sugar and are more acidic than those exposed to sunlight. Light also affects bud fertility, so one of the key viticultural goals is therefore to encourage the vine to produce an open canopy, without dense, vigorous growth that could produce shading. This is covered in Chapter 3.

DIFFERENT VARIETIES, DIFFERENT CLONES

In taxonomic terms, *Vitis* is what is known as a genus, the taxon above species. Way back in evolutionary time, this genus split into three lineages. The Eurasian vine *Vitis vinifera* is a single species and is responsible for almost all wine made today. In contrast, there are dozens of different American species of *Vitis*, whose main importance in modern viticulture is to provide phylloxera-resistant rootstocks for grafting *Vitis vinifera* onto. Finally, there are a few Asian species of *Vitis*, of little importance for wine.

Vitis vinifera may be just one species, but through evolution and numerous cross-fertilizations it has produced thousands of different varieties that are used in winegrowing today. The effect of domestication has been largely to improve the fruitfulness of these various varieties and to distribute them widely.

So what, exactly is a grape variety? I posed the question to researcher José Vouillamoz. "A grape variety is the result of a seed that has grown into a plant that has been selected by humans, propagated over centuries or millennia by layering or cuttings, accumulating mutations along the way," he responds. "The older the grape variety is, the more diverse it is in terms of shape, color, qualities, and characteristics. All of these differences are what we call clones, and all the clones together make up the grape variety. The starting point is one father and one mother." Vouillamoz points out that a new variety cannot be made by a series of accumulated mutations. Instead, sexual reproduction is necessary. "It is just like a human being: every grape variety has a father and a mother. You can't say that by accumulating mutations you develop a new variety. Otherwise, where do you put the limit?"

By this definition, Pinot Noir, Pinot Gris, and Pinot Blanc are not separate varieties. "Pinot is one single grape variety that is very old and that has had many 'accidents' in its life—many mutations," says Vouillamoz. "Some of them were spectacular because they touched the color of the grape varieties and almost nothing else. When I do the DNA profiling of Pinot Noir, Pinot Gris, and Pinot Blanc, since I look at, say 10, or 12 different DNA regions, I don't see where the color mutation has happened. They are all the same in the DNA profile."

Vine propagation occurs vegetatively, meaning that cuttings taken from vines are used to produce new plants, which are genetic clones of the parent variety. Attempts to grow vines from seeds are almost certainly doomed to failure because the genetic reassortment that takes place usually means the loss of positive features of the variety. Generally, growers are happy with the varieties they have and just want to improve them in ways that don't affect the expression of varietal character.

Within each of the different varieties a range of clones exist. Some of these differ through spontaneous bud mutations, which then result in genetically altered shoots. Almost always, such mutations are deleterious. Sometimes, however, they are positive, and can be propagated by cuttings taken from such an affected shoot, resulting in a new clone of the variety. Over enough time, a range of such clones might be developed. Other times, the clonal differences reflect nothing more than differing levels of virus infection, or perhaps epigenetic differences (heritable changes that aren't based on DNA-sequence changes).

Vouillamoz adds, "The definition of clone is not really clear, either. For me, a clone is one or several mutations that are spectacular enough to the human eye for them to be selected. It is very subjective. For the producer, the color of the grape is important, or the size of the bunch is important. But maybe for a biochemist, a mutation in a biochemical synthetic pathway could be more important, but it is hidden."

Another mechanism underlying differences between varieties and clones is a process called chimerism. This is where separate portions of the plant have different genotypes. Mark Thomas and colleagues have demonstrated that the phenotype of the variety Pinot Meunier results from the interaction of two genetically distinct cell layers. They separated the two by tissue culture and showed that one layer is the same as Pinot Noir, while the other is a mutant that is insensitive to the plant hormone gibberellin, and produces a short, stubby vine with a fruit cluster at every node, instead of the more usual mix of fruit clusters and tendrils. I asked Thomas how common this sort of grapevine chimerism is. "Since our work, other research groups from various countries have looked at other varieties and found similar results. So I would guess that most, if not all, old varieties would have accumulated somatic mutations (one occurring in any cell in the plant except the germ cells, the pollen, or ovule) and a chimeric situation would be very

Left Wine scientist Dr. José Vouillamoz, who has done important work on grape varieties using modern molecular biology techniques.

common due to somatic mutations arising from one cell in a specific cell layer, and that mutated cell eventually taking over the whole layer."

Modern molecular biological techniques have provided some important new insights into the relationships among different grape varieties. Dr. Carole Meredith and her colleagues were the first to use microsatellite markers (also known as simple sequence repeats, a feature of complex genomes such as that of the grape that can be used as a molecular fingerprinting device) to sort out relationships between grape varieties that weren't apparent from the traditional chromosome studies. In 1997 they found that the parents of Cabernet Sauvignon were Cabernet Franc and Sauvignon Blanc—the first successful vine paternity test. Among other discoveries they have shown that Zinfandel is the same grape variety as Primitivo (grown in Italy) and Crljenak (from Croatia, also called Pribidrag there), and Chardonnay are the result of what was likely an accidental cross between Pinot Noir and an undistinguished white variety called Gouais Blanc.

"Microsatellites are pieces of DNA where the bases of the DNA repeat themselves. When you have coding DNA, the bases A, T, G, C are in a certain order that would translate into a protein," explains Vouillamoz, who has collaborated with Meredith in these studies. "At one point the DNA starts to stutter and to repeat, like GAGAGA 100 times, for example. This part of the sequence does not contain any genetic message. We don't know what the use of these microsatellites is, but they exist in all organisms. The first simple sequence repeats (SSRs) were spotted with fluorescence at the top of the chromosomes, so they were called microsatellites because they looked like small satellites orbiting around the chromosome centers. Later on we found them all along the chromosomes, and their name is somehow obsolete."

Vouillamoz continues, "These pieces of DNA are repetitions of bases that exist in every organism. In one single individual—either a grape variety or a human being—they are stable, and they differ between two individuals. If you look at, say, 10 different microsatellites, you get a DNA profile of your individual or grape variety. It remains the same within the grape variety, and it differs from one to the other. The first use of microsatellites [in grapes] was in 1993. The idea was to identify the grape varieties in the collections. Australians did it, and a few years later John Bowers at UC Davis, California, started to analyze the collection there. The idea was just to sort out which were duplicates, because if two varieties are the same you can eliminate one from the collection and save money. He was comparing the DNA profiles and he noticed that Cabernet Sauvignon had a profile that was very similar to Cabernet Franc and Sauvignon Blanc. So he looked into the human genetics, to see how parentage testing was done. He then did the same with grape varieties. He realized that these markers were completely consistent with Cabernet Sauvignon being the progeny of Cabernet Franc and Sauvignon Blanc. He went on with the analysis and statistics, and 1997 was the first publication of grape parentage for Cabernet Sauvignon. This was groundbreaking news in the wine world, because everyone thought that all these varieties, especially in Bordeaux, had always existed. No: it was born sometime, somewhere, and we know the parents. Most of the time the parents have disappeared or we know only one. Here we know both of them. No one would have thought that a white grape variety could be a parent of Cabernet Sauvignon, which is black-berried. This was a breakthrough, and this is the technique I have used since then."

Vouillamoz says that this technique is quick and inexpensive, and he charges about $130 per sample. "When I was at UC Davis, one producer wanted to plant 100 hectares [about 250 acres] of Syrah, and he wasn't sure of the nursery. He sent us a sample and we analyzed it, and found out it wasn't Syrah, but Peloursin. It is in the same family as Syrah, but a very rare variety. It is a parent of Durif, which is a cross between Peloursin and Syrah, and is often called Petite Syrah. Carole Meredith analyzed the vineyards in California and nine out of ten times what they call Petite Syrah is Durif, and one in ten times it is Peloursin."

A famous example of mistaken identity was with the Australian Albariño, which was recently found to be Savagnin, which has been shown by DNA analysis to be the same as Traminer, the nonaromatic version of Gewürztraminer. This mistake was revealed when France's top ampelographer, Jean-Michel Boursiquot, toured Australian vineyards and suggested that the vine they were calling Albariño was in fact Savagnin. The Australian authorities compared samples of Spanish Albariño and French Savagnin with what they were selling as Albariño, and then realized their mistake. So, from the 2009 vintage, Australia's "Albariño" producers have had to rename their wines Savagnin Blanc or Traminer.

FOUNDER VARIETIES

Savagnin turns out to be very important, because it is what is known as a founder grape variety. "We realized that a small number of grape varieties have given birth to all the diversity that we observe today," says Vouillamoz. "This is a new concept; we are the first ones to propose this hypothesis, because when you read other books or even scientific papers, they give the impression that all the important grape varieties that we have today have been introduced a long time ago from different places—from Egypt, the Near East, the Middle East, and so on. I do think that we had a limited number of ancient introductions of the ancestors of what we have called the founder grape varieties, among which we find Savagnin Blanc, Pinot, and Gouais Blanc [the parent of no fewer than 80 varieties that are cultivated today in Western Europe]. We have established so far a total of 13 founder varieties that gave birth to most of the diversity that we observe today in western Europe, but many more remain to be discovered in other regions were grape family trees have not been studied yet."

RETROTRANSPOSONS AND GRAPE COLOR

Have you ever wondered what makes some grapes white and others red? Now we know. Scientists from Japan have uncovered the molecular basis of grape-skin color, identifying the genetic mutation responsible for white-skinned grapes. It is thought that ancestral wild grape species were all dark-skinned, and that white grapes arose by mutation, but the precise details were a mystery. A research group led by Shozo Kobayashi discovered the gene mutation that is thought to underlie the emergence of white grape-skin color, in research published in the journal *Science* in 2004.

Black- or red-skinned grapes owe their color to a group of red pigments known as anthocyanins. The synthesis of these pigments is controlled by a specific set of genes. Kobayashi's group has shown that a specific sequence of DNA, known as a retrotransposon, is responsible for turning off the expression of one of these genes and thus switching off pigment production in white grapes. Retrotransposons are mobile pieces of genetic information, that are able to move genes around the genome. They are especially abundant in plants and are able to induce mutations by inserting next to, or actually in, other genes.

"We hypothesize that Gret1 [the specific retrotransposon] originally inserted upstream of one of an anthocyanin-regulating genes of a black-skinned ancestor, and that subsequently a white-skinned grape was produced by spontaneous crossing," says Kobayashi. They also think that this was an ancient mutation, taking place before grapevines were first cultivated.

Kobayashi's group have also shown that red pigmentation can be induced in the skins of white grapes by the insertion of the anthocyanin-controlling gene, and that white grape varieties have mutated to red by the spontaneous removal of the retrotransposon from this gene.

"It makes sense," says Professor Andrew Walker of UC Davis' Viticulture department, "and backs up observations by breeders, geneticists, and viticulturists with some varieties, particularly Pinot Noir and its color morphs Pinot Blanc and Pinot Gris." These results explain how a black grape, Cabernet Sauvignon, can have both a red grape (Cabernet Franc) and a white grape (Sauvignon Blanc) as its ancestors.

Renowned grapevine geneticist Carole Meredith thinks that the work is significant. "It adds to the growing evidence that retrotransposons have long been active in grape and may well have contributed to characteristics that are important in wine composition and flavor." She adds that there is ongoing research into the role of retrotransposons in the development of clonal variation within varieties. "I think we'll find that retrotransposons are very important in wine grapes," says Meredith. "It adds to previous evidence that white fruit color in many varieties may have a single common basis. Meredith concludes, "This is fascinating work, but a lot of genetic analysis is needed before it can be concluded that this retrotransposon insertion is solely responsible for white fruit color in grapes."

Ancient wine: the new field of molecular archeology

The science of archeology must be a frustrating one at times. Detective work of the highest order is necessary to meld together the few remaining pieces of surviving evidence of a bygone era into some sort of coherent story. Patrick McGovern, an archeologist at the University of Pennsylvania, is probably the leading expert on the ancient origins of wine. But rather than just rely on old fragments of pottery and a few vine seeds, McGovern has turned to advanced molecular biological techniques to provide new evidence to shed light on the origins of wine in ancient civilizations. This new avenue of research has been dubbed "molecular archeology."

McGovern has collaborated with grapevine molecular biologists to study the DNA of ancient grape relics, such as seeds. Using similar techniques as those employed by Carole Meredith and her colleagues to assess relationships among modern-day varieties, McGovern is using microsatellite repeats in the ancient DNA to identify the grape variety and its relationship to modern vines. It's a work in progress as he and his collaborators continue to fine-tune the complex process of extracting useful DNA from ancient plant tissue and then making sense of the results. As yet, more concrete results have come from the array of chemical techniques that have been used to study residues present on archeological samples, including infrared and UV spectrometry, gas chromatography/mass spectrometry, and liquid chromatography/mass spectrometry. Together, these are powerful tools for providing scientifically reliable answers to questions that were previously just a matter of conjecture.

McGovern's DNA search for the site of domestication of the Eurasian grapevine, dubbed the Noah Hypothesis, has supported the thesis that it first happened in the Near East. As well as genetic evidence, linguistics and archeology point toward an area in what is known as the Fertile Crescent, in eastern Anatolia, as being the site of the first domestication of *Vitis vinifera*. However, Transcaucasia also remains a possibility.

I asked one of his collaborators, José Vouillamoz, how many domestication events there were. "We don't know how many," he says. "But our belief is that there were a small number of primary domestications somewhere in the Near East, most likely in southeastern Anatolia. It wasn't a single plant; we don't believe in an 'Eve' hypothesis, like we have for humans. I think we had a large population of wild grapevines, and from this large population a few hermaphroditic individuals gave birth to all the varieties that we have today. In a natural wild grapevine population, 2–3% of the plants are hermaphroditic. It is a small number. It has created a kind of bottleneck, genetically speaking, because they selected only the hermaphroditic plants, which was the starting point of grape domestication. I believe there were several primary domestications. I am still not convinced about secondary domestication centers in Western Europe, perhaps in Greece or Italy. I am open-minded about this, though. I would be more in favor of a secondary domestication center in the Iberian Peninsula, because they have a lot of wild grapevines."

Left Do old vines produce better wines? This Shiraz vine, planted on its own roots, is more than 100 years old and is from Wendouree in the Clare Valley, South Australia.

2 Terroir: How do soils and climate shape wines?

THE UNIFYING THEORY OF WINE

"Terroir" is a concept that is rapidly emerging as the unifying theory of fine wine. Once almost exclusively the preserve of the Old World, it's now a talking point in the New World, too. The traditional, Old World definition of terroir is quite a tricky one to tie down, but it can probably best be summed up as the way that the environment of the vineyard shapes the quality of the wine. It's a local flavor, the possession by a wine of a sense of place or "somewhereness." That is, a wine from a particular patch of ground expresses characteristics related to the physical environment in which the grapes were grown.

The goal of this chapter will be to give a broad introduction to this hotly debated topic, examining why it is still a controversial issue. Then I'll focus on the scientific underpinnings of this concept, concentrating on teasing out the relationship between vineyard characteristics and wine flavor. It's unfortunate that just the mention of the word "terroir" rouses such strong negative feelings in some, because it really is one of the most absorbing topics in the study of wine, something that I'm hoping to demonstrate in the following pages.

DEFINITIONS

One of the problems with many discussions of terroir is that this word means different things to different people. Indeed, defining terroir in precise terms is quite difficult, partly because it is a word used in three rather different ways.

A SENSE OF PLACE

The primary definition is that terroir is the possession by a wine of a sense of place. That is, the wine expresses flavor characteristics influenced by the properties of the vineyard or region it hails from. Immediately we see that

scale is an issue here; environmental variation affects wine flavor, but this variation operates on a number of scales. Wines made from grapes harvested from different parts of the same vineyard may well taste different. On the other hand, there might be characteristics held in common by wines made from larger geographic regions that are evident when these wines are compared with those of other regions: for example, Burgundian Pinot Noirs compared with Californian Pinot Noirs. And which factors should be included in the definition of terroir? It's clear that human intervention in terms of viticultural and winemaking practices may also confer a sense of place to a wine, but most people wouldn't count the human element as part of terroir.

This raises the question of the differences between terroir and type. Most definitions of terroir rule out human intervention as part of the equation. But could winemaking play a role in maintaining type? Certainly, in the classic Old World regions where terroir is so precisely delineated, the fact that winemakers commonly use similar techniques could help lend a distinctive regional style. Winemakers could also be adapting their techniques to best exhibit regional differences. This type, owing more to human intervention than classical definitions of terroir, is still of merit because it helps maintain the sort of stylistic regional diversity that makes wine so interesting. In general, though, terroir is easier to conceptualize and is a more useful concept if winemaking is excluded. "I don't see winemaking as part of terroir," says Jeffrey Grosset of Australia's Clare Valley, "but rather that poor winemaking can interfere with its expression and good winemaking can allow pure expression."

At Felton Road, the celebrated Central Otago (New Zealand) winery, the concept of terroir is seen as a partnership. Owner Nigel Greening explains that when it comes to blending the single-vineyard wines, Cornish Point and Calvert, only about 30–40% of the total production will go to a single-vineyard bottling, with the balance going to the winery's Bannockburn label (the wines from other two single vineyards, blocks 3 and 5, pretty much all end up in the single-vineyard bottlings). "So we have a dilemma," explains Greening. "We have to choose which 30–40% to use. There will typically be eight lots from each of the vineyards, and we will taste them about three times, blind. We score them not for their quality as a wine but for their expression of site. The 'Calvertiest' ones become Calvert, and those that show the least Calvertness go into Bannockburn. The same applies to Cornish Point." My next question is, how do they decide which lots are most reflective of the respective single vineyard sites? "Calvert is the more elegant, tighter, more linear wine," says Greening, "while Cornish Point is voluptuous, perfumed. This is naturally what this vineyard does and we want to show that expression of site as clearly as we can."

THE VINEYARD SITE

The next use of terroir is in describing the vineyard site itself: the combination of soils, subsoils, and climatic factors that affect the way that grapes grow on the vine, and thus influence the taste of the wine made from them. This is probably the least controversial use of the term, because it is purely descriptive.

GOÛT DE TERROIR

Finally, there is a third use of this phrase. The term *goût de terroir* is sometimes used to describe flavors that are presumed to be imparted by the vineyard site itself. Thus, someone might say that they taste notes of "terroir" in the wine. This is the most confusing use of this word, and in scientific terms it is hard to defend because it makes assumptions about mechanisms that can't be demonstrated, as we will see later.

TERROIR IN PRACTICE

Let's try to explain terroir in practical terms. Take a property with three different vineyard sites, one flat, one on a south-facing hillside,

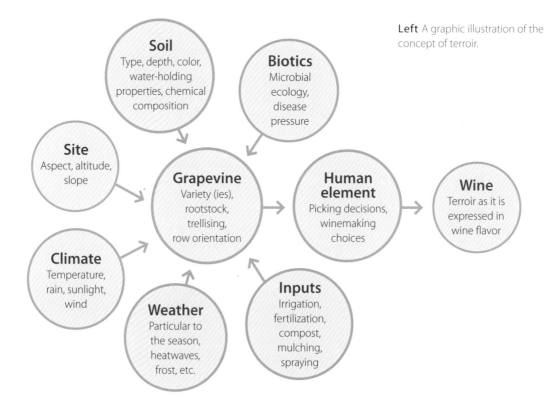

Left A graphic illustration of the concept of terroir.

and one on a north-facing slope. We're assuming for the sake of simplicity that the same grape varieties and clones are used in all three, that the vineyards share the same geology, and that they are farmed the same way. Three wines are then made, one from each vineyard, in identical fashion. They will likely all taste different. That's terroir in action. Typically, vineyard sites will differ not only in one variable, as in this example, but in several. And such factors as differences in slope, orientation, or soil type will also influence decisions about which varieties are planted where, bringing further variables into play.

TERROIR IN THE NEW WORLD

One interesting question surrounds why there has recently been a transformation in attitude among New World winemakers, where for so long the job of grape-growing was seen merely as a mundane prelude to the work of the all-powerful winemaker. Even fairly recently, the New World response to "terroir" was typically that it was a last-ditch marketing ploy by European winegrowers who were panicking about their increasing loss of market share. This turnaround has occurred for two reasons. First, because New World winegrowers have realized that one of the keys to wine quality is starting with grapes that show homogeneous (even) levels of ripeness, and the recognition of the role that natural variation plays within and between their vineyards. With the increasing adoption of a technique known as precision viticulture, vineyards are commonly broken up into subplots sharing similar characteristics (known as natural or basic terroir units), so that vineyard interventions can be precisely targeted to where they are needed. The second reason is that New World winemakers have realized that regionality is the way forward with fine wine. But you don't have to look too far below the surface to see that there are subtle but important differences between Old World and New World notions of terroir. Speaking generally, in the Old World terroirists aim to make wines that express the typicity of the specific vineyard site, whereas in the more pragmatic New World, understanding terroir is seen as a route to improved quality. Of course, there are exceptions to this generalization.

WINEMAKERS' VIEWS ON TERROIR

Although the idea that soils and climate influence wine flavor seems to be rather obvious, terroir is in fact a highly controversial concept. It doesn't help that it has long been regarded as the exclusive preserve of the French. "The French feel they have ownership of 'terroir'," says Barossa winegrower Charlie Melton, "but in good winemaking the idea is universal." Melton prefers not to use the "T word" itself, but instead talks of characterization of distinct vineyard blocks. "In the Barossa the subregions have their own character," he says. "Wines from the southern end have finer, slightly sweeter aromatics, and those from the northern end have a *garrigue*-like earthiness rather than sweetness." He points out that the humidity is on average 3% higher in the southern end, and that the soils in the Barossa vary quite widely, with adjacent blocks sometimes making quite different wines. "Call it terroir if you like," he adds.

In fact, one Australian winemaker, Jeffrey Grosset, suggests that the French weren't the first to come up with the concept. "Terroir is the French word for what some have known in Australia for thousands as years as *pangkarra*." He explains that this is an aboriginal word that encompasses the characteristics of a specific place, including the climate, geology, and soil–water relations.

TERROIR IN ARGENTINA

There's also a word in Spanish that's equivalent to "terroir." According to Santiago Achaval, president of premium Argentinean winery Achaval-Ferrer, this term is *terruño*. Achaval is an eloquent proponent of this concept. "Our word has the same nuances as the word "terroir", plus an additional one: it's the land a man belongs to, not the land that belongs to a man. It describes a man's bond with the land where he was born," he says. "We think the concept of terroir is of the highest importance. Terroir is for us the only source of originality and personality of a wine. It is also a source of never-ending wonder: how small distances and slight differences in soil composition, exposure, and even surrounding plant life result in very noticeable differences in the wines."

He refutes the idea that terroir is confined to classic Old World regions. "Argentina does have terroirs in the same way as France and Italy do.

The difference with those other countries is that the discovery of our terroirs is just now beginning. Both France and Italy have been perfecting their knowledge of their soils and microclimates since the early Middle Ages. Argentina started a century ago, with a hiatus during the turbulent economic times between the 1970s and the 1980s. So there's a lot of exploration to be done until we can really say that we know our terroirs, and that we can design their hierarchy; not every vineyard is capable of expressing a powerful personality through its wine. And as in the rest of the world there are differences in quality of the wine that are driven only by location." I asked him whether terroir influences the way he works. "Yes it does, and strongly so."

OBJECTIONS TO TERROIR

But one New World *vigneron* who objects to the notion of terroir is Sean Thackrey, a Californian winemaker famous for his single-vineyard Orion wine. "My objection is simply that it's so ruthlessly misused, and with such horrifying hypocrisy," says Thackrey. "It's very true that fruit grown in different places tastes different. In fact, it's a banality, so why exactly all this excess insistence?" Thackrey himself allows terroir to influence his work—"I don't know how it will be possible to observe the delicacies of change in a particular vineyard more attentively than I do in making the Orion"—but he feels that the French overemphasize terroir for largely economic motives. He describes it as "an intensely desirable

and bankable proposition because their property can then be sold, transferred, and inherited with the full value of the wine produced from its grapes attributable to the property itself." Thus the work of the winemaker and viticulturalist is played down and the role of the vineyard site talked up. "The immense psychological imperative to have wine borne from the earth without human intervention other than caretaking—which may in itself be why there is no French word for "winemaker"—would make a long and complex book in itself," Thackrey suggests. "Personally, I believe that the quality of French wine is due to a French genius for viticulture and winemaking, just as I believe that the quality of French cuisine is due to a French genius for gastronomy."

A KEY TO SUCCESSFUL BLENDING

Portugal's Dirk Niepoort, well-known for his trailblazing Douro table wines, adds a slightly different complexion to the debate: "As a rule, I believe blended wines to be better than single-vineyard wines." This doesn't mean he's a nonbeliever, though. "I believe that terroir is essential," he says. "I think that a good blender has to be someone who understands, knows, and respects different terroirs." Niepoort maintains that the search for great terroirs is very important to him. "But it is not only finding the terroirs that is important; it is also important to understand them and then adapt your winemaking to them."

CLIMATE VERSUS SOILS

Climate is clearly vital to the concept of terroir. The average climate of a vineyard site determines which grapes it can effectively grow. In large vine repositories, such as the ampeographic collection at the École Nationale Supérieur d'Agronomie de Montpellier (National Superior School of Agronomy of Montpellier), in France, the earliest and latest varieties ripen some two months apart. Remember: these are all varieties of the same species. Vines are indeed incredibly fussy about climate, and only perform well if fairly narrow climate parameters are met.

But while climate determines which varieties can be grown successfully at a particular location, it is the soils that are of real interest to wine-science types. If the climate is just right for your variety, this is but part of the story, as one of France's most famous wine region, Burgundy, illustrates perfectly.

There aren't that many places where Pinot Noir can be grown successfully, because among all grape varieties, it is the fussiest of the fussy. It has taken wine producers in the New World decades to find sites where they can get it to perform properly. Even within Burgundy, though, some vineyards do brilliantly with it, while others just a few yards away make mediocre wine. The climate isn't changing much as we progress from regional to village to *premier cru* to *grand cru* vineyard; it is the soils and geology that are making the difference.

Burgundy is therefore seen as the test case of terroir. It is a region divided into a patchwork of vineyards based on long experience. Over hundreds of years, people observed that certain vineyards did consistently better than others year after year. This resulted in a hierarchical classification and structuring of vineyards based on differences in wine characteristics, and these characteristics have since been found to have their origin in the vineyard's physical properties. When the boundaries for the Burgundy vineyards were put in place, no one knew much about geology and soil maps; they just recognized that some vineyards, rather inexplicably, were privileged, and produced better wines. But now geologists can show that the hierarchy of Burgundy vineyards reflects changes in the subsoil properties that influence grape and thus wine quality.

Interestingly, the division of Burgundy vineyards into white (Chardonnay) and red (Pinot Noir) doesn't seem to be perfect. Noted Burgundy expert Jasper Morris points out in his famous book on the region that the soils of some red vineyards are better-suited to white-wine production, instead of red. The choice of variety is sometimes governed by economic factors.

MECHANISMS OF TERROIR: A TASTE OF THE SOIL?

The notion of terroir is fundamental to the wine industries of such Old World countries as France, Italy, and Germany. It's a philosophical framework within which winegrowers work. Local wine laws are built around the concepts of appellations, which lend official sanction to the idea that a combination of certain vineyard sites and grape varieties creates unique wines that faithfully express their geographical origins. Correspondingly, many Old World growers feel they have a duty to make wines faithful to the vineyard sites they are working with. These growers will commonly make associations between properties of the wines and the soil types the grapes are grown on. In some cases these putative associations are quite specific: people will talk about mineral characters in wines and associate them with the minerals in the vineyard, taken up by the roots of the vines. Do chalk, flint, or slate soils impart chalky, flinty, or slate-like characters to wine? As a scientist who has a working knowledge of plant physiology, I find this notion, which I call the "literalist" theory of terroir, implausible. Yet I can't get away from the fact that an overwhelming majority of the world's most compelling and complex wines are made by people who hold the notion of terroir as being critical to wine quality. Thus, one of the goals of this chapter is to explore the mechanisms of terroir, focusing specifically on soils. Just how do soils affect wine quality? Is it a direct or indirect relationship? What are the scientific explanations for terroir effects?

If we are going to frame terroir in scientific language, then we'll need to start with some plant physiology (see Chapter 1 for a more in-depth account of this). The miracle of the plant kingdom is that these complex organisms build themselves from virtually nothing: all that a plant needs to grow is some water, sunlight, air, and a mixture of trace elements and nutrients. All the complex structure and chemistry of an oak tree, a daffodil, or a grapevine is fashioned from these very basic starting ingredients. What do vine roots take up from the soil? Primarily water, along with dissolved mineral ions. It seems implausible that such a complex structure as a vine is created from virtually nothing by photosynthesis: the capture of light by specialized organelles called chloroplasts, which turn light energy into chemical energy that the plant can use. But that's the way it is. As Richard Smart emphasizes, "All flavor compounds are synthesized in the vine, made from organic molecules derived from photosynthesis ultimately, and inorganic ions taken up from the soil." Professor Jean-Claude Davidian, of the École Nationale Supérieure d'Agronomique in Montpellier, echoes these sentiments. "Nobody has been objectively able to show any links between the soil mineral composition and the flavor or fragrance of the wines," he says. Davidian adds, "Those who claim to have shown these links are not scientifically reliable."

TREATING VINES MEAN TO KEEP THEM KEEN

It is helpful to think about plants as sophisticated environmental computers. In the same way that we sense the world around us and then use this information to guide our actions, so do plants, but over a longer timescale. Literally rooted to the spot, they adapt their growth form to best suit the local conditions. This extends to their reproductive strategies. Generally (and simplistically) speaking, if conditions are good, then plants opt for vegetative growth; if they are bad, they choose to reproduce sexually, which means fruit production. So viticulturalists want to treat their vines mean enough that they focus on fruit production, while giving them just enough of what they need so that they don't suffer from water or mineral deficit, which would hamper their efforts at producing ripe fruit. Thus many viticultural interventions aim at encouraging the vine to partition nutrients to the grapes so that they ripen properly, rather than concentrating on growing more leaves and stems (vegetative growth or "vigor").

An example of this "environmental computing" is seen in the growth of plant roots. Root growth is determined by interplay between the developmental program of the plant and the distribution of mineral nutrients in the soil. The roots grow to seek out the water and nutrients in the soil; to do this it appears that they sense where the various nutrients are and then preferentially send out lateral shoots into these areas. Low levels of nutrients in the upper layers of the soil results in the roots growing to a greater depth, which is likely to improve the regularity of water supply to the vine. "Vines have roots which can reach up to 3 meters [10 feet] in depth," reports Davidian. "These deep roots can actively take up water and minerals, even though most mineral ions are more abundant at the root surface."

A popular notion is that very old vines with deep roots express terroir better. "The claims often made regarding the importance of deep-rooted vines are based on the assumption that the roots are then better able to exploit the underlying geology," says Dawid Saayman. "In turn, this is considered by some to contribute certain minerals and thus impart a certain character to the wine. There is no scientific proof for this."

It's also worth mentioning here the existence of mycorrhizae. Many plant roots form an association with specific soil fungi where the fungi hitch a ride on the roots, gaining energy from the plant, while the plant root gains an enhanced absorptive area and ability to extract mineral nutrients from the soil. This is known as a "mutalistic symbiosis," because both partners benefit. Some people have claimed that mycorrhizae are important for terroir expression, but this is not clear from the scientific literature. Dawid Saayman points out, however, that grapevine mycorrhizae mainly assist in phosphorus uptake, an element that vines usually don't have problems getting enough of. "It is highly unlikely that mycorrhizal associations are prominent enough to contribute to a terroir effect," he concludes.

HOW SOILS HAVE THEIR EFFECT

Soils differ in their chemical and physical properties. According to Victoria Carey, a lecturer in viticulture at Stellenbosch University in South Africa, who specializes in terroir, the latter are more important for terroir effects. "The most convincing indications in the scientific literature are that the effect of soil type is through its physical properties, and more specifically, through the water supply to the grapevine," she suggests. This is a position with which Richard Smart agrees: he cites the pioneering work of French scientist Gérard Seguin, who conducted a survey of the properties of the soils in the Bordeaux region. Seguin couldn't find any reliable link between the chemical composition of the soil and wine character or quality, and maintained that it was the drainage properties of the soil affecting the availability of water that mattered. He concluded that it is "impossible to establish any correlation between the quality of the wine and the soil content of any nutritive element, be it potassium, phosphorus, or any other oligo-element." The verdict was that it was the physical properties of the soils, regulating the water supply to the vine, that were all-important in determining wine quality. The best terroirs were the ones where the soils are free-draining, with the water tables high enough to ensure a regular supply of water to the vine roots, which then recedes on *veraison* (when the berries change color) so that vegetative growth stops and the vine concentrates its energies on fruit-ripening.

The consensus among the viticulture experts I consulted seems to be that the chemical composition of the soil—that is, nutrient availability—is only important when there is excess nitrogen, leading to excess vigor, or when there is a serious deficiency. "Nutrition can be instrumental to the specific growth pattern of the vine and thus can cause a specific canopy architecture and therefore ripening pattern," says Dawid Saayman. "The plant performance therefore modifies the vineyard climate, creating a specific microclimate in the bunch zone, and in this way it can greatly determine the character of the wine," he says. "Overall, nutrient effects are minimal," adds Smart.

SOIL-CHEMISTRY EFFECTS

But before we give up on soil chemistry as an important factor in terroir, it's worth taking a look at recent research on the effects of mineral nutrition on plant physiology. I spoke to a number of researchers who are actively working on plant mineral nutrition, to see whether their work might shed some light on the mechanisms of terroir transduction. "I wouldn't be at all surprised if soil chemistry had an effect on the expression of genes that are involved in the production of the compounds that determine flavor," says Professor Brian Forde of Lancaster University (UK). "There is certainly plenty of evidence that plants are tuned to detect and respond to soil nutrients," he adds. "The balance between the nutrients [nitrogen, phosphorus, potassium, sulfur, and calcium, and even the micronutrients] is likely to be important and the plant stress responses elicited by limiting amounts of one nutrient would probably be subtly different from the stress responses elicited if another nutrient is limiting." Forde referred me to some publications showing that the levels of various plant metabolites were significantly altered under different nutrient regimes. At a more detailed level, it is now clear that patterns of gene expression in plants are altered by the presence and absence of various nutrients.

I spoke to Professor Malcolm Bennett and Dr. Martin Broadley of Nottingham University in the UK, who have studied the effects of phosphate deficiency on plant gene expression. Broadley feels that it won't be too long before we have a much better idea about the influence of soils on wine flavor. "There is a large amount of work underway to understand the molecular biology of grapes, and scientists are identifying genes that influence wine flavor," he explained. "As more grape molecular biology is known, the easier it will be to understand mechanisms of terroir on wine taste. When genes encoding for proteins that influence wine taste are identified, then the effects of different components of terroir (e.g. the availability of different minerals, soil pH, soil water content) on specific biochemical pathways can be identified and tested. This research may allow current agronomical practices to be improved to enable better-tasting grapes to be produced, or it might even allow varieties of grapes to be selected or bred more effectively."

Even if science leaves us with what currently looks like rather an emasculated version of terroir, I don't think that this necessarily diminishes the importance of this cherished concept. Winegrowers who use terroir as their guiding

Above The schist soils of the Douro Valley in northern Portugal are inhospitable to plants, but ideal for high-quality viticulture. The schist fractures vertically, allowing deep penetration by roots, and ensuring a slow but steady water supply, even in the very hot, dry summers that the region often experiences. They are also low in fertility, leading to low vine vigor, which is good for red-grape quality.

Above Modifying terroir: this is an experimental row in a vineyard (Ata Rangi) in Martinborough, New Zealand. The soils have been covered by crushed glass in a bid to reflect light up to the grapes. In a cool climate such as this, extra light can be good for red-grape quality.

philosophical framework and focus on the importance of the soil are responsible for a disproportionately large share of the world's most interesting wines. Even though it seems that there is unlikely to be a direct link between soils and wine flavor, by framing their activities within the context of a soil-focused worldview and trying to get a touch of "somewhereness" and minerality into their wines, winegrowers might be vastly increasing their chances of making interesting wine. And that's something the world needs more of.

As with so many other areas of viticulture and winemaking, it seems that more research is needed. While the current consensus is that it's the physical rather than the chemical properties of the soil that are all-important in separating good terroirs from bad ones, there's still the possibility that new data could emerge demonstrating that soil chemical properties can modulate grape characteristics significantly through altering gene expression.

What's the use of this research, besides satisfying our intellectual curiosity? In an ideal world we could all drink fine wines, such as first-

growth Bordeaux and *grand cru* Burgundy, every day. We can't because there isn't much made and they are expensive. It's all very well if you own some of Montrachet, or have vines on the hill at Hermitage, or preside over a Médoc first growth. You are blessed. Winegrowers with less auspicious terroirs, however, would understandably like to make better wine, and if scientists uncover precisely what it is that makes a great vineyard as opposed to an ordinary one, then they might be able to implement better management strategies that would improve their wines, or engineer vines that perform superbly on indifferent terroirs. After all, there's nothing magical about great terroirs; they are just patches of ground that naturally possess the conditions that encourage grapevines to produce grapes that, when handled correctly in the winery, can make outstanding wines. Wouldn't it be great if top-class wines were affordable for all? What a great scientific objective!

In the next chapter, I'll be taking a closer look at the topic of soils, seeing how they might be important in shaping wine quality. I'll also be addressing the issue of minerality, which is currently a hot topic in the world of wine.

3 Soils and vines

To the ancients, the idea that plants are formed from the soil would seem self-evident. The communion between the roots and the earth suggests that the composition of plants and, by extension, the fruit they produce, is determined largely by the composition of the soil. Modern science, however, paints a rather different picture. The fact that you can produce perfectly delicious-tasting fruit using hydroponics, where the plant is supplied with just water, light, and a solution of 16 trace elements, demonstrates that the intuitive notion that the soil makes the plant is quite mistaken.

Plants use light, water, air, and trace elements to synthesize everything they need. Think of them as remarkable chemical factories, taking very basic raw ingredients and synthesizing complexity from them. Moving to viticulture, specifically, grapes—the starting place for wine production—are made entirely through the process of photosynthesis and the subsequent biochemistry that builds and modifies the building blocks of sugars into complex biology. The soil? It's merely supplying water and dissolved mineral ions. These nutrient minerals are derived from the vineyard geology, but they are only needed in only tiny quantities by the vine, and have little, if any, aroma or taste.

But let's leave scientific fundamentalism aside for a second, and consider the experience of winegrowers worldwide, over many centuries. This experience testifies that soils are actually vital for wine. You can take a trip to a vineyard region such as Burgundy and discover that when vineyard boundaries are coupled with changes in underlying soil structure, two neighboring vineyards can differ significantly in the quality of wine produced.

That underlying geology impacts wine so strongly is undoubted. The cost differential between a *grand cru* Burgundy and a lowly generic Bourgogne, or even a respectable village-level wine, is such that there is a significant financial incentive for winegrowers to do all

they can to improve the quality of their wine. But even where great care is taken in the vineyard, yields are dropped, and the highest level of winemaking is practiced, there seems to be a quality ceiling that is imposed by the vineyard.

So we have a dilemma to solve: how is it that soils seem to be so important for wine quality, when science indicates that they are only playing a limited role in influencing the flavor of grapes?

ROCKS, STONES, AND SOILS

Those of us in the wine trade do tend to talk quite a lot about the geology of vineyards, so it's helpful to explore some of the science behind what we're talking about, lest we practice the folly of talking nonsense. So how does a rock become soil? Very slowly is the simple answer. But in geology, time is not of the essence. Slow is fast enough. The two processes that need to occur are weathering and the incorporation of organic material.

Consider a rock exposed to the atmosphere. It is going to get rained on from time to time, and it will experience temperature changes. Rain is very slightly acidic, because atmospheric carbon dioxide is dissolved in it. And, if any water should get into cracks or microfissures in the rock, and then freeze, because water expands by 9% on freezing, it will exert pressure on the rock. These forces may be small, but over extreme lengths of time, they will have an effect.

Organisms also play a role. Primary colonizers, such as bacteria, algae, and lichens, dramatically accelerate the process of weathering. Lichens are amazing. They are a fusion of two quite different organisms, a fungus and an alga. The fungus is the host: it breaks down the surface of the rock by acidic secretions, releasing minerals that can then be utilized by the alga for photosynthesis, which produces carbohydrates that are then shared by the fungus. Lichens are thus able to grow where very little else can, such as on the surface of a rock.

Lichens are resistant to drying out, and actually only grow when they are in the process of being wetted or are drying. Mosses are also effective colonizers, able to survive repeated cycles of wetting and drying, and needing very little mineral input to grow. As the first colonizers of rock become established, the basic elements of soil begin to be developed, albeit on a basic scale. Organic material from the decomposition of organisms begins to be incorporated, with small particles of rock produced by mechanical and chemical weathering, and a succession of plant species then begins.

So, this is the simplest sort of soil: parental bedrock undergoes weathering, and organic material is combined with the weathered particles. But soils can also move around a little, especially when water is involved. There's the action of glaciers, carving through the landscape and scattering material around. Rivers also move material, depositing it. For example, the course of rivers might change quite a lot, creating alluvial fans where the deposits show an irregular distribution on a valley floor. Wind can also shift soils around: "loess" is the term used for wind-blown soil.

Of course, there are also the more ancient geological movements that need to be considered, such as continental drift. Mountain ranges getting pushed up, sea beds rising (or sea levels falling), crumpling and folding of the Earth's surface, volcanic activity—all these will alter the way soils develop in complex ways. Look at any geological map of Western Europe, for example, and you'll likely see all sorts of colors, representing different types of geological activity and parental bedrock.

Generally speaking, soils themselves consist of particles of different sizes. There are rocks and stones, but the interesting parts are the three different particle sizes known as sand, silt, and clay. The size limits for these differ with classification type but, generally speaking, coarse sand is 2–0.2 mm, fine sand is 0.2–0.02 mm, silt is 20–2 μm (micrometer), and clay is less than 2 μm.

Add into this mix the organic material known as humus, and you have your soil. (Humus is organic, i.e. mostly plant, residues that have been mineralized and recycled by soil organisms. But as well as this, there is ongoing biological activity: soils are teeming with life.

The organic material in soil is vital for its structure and texture, both of which are important soil properties. The term "structure" describes the way that the different particles are bonded together into what are known as aggregates. This structure is a vital property, because if the aggregates are unstable, then the soil is easily compacted, and then this can lead to poor water exchange with the roots and poor gas exchange (air is an important component of healthy soils). The stability of soil aggregates can only be maintained if the organic material—humus—that helps hold them together, is replenished. Vine roots need at least 10% air-filled space in the soil, and preferably 15%, otherwise waterlogging is a problem. Dormant vine roots can withstand several weeks of waterlogging, but during the growing season even five days of waterlogging can start to kill them.

Crusting can occur at the surface of the soil if the aggregates there are broken down. This leads to small particles being released and then washed into the pores at the surface of the soil, making water penetration of the soil difficult. Structure is very important to soil because this will govern how much space there is in the soil for air, as well as its ability to drain and retain water. Air is important in soil for encouraging microbial life. Without structure, soils easily become compacted, and thus the flow of air is reduced. They are also more likely to become waterlogged. Soil microbes have been shown to be highly important in helping to form soil structure, in particular because of their role in aggregate formation and stability.

The role of microorganisms in the stability of soils has been studied in depth by Claire Chenu, one of the leading researchers on soil microbes and their role in soil quality, and her colleagues. A paper they published in 2011 reports on seven long-term experiments on different soil plots. They found a rapid increase in soil aggregate stability when practices that increase soil carbon (organic content) were implemented, such as no-till agriculture, the permanent cover of the soil by plants, and repeated compost additions. Follow-up laboratory experiments showed that this increase in stability was an indirect effect, occurring through the stimulation of organisms that decompose organic material, which helped aggregate the soil.

Another vital aspect of soils is known as "colloids." Clays and humus represent these, and they are able to act as chemical stores, soaking up and releasing mineral ions. Clays have an enormous surface area to volume ratio, which makes them very good at this.

There are also different sorts of clays, and the type of clay can be significant for viticulture. Clays are composed of layers of silica and aluminum, but the actual structure can vary. In simplified form, clays can be separated into two different groups, called kaolinite/illite and smectite/ montmorillonite. Kaolinite/illite clays don't swell much when they are wetted, and have a low "cation exchange" capacity (meaning, they have limited ability to hold nutrients and exchange them with plant roots). Smectite/montmorillonite clays swell more when wetted and then crack when they dry out. They are stickier and more plastic, and have a high cation exchange capacity.

A "clay" soil isn't composed solely of clay, but consists of different proportions of clay combined with other particles. The proportion of clay is important: a high clay content will allow the soil to store more mineral ions, and then exchange them with the roots, especially if the clay is smectite rather than kaolinite/illite. We'll return to clays later, because they are very interesting from a viticultural point of view.

WHAT DO VINES NEED FROM THE SOIL?

So what do plants—and more specifically, vines — require of soils to be able to function? Surprisingly little, it seems. The key factor, and this can be a deal-breaker, is, of course, water. Water availability and grape quality are inextricably linked, to the point that some have suggested that the main significance of vineyard soils is in the water supply to the vine. More on this later.

Aside from water, vines need mineral nutrients, which are divided into macro- and micronutrients, depending on how much is required by the vine. (Bear in mind that they are able to get carbon dioxide and oxygen from the air, and hydrogen from water: these are, of course, essential.)

Nitrogen is the key nutrient, needed in relatively large quantities. But while it is vital, too much nitrogen is actually a problem for grape quality, because it promotes vigor—the excessive growth of the vine and canopy. While the vine likes nitrogen, if it gets too much of it forgets to do what we really want it to do, which is to produce top-quality grapes. And as an aside, it is worth mentioning that plants generally have a reproductive strategy where they favor vegetative growth when things are going well (they are clearly in a good place), and sexual reproduction when things are tricky (time to get out of here).

The extra shading of the lush canopy prevents exposure of the base of the canes (the part of the woody stem that the leaves and grapes form along) to light, which compromises fruitfulness for the following season. This is where next year's buds are being formed, and the vine doesn't want to produce fruit inside a dense canopy (remember, the vine is designed as a woodland climber), and this is a way of making sure that the fruit is produced near the edge of the canopy, where birds are more likely to find it. Overly vigorous vines with dense canopies also have more problems with disease, and if the grapes are shaded too much they can struggle to ripen. It also increases the amount of vineyard labor trying to keep the canopies in check. Vines actually prefer less fertile soils, with lower nitrogen content, especially in cooler climates. In warmer climates, water stress can reduce vigor naturally, so in those cases more fertile soils might not be as bad. If soils are too infertile and contain very little nitrogen, then this can cause problems for fermentation, because nitrogen in a form that yeasts can use (known as YAN, for yeast-available nitrogen) is needed for fermentation to carry on successfully.

It is becoming apparent that white and red grape varieties might differ in their requirements for nitrogen. In an interesting study for his PhD thesis, Jean-Sebastien Reynaud looked at the influence of different soil types on the Doral grape variety grown in 13 different vineyards in Vaud, Switzerland, where the soil was reduced to the only variable. He found that soil nitrogen was a critical variable. The 13 vineyards could be grouped into two types of soil, and one type was much lower in nitrogen than the other. Low vine nitrogen resulted in higher soluble solids content in the wine, low malic acid, higher pH, and smaller berries. The lower vine and juice nitrogen resulted in the lowest-quality wines, with reduced aromatic complexity. However, Doral is a white grape variety, and it seems that many of the factors that are associated with low soil nitrogen might actually be correlated with higher red-wine quality.

Potassium is another important nutrient. A deficiency can reduce yields and fruit quality, as well as increasing the susceptibility of the vine to disease. Too much potassium can be a problem, though, especially for red wines. This is because most of the potassium is in the skins, and during red-wine fermentation will find its way into the wine. This can cause loss of acidity, because the potassium ions bind with tartaric acid, the main grape acid, and the combination precipitates as potassium bitartarate. Thus acidity is lost, and that is almost always a bad thing in winemaking.

Calcium and magnesium matter, too. Apparently, the ratio of calcium to magnesium determines the tightness of structure of a soil. More calcium makes the soil structure looser. This is a good thing, because the soil will have greater air penetration (encouraging the life of the soil) and will drain more easily.

VINE ROOTS IN ACTION

"The extent, health, and physical and chemical environment of the roots must be a major key to the best ripening and terroir expression."
John Gladstones, *Wine, Terroir and Climate Change (Wakefield Press, 2011).*

Vine roots respond to the conditions of the soils they are growing in. First of all, a large permanent framework of roots is established, followed by a network of finer lateral roots, and finally even finer tertiary roots, which are vital for uptake of water and nutrients. Nutrient uptake by the roots can be both passive and active. As the vine takes up water, it will usually take up whatever is dissolved in that water. But if it lacks specific nutrients, it can take them up actively, if they are present in the soil. There are some situations where the vine is fooled, though, by mineral ions that look quite similar, such that a deficiency of one can occur when there's an abundance of another. And, for example, in soils with a lot of limestone, chlorosis can be a problem. This is because in limestone-rich soils, vines find it very difficult to take up iron, which is needed for photosynthesis, as a vital component of

the green-colored chorophyll pigment. As a result, the leaves turn a yellow color and are diminished in their ability to carry out photosynthesis, the process of transforming light into energy.

A special layer of material, called the Casparian strip, surrounds the root endodermis (the layer of cells that circle the vascular tissue). This strip contains suberin, a waxy, rubbery material that is impermeable to water. Thus water and solutes entering the roots have to pass through plasmodesmata (pores in the cell walls) and therefore through the cytoplasm of root cells, before they can be transmitted to the rest of the plant. This gives the vine a level of control over what is taken up. The plasmodesmata are significant because they allow direct communication between the cytoplasm of adjacent plant cells, through the otherwise rigid cellulose cell walls.

How do vine roots take nutrients from the soil? One of the key concepts here is cation exchange. Roots are able to exchange hydrogen ions, which they pump out, for the cations attached to the negatively charged soil particles, such as clay and humus. Clay often carries a negative charge, whereas humus—decayed organic material—can carry both negative and positive charges, and so can hold both cations and anions. Cation exchange capacity (CEC) refers to the number of positive ions (such as calcium, magnesium, iron and the nitrogen-containing ammonium ion) that the soils can hold. When clay and humus have a negative electrical charge they are able to hold on to positively charged ions. Generally speaking, CEC correlates positively with soil fertility, because it determines how many plant nutrients the soils can hang on to. Soil pH also affects CEC: more acid soils (lower pH) have a lower CEC than more alkaline soils (high pH). One way to increase CEC is to increase the organic content of soils. This has the benefit of both increasing CEC, and thus fertility, and also increasing soil texture. Without organic material or clay, soils find it hard to retain nutrients. For example, an excessively sandy or gravely soil will allow mineral ions to be leached rapidly from the soil by rainfall.

So where do the mineral ions (nutrients) come from in the first place? It is not really from the bedrock, which would be the intuitive assumption on many. Some mineral ions might be bedrock-derived, but these would largely be in the subsoil. Low levels can come from rain, and some can come from the weathering of larger soil particles, such as stones and rocks. However, the bulk of soil nutrition will come from decaying organic material.

To get a better handle on this I spoke with Tim Carlisle, who has studied soil science, but also has a deep understanding of wine from his current employment as a wine merchant. "We need to then look at the microbial activity in soil," he states. "This affects the speed and ability of soil to break down organic matter into mineral ions that can be used by plant—and also aids the uptake of ions by plants. Because of this, no discussion about soil should exclude them, because whatever the terroir is, the level of microbial activity is an important and always overlooked element."

Carlisle points out that there are many factors that influence this microbial activity, but primarily water, food, and oxygen. "Oxygen is more available in a loose, uncompacted soil," he says. "A soil that is overly compacted has little oxygen and so little microbial activity—the same is true of a waterlogged soil—which is one reason why porous bedrock and/or slopes are important, not just because of vine stress but because the microflora and fauna don't get drowned." The term "microflora" refers to the bacteria and fungi in the soil. Their existence also governs the level of microfauna, which refers to soil organisms ranging from single-celled protista to small arthropods and insects, through to nematodes and earthworms.

"The food they need is organic matter. If you visited a conventional agriculture wheat field you'd find that there was very little organic matter in the soil, and as a result very little microbial activity, which is further diminished by crop spraying—hence the vicious cycle of needing to use tons of fertilizer."

During his studies of soil science, Carlisle looked at the effect of fungicides, herbicides, and insecticides on the soil microbes. He did this two different ways. First, he took microbes from the soil, grew them in culture, and then studied the effects of dilute agrochemicals on their growth. "What I found in this was that fungicides over herbicides and insecticides kill off not just

Nutrients needed for plant growth

Macronutrients	Micronutrients
Nitrogen	Boron
Phosphorus	Chlorine
Potassium	Iron
Calcium	Manganese
Sulfur	Zinc
Magnesium	Copper
Silicon	Molybdenum
	Nickel
	Selenium
	Sodium

fungi (which includes yeasts and molds)," says Carlisle, "but also a high proportion of bacteria, and actinomycetes (I didn't do anything with algae), but also that herbicides and insecticides also killed off a proportion of all types of microbe, and restricted growth of others."

He also studied the overall microbial activity in the soil. "What this showed up was that untreated soil was healthier than anything with any kind of treatment, including one that was sprayed with fertilizer," reports Carlisle. "If you think about it, minerals are essentially the excretion of microbes. Too much excretion to soil will poison them and so spraying with fertilizer actually caused a check in microbial activity—it continued but at a lesser rate." He adds, "The thing that was by far the most interesting from a viticulture perspective was that one of the samples was sprayed with copper sulfate, which is permitted in organic viticulture. This sample was the one in which microbial activity was reduced by the most."

In 2010, Elda Vitanovi and colleagues published a study looking at the level of copper in vineyard soils in Croatia, which routinely receive about 3–5 kg per hectare (6.5–11 lb per 2.5 acres) of copper fungicides each year. They found that copper levels were significantly higher in vineyard soils than in other soils they looked at, and that 17 out of 20 vineyards studied officially fell into the category of copper contamination. A French

1 Perhaps one of the most famous vineyard soils of all: Romanée Conti *grand cru* in Vosne Romanée, Burgundy. The soils have limestone and clay.

2 This is the famous soil of the Médoc in Bordeaux: this is a vineyard of a first-growth property in Pauillac. The alluvial pebbles offer excellent drainage, with the water supply to the vine slowing down at *veraison*, allowing the grapes to ripen perfectly.

3 Granite soils from the hill of Hermitage, in the northern Rhône, France—the home of Syrah.

4 The famous galets (large river pebbles) of Châteauneuf-du-Pape in France's southern Rhône.

5 Slate soils in the Mosel, Germany. This is the Erdener Prälat vineyard.

6 Schist soils from Portugal's Douro Valley. This is Quinta de Roeda.

research group reports that copper levels in vineyard soils typically range from 100–1,500 mg per kg (2.2 lb). They reckon that across the world, some 8 million hectares (nearly 20 million acres) are affected by this copper pollution, with half of them in Europe and one-eighth of them in France. It's clearly an issue that needs addressing. "It's not a problem for the quality of wine directly," says Claire Chenu, "because copper is little assimilated by vines. But it does lead to a decrease in the diversity and abundance of soil microorganisms, and a decrease in diversity and abundance of soil fauna (including earthworms)." Correspondingly, says Chenu, there will be a decrease in the soil properties and functions that these organisms help to promote, such as aggregate stability.

Franz Weninger, a biodynamic winegrower in Austria and Hungary, points out that some of the elevated copper levels in vineyard soils date back to the time when up to 30 kg (66 lb) of copper per hectare (per 2.5 acres) was applied annually. The current situation in Austria is that the maximum amount allowed is 6 kg/ha (13 lb per 2.5 acres) and organic/biodynamic winegrowers can use only 3 kg/ha (6.5 lb per 2.5 acres). "We manage with tunnel sprayers to come down to 1–1.5 kg/ha [2–3 lb per 2.5 acres]," he states. "This is about 0.15 g per square meter [.005 ounces per square

7 Schist soils from Priorat in Catalonia, Spain.

8 The characteristic sandy granitic soils of the Dão region in northern Portugal.

9 The distinctive sandy soils of the Terras do Sado region in central Portugal.

10 The alberese soils in Chianti Classico, Tuscany, Italy, consisting of compact clay and limestone.

11 The famous Gimblett Gravels, a special subdistrict of the Hawke's Bay region in New Zealand with free-draining, alluvially deposited gravel soils.

12 Terroir expert Pedro Parra.

yard]." Weninger points out that while copper has a very toxic effect on soil life in the lab, copper in the field is about six times less aggressive to soil life. "I hope that in the long-term organic and biodynamic winemakers can get rid of copper," he says, "but believe me, glyphosate and the systemic fungicides are far more of a problem for the soil life."

RHIZOSPHERE INTERACTIONS

The root environment is relatively neglected because it is out of sight, hidden underground. But this space, known as the rhizosphere, is nevertheless home to some complex physical and biological interactions among the roots and soil-dwelling microorganisms.

The roots have been described as "rhizosphere ambassadors," facilitating communication between the plant and other organisms in the soil. Plants secrete a wide range of compounds into the soil from their roots. These include amino acids, organic acids, vitamins, nucleosides, ions, gases, and proteins. Such secretions are important in maintaining the rhizosphere biology. This is illustrated by an interesting experiment in a Swedish forest, in which the photosynthates produced in the leaves were prevented from being transported to the roots by the stripping away of a ring of bark all around the diameter of the trunk. This removes the phloem layer of vascular cells, just under the bark, and would normally kill the tree. (As an aside, cork trees are different: fortunately for the wine industry, the bark of a cork tree can be removed without damaging the tree, and then will grow back again.) It resulted in a massively reduced amount of life in the soil within just a few days. The trees were clearly supplying the microbes and other soil organisms with nutrients.

Why do plants take part in this energetically costly charity work? Because eventually it benefits them. Some of the chemicals released help with nutrient uptake. This can be an indirect effect, through reducing the soil pH (which can help with making some nutrients more available), or promoting the life of soil microbes that take part in nutrient recycling. Some root exudates can act as "chelators" of metal ions that are needed as nutrients. For example, iron is frequently abundant in the soil but is often present in a form that is hard for plants to access. Various root extracts can chelate (form chemical complexes) with iron that is then made available for uptake by the roots. Phosphorus is also often present in an unavailable form, but organic acids released by the roots can help make it more available.

Root secretions can also play a role in signaling: there's a chemical conversation going on between plants and other plants, and plants and microbes, for example. This can be both positive and negative. For example, isoflavone secretion by soybean roots attracts both a mutualistic fungus (good) and a fungal pathogen (bad). There's a phenomenon known as allelopathy, which is when one plant secretes compounds that discourage the growth of other plants nearby. Sometimes they even discourage the growth of the same species nearby (autotoxicity). Plants under attack by a pathogen can signal through their roots to their neighbors of the same species and thereby induce resistance in the plants that have not yet been attacked. This induced herbivore resistance can be direct (turning on defense strategies), or indirect (sending out volatile chemical signals that then attract predators of the insect pest, and airborne SOS signal). An example of this would be in lima bean, where attack by spider mites causes root exudates from the plant under attack that cause neighboring plants to release a volatile signal that attracts predatory mites that feed on spider mites.

One well-known example of a beneficial rhizosphere interaction is that of the mycorrhizal association. Some 80% of terrestrial plants have these mutually beneficial symbiotic relationships in which certain fungi become associated intimately with the roots, and help with nutrient uptake (through massively increasing the surface area of root to soil contact), in exchange for lipids and carbohydrates from the plants. In legumes, a very special relationship occurs in which signals from the roots begin a series of interactions with nitrogen-fixing bacteria, which then cause the root morphology to change so that these bacteria become enclosed in specialized structures called nodules. The plant root secretions cause the bacteria to arrive and release chemicals that then modify the way the root cells grow. It's a brilliant arrangement: they get a home and nutrients; the plant gets a nitrogen supply.

While the importance of rhizosphere interactions in grapevine growth hasn't been studied in detail, the take-home message here is that what happens underground is complex and important. The biology occurring in the soils has been a neglected area of viticulture research, and there is more happening here than we are aware of. How we manage vineyard soils matters. There could be unintended consequences to the use of fungicides and herbicides. As discussed earlier, adding copper to the soil clearly isn't a great idea, but copper-containing fungicides are still widely used in organic and biodynamic viticulture.

ROOTS AND HORMONE SIGNALING

Vine roots are important to grape quality in a number of ways. Significantly, root growth causes hormonal signals to be sent to the portion of the plant above ground, and these act like instructions to tell the vine to modify its growth. Perhaps the best summary of what is happening here comes from the Western Australian plant biologist John Gladstones, in his book *Wine, Terroir and Climate Change* (Wakefield Press, 2011), in which he brings together the existing literature and adds some theories of his own.

Gladstones points out that gibberellins (one of the major plant hormones) promote shoot internode growth through cell extension (the nodes are the parts of the stem between two buds), and these gibberellins are formed in the region of cell division behind root tips. They also promote the formation of tendrils rather than fruit clusters in the newly forming lateral buds. Presumably, if conditions are good for root growth, then the vine is likely to favor vegetative growth, and gibberellins are sending up these instructions from the roots.

Cytokinins (another major plant hormone), produced in the root tips, promote new node and leaf formation, the branching of shoots, the development of existing fruit clusters, and the fruitfulness of newly forming lateral buds. Gladstones observes that warm spring soils promote cytokinin dominance; cool soils gibberellins. Warm spring soils are therefore a good thing for grape production.

Abscisic acid (ABA) is probably the most interesting of the plant hormones for wine quality, though. Roots signal to the vine that water stress is coming using ABA, so that the leaves can respond appropriately, closing their gas-exchange pores. ABA also acts in tension with another group of plant hormones, auxins, in controlling ripening. There's a kind of tug-of-war going on here. The auxins, produced by developing seed, slow down berry development, prolonging the pre-*veraison* phase. ABA is pulling for berries to develop faster. In conditions of water stress, ABA is signaling to the berries to develop faster. A massive transfer of ABA to fruit coincides with *veraison*, possibly because of declining berry auxins at this stage. Therefore, root moisture stress is contributing to berry ripeness. Gladstones concludes that ABA is the primary hormone imported into the grape clusters that both triggers and continues to stimulate ripening.

A recent study by Dr. Hendrik Poorter and colleagues in Germany used magnetic resonance imaging (MRI) to look at root growth in potted plants. They were interested in finding out how big the pots have to be for experimental work, looking at a wide range of different pot-grown plants. The results emphasized how important root signaling is for the growth of the portion of the plant above ground. The pot size restricted growth for a wide range of species, and doubling the size of the pot increased growth by 43%. The MRI results indicated that the plants were

using their roots to "sense" the size of the pot, and then signaling this to the rest of the plant.

Gladstones' assessment of the literature on vine physiology agrees with this idea. Roots are signaling by means of hormones to the portion of the plant above ground. The root structure is determined by soil conditions; deep soils with ample water supply prolong the phase of root development, and signal the above-ground plant to keep growing; shallower soils with limited water or a texture that obstructs root growth reduce vegetative vigor. Low vigor is usually best for wine quality, especially for red wines.

WATER RELATIONS AND WINE QUALITY

One of the most important properties of vineyard soils is how they control water supply to the vine. Vine roots are designed to be particularly effective in taking up water and nutrients from the soil, because owing to their climbing habit in the wild, vines are establishing themselves in soils already colonized by other plants. Supply vines with too much water and they will grow big, lush canopies and not put much effort into grape production.

Researcher Gérard Seguin carried out a famous study on the soils of the Médoc (Bordeaux) in the 1980s. He found that many of the best vineyard sites had poor levels of soil nutrients, but that this was compensated for by deep root systems. These top sites were frequently acidic gravels, and showed magnesium deficiency due to high potassium levels, as well as low levels of nitrogen. But aside from this, it was hard to correlate potential wine quality and soil nutrient levels. Seguin stated that, "It is impossible to establish any correlation between quality of wine and the soil content of any nutritive element." He adds that if there were to be such a correlation, you could give yourself a good chance of making great wine simply with the assistance of chemical additives to the soil.

Seguin's major conclusion was that the vital way in which the soils affected grape quality was through regulating water availability during the vegetative cycle of the vine. Moderate water deficit has been shown to reduce shoot growth (vigor), berry weight and yield, and increases berry anthocyanin and tannin content—ideal for high-quality red-wine production. Vine water status is dependent on soil and climate characteristics, and soil influences vine water status through its water-holding capacity. Seguin showed that in the Bordeaux vineyards, which are not irrigated, berry size is decreased and total phenolics are increased when vines face water deficits, resulting in higher grape-quality potential but lower yields.

A more recent study by Bordeaux-based terroir researcher Kees van Leeuwin, who has worked with Seguin in the past, examined this in more detail. Van Leeuwin and colleagues looked at 32 vintages in Bordeaux from 1974–2005 and found a correlation between vine-water deficit stress index and vintage-quality ratings. "The quality of red Bordeaux wine can be better correlated to the dryness of the vintage than to the sum of active temperatures," he concludes. In none of these vintages included in the study did the quality suffer because of excessive water deficit. Vintage quality ratings don't correlate well with the average growing season temperatures, which is surprising. However, the story isn't a simple one; some vintages, such as 1982, were excellent with no real water deficit.

In Bordeaux, the vineyards that rarely experience deficit are either planted with Merlot, or with white varieties. Closer planting, growing higher canopies, and using rootstocks that only partially use soil-water reserves (such as Riparia Gloire de Montpellier) are vineyard interventions that can be used on sites that usually lack natural water deficit.

But water deficit is not always associated with higher wine quality. As part of his PhD studies, Jean-Sebastien Reynaud looked at the effect of soil water-holding capacity on wine quality in 23 different Vaud vineyards, in Switzerland. These were planted with Gamaret, a Swiss variety that's a cross between Gamay and Reichensteiner, and is popular in Switzerland because of its resistance to botrytis. Many studies of water deficit have involved irrigated vineyards, but Reynaud looked at unirrigated sites over three vintages, 2007–2009. Water stress can have both positive and negative effects on vines. In drying soils, plant roots synthesize abscisic acid, which signals to the portion of the plant above ground and encourages grape ripening, and the partitioning of carbon resources to the fruit rather than the canopy. But it also causes the pores in the leaf (known as the

stomata) to close, reducing carbon assimilation. If the stress is severe then leaves are lost. So stress tends to increase soluble solids in the grape (generally good for wine quality, especially in reds) until it reaches a certain point and soluble solids decrease when the stress is too much.

Reynaud found that water deficit improved wine color, but there was no clear relation between water stress and the sensory attributes of the experimental wines he made. "In the Vaud conditions, vine-water stress was not the major parameter responsible for differences in wine quality," he concluded.

Perhaps Seguin's viewpoint that water-holding capacity is the key soil factor for wine quality—widely accepted around the wine industry, particularly for red wines—has overshadowed the potential importance of soil chemistry. The soils he studied were of a certain type: what Chilean terroir expert Pedro Parra describes as "geomorphic" soils—those where the soil isn't formed directly from the underlying bedrock. Parra suggests that "geological" soils, formed from weathered bedrock, may be quite different in this respect than geomorphic soils. Examples of gemorphic soils would include alluvial soils such as those of Bordeaux, or wind-blown loess. In contrast, the soils of Burgundy and the Rhône would be geological.

THE IMPORTANCE OF CLAY

Earlier I touched on the different sorts of clays, and the implications of these for viticulture. Clays are made of the smallest of all soil particles, and because of their structure present an extremely large surface to volume ratio. They can hold water and nutrient ions very effectively. "I'd always thought that clay was an unlikely viticultural soil," says John Atkinson MW, who has a vineyard in the UK and is the author of an interesting research paper on terroir in Burgundy's Côte de Nuits. "It just seemed so charged with minerals and water, and therefore overly invigorating in temperate climates. I was therefore surprised to read in Denis Dubordieu's two-volume work on oenology that Petrus' smectite clay soil hydrically stressed the property's Merlot vines.

"I read around the subject," continues

Atkinson, "and came across a report on Geelong's [Victoria, Australia] soils. The article makes the important distinctions between soil water capacity, availability, and extractability." Atkinson points out that clay soils hold on to their water, and their density makes rooting difficult. "Clays might appear humid, but extractability can limit availability," he adds. "This tied in with a paper I'd read by Kees van Leeuwen, which reported hydric stress occurring more rapidly in vines grown on a clay-based Bordeaux soil than they did on a more typical Medoc gravel soil."

Apparently, Petrus, the famous Pomerol estate, has more smectite clay than any other estate in Pomerol. "Smectite is a volcanic mineral that increases the internal surface areas of clays, and exaggerates their shrink–swell properties," says Atkinson. "It is part of the montmorillonite group of clays. Expansion of the clays closes the pores in the soil, and makes conditions too anaerobic for root growth while impeding existing root function. Consequently, they impose a limit on the extractability of water by limiting root development. Conversely, as smectite/momtmorillonite clays dry out, they shrink and crack, allowing rootlets to populate the developing capillaries. By contrast, kaolinite and illite clays expand and contract very little.

"The relevance of all this is that one could model a viticultural regime based upon montmorillonite clay, warm summers, and irregular rainfall in which vines are nearly always under stress; even heavy rainfall wouldn't penetrate the soil, because the clays expand and seal at their surfaces. Petrus would be the paradigm of this sort of interaction."

The main focus of Atkinson's work on terroir has been in Burgundy. The famed vineyards of the Côte d'Or are on the side of a rift valley and, because of this, soil can vary over just a few yards. Atkinson says that the best vineyards are those where there is limestone bedrock and plenty of active calcium carbonate, which helps create an open soil structure. The flocculated clays in these soils have the physical drainage properties of sand, but can hold nutrients. The porous limestone soils can help with drainage because of their effect on the structure of the soil, but can also act as a water reserve. The clay content in the soil is able to extract the water from the limestone reserve, and make it available in limited quantities through the growing season. The red grand crus typically have smectite (swelling clays), while those with kaolinite clays are better-suited to white wines.

Atkinson cites the work of Frank Wittendal, who used a statistical technique called principal components analysis (PCA) to look in detail at 2,816 specific climats in Burgundy. The climat is the Burgundian terroir unit, and represents a single patch of presumably homogeneous terroir. This may be a vineyard in its entirety, or may just be a part of a larger vineyard. For example, Clos de Vougeot is a 50-hectare (123-acre) grand cru vineyard with 16 climats, while Corton, another grand cru vineyard, has 24 climats.

Wittendal described each climat in terms of 14 soil-description variables and four landscape/climatic variables, which were then fed into the PCA. The great significance of this work is that by using statistics, he was able to show that factors that would be assumed to be important—such as altitude, aspect, parent rock, and gradient—weren't significant in separating the different hierarchical levels in Burgundy's vineyards.

The PCA work shows that in terms of vineyard classification, the soil properties are really what matter, and these have precedence over altitude and slope. There is some evidence that east-facing vineyards are favored, but this is because facing east correlates well with interesting soil types, rather than the angle of the sun's illumination.

Wittendal's analysis was able to split the different climats up into three quite separate groups. Group 1 consists of colluvial soils. These are formed by the accumulation of fallen, eroded soils, which are retained on shallow slopes. Colluvium is made of fragmented rocks. Group 2 is noncolluvial compact limestone. This is able to retain water and possibly sequester it from deeper sources. Group 3 is alluvial soils, at the lower end of the altitude and slope indicators. There is just one grand cru climat with an alluvial soil, a Bâtard-Montrachet. Root growth in limestone and colluvial soils is quite different. In the limestone soils, roots go mainly down, searching for water. In colluvial soils, made of fragmented rocks, roots travel in all directions.

If just the red Grand Cru vineyards are included, then 80–90% of the clayey limestone soils with active carbonate are eliminated, and the majority soil type is colluvium. That is, the *grand cru* vineyards had a high proportion of gravely, colluvial hill-wash in their subsoils. Atkinson suggests that the significance of this is that the best vineyards put the vines into the optimal deficit zone, and it's the hydrology properties of the vineyard that determine its quality potential. This is similar to the findings of Seguin and others in Bordeaux, where the alluvial sand and gravel soils, coupled with a retreating water table in the summer, put the vines into deficit. In the best Bordeaux vineyards clay lenses (a lens-shaped block of clay) run through the soil, helping to buffer water availability, keeping the vines supplied with just enough water but letting them experience mild deficit.

MINERALITY IN WINE: CAN WE TASTE THE SOIL?

This brings us on to the interesting question of minerality. "Minerality is the perception of the rocks in the soil, by the palate," claims soil scientist Lydia Bourguignon who, with her husband, Claude, forms one of the most respected and influential consultancies on the role of vineyard soils. "We hold geosensorial tastings," she continues. "It is a sort of taste training in which you actually touch, even taste, different rocks to then be able to find the same sensations on the palate with the wine. For example, touching granite gives a cold impression, while limestone seems warm."

What is minerality as applied to wine? Can we define it as a tasting term? Do we even mean the same thing when we each talk about "mineral" wines? And can we try to link minerality to chemicals present in the wine? These are all complicated questions, and for many people the whole concept of minerality is so muddled and confusing they'd rather avoid it all together.

"Minerality" is a really useful descriptor; many of us use it frequently in our tasting notes. Yet it's also a term that means different things to different people. I know what I mean when I encounter some characteristic in a wine that makes me think "mineral," but I can't be sure that when others use the term they are referring to the same thing.

And minerality may well be a sort of syndrome, like some medical conditions, when different underlying factors cause symptoms that look quite similar. I also suspect that it's sometimes used as a way of praising a deliciously complex wine, in the same way that "long" is often thrown into a tasting note when people really like a wine but have run out of more concrete descriptors.

HOW EXPERTS USE THE TERM

What do different tasters mean by "minerality"? "'Mineral'" is interesting," says well-known English wine writer and contributing editor for *Decanter* magazine Stephen Spurrier, "but it did not exist as a wine-tasting term until the mid-1980s. During most of my time in Paris I don't think I ever used the word," he recalls. "I think this was because most French vineyards were overproducing, chaptalizing and doing all those things, which means minerality, which has to come from the soil and nothing else, was not looked for and not present." Spurrier does use the term quite a lot in his tasting notes. "I probably associate minerality with stoniness, but then stones are hard and minerality is generally 'lifted.' As a taste, it just comes into my mind and I very often find myself writing 'nice minerality on the finish.' I suppose it is easier to describe what it is not—that is, it is not fruit, nor acidity, nor tannins, nor oak, nor richness, nor fleshiness. It is not really a texture, either, for texture is in the middle of the palate and minerality is at the end. I think it is just there, a sort of lifted and lively stoniness that brings a sense of grip and also a sense of depth, but it is neither grippy (which is tannin) nor deep (which is fruit)." Spurrier adds, "No wonder we are all a little confused."

I asked another well-known wine writer, Jancis Robinson, about her use of "minerality" in tasting notes. "I am very wary of using it because I know how sloppily it has been applied," she replied. "In general I try to use it as little as possible and be a bit more specific. 'Wet stones' is a favorite tasting note of mine but there is sometimes something 'slatey' about some Mosel wine and 'schistous' (grainier) about some St-Chinian and Catalan wines from both sides of the Pyrenees, I think. But when wine definitely doesn't smell of anything fruity, vegetal, or animal, I might use it."

Noted French critic Michel Bettane describes minerality as "a fashionable word never employed in the 1970s and 1980s," agreeing here with Spurrier that it is a fairly recent invention as a tasting-note term. Bettane says that for many tasters it is a "politically correct" term, describing a "wine nonmanipulated by the winemaker and from organic viticulture, aromas and fruit being the sign of manipulation or the lack of expression of origin." He continues, "For me the only no-nonsense use is to describe a wine marked by salty and mineral undertones balancing (and not hiding) the fruit, more often a white wine rich in calcium and magnesium as many mineral waters are. For a red wine I have no idea, with the exception of some metallic undertones (iron in Château Latour or copper in Nuits-St-Georges les Pruliers.)"

Jordi Ballester, a researcher from the Centre des Sciences du Gout in Dijon, in Burgundy, in France, has been studying the use of the term minerality by lexical analysis. He has noted that it's a term being increasingly used, but without a clear definition. He and his colleagues compared the ratings of 34 wine experts and a trained panel, looking at the sorts of tastes and smells that people describe as "mineral." They found widespread differences among the tasters, and while there was some agreement about definitions when people were quizzed, the use of the term differed in practice. Interestingly, though, some of the subgroups used the term in similar ways, suggesting that there is a cultural basis for its use. The Sancerre winegrowers, for example, agreed on what they thought minerality was. So maybe minerality is in part a local concept?

TAKING MINERALITY LITERALLY

It is interesting that Bettane brings up mineral water. Frequently, you will see on the side of a bottle of mineral water the different levels of mineral ions present in that water. These depend on the source of the water, and different mineral waters tend to have subtly different flavors if they are compared side by side. Presumably, these differences in flavor, subtle as they are, are down to the mineral composition, although there are plenty who will contend that most mineral ions don't taste of anything. This is the first definition of minerality we are going to explore:

the literal one. In this defintion, minerality in a wine context is a result of mineral ions, present in the soil, which find their way into grapes, and then affect the flavor of the wine.

If a soil has mineral ions in it, the roots will take these up passively, along with the water that the roots sequester. However, roots are also able to take up mineral ions selectively in certain circumstances. Some people object that the typical differences in mineral-ion concentration, such as those found in different mineral waters, would not be noticeable against the backdrop of the other flavors present in wine. However, the levels present in wine seem to be considerably higher.

"Minerals can be detected while tasting a wine," says Olivier Humbrecht, one of Alsace's leading winegrowers, famous for his advocacy of biodynamic viticulture. "It is the fraction on the palate that makes the wine taste more saline or salty. High acids or high tannins do not mean that the wine has lots of minerality. High salt contents make the acidity more 'savory' and therefore less aggressive. Good minerality makes one salivate and want to have another sip or glass or bottle."

There is a very interesting pair of scientific studies in support of this notion of minerality in wine. Back in 2000, a plant researcher from Germany, Andreas Peuke, grew Riesling vines in pots containing three different soils from Franconian vineyards: loess, muschelkalk (seashell lime), and keuper. He collected sap from the vines and analyzed its chemical composition, and found differences among the different soil types. In muschelkalk soil, carbon, nitrogen, and calcium were present in the greatest concentrations. Sulfur, boron, magnesium, sodium, and potassium were greatest in keuper, and the concentrations in loess soil were intermediate. Aqueous extraction of the soils resulted in a two-fold greater concentration of total solutes in keuper extract compared with muschelkalk, and more than threefold than in loess.

A few years ago, Californian winegrower Randall Grahm carried out some interesting experiments. In his quest to try to understand minerality better, he actually put some rocks into tanks of wine. However, in this case the rocks were in a wine environment at low pH, and are therefore likely to release more "minerals" than if they had been in the ground. And the rocks also had the side effect of raising the pH of the wine, which can change the flavor. "Our experiments were incredibly simplistic and gross in comparison to the very subtle chemistry that occurs in mineral extraction in real soils," Grahm recalls. "We simply took interesting rocks, washed them very well, smashed them up and immersed them in a barrel of wine for a certain period, until we felt that the wine had extracted some interesting flavors and we were able to discern significant differences between the various types."

Grahm saw major changes in the texture and mouthfeel of the wine, as well as dramatic differences in aromatics, length, and persistence of flavor. "In every case, low doses of minerals added far more complexity and greater persistence on the palate. It is my personal belief that wines richer in minerals just present way differently." He adds that, "They seem to have a certain sort of nucleus or density around their center; they are gathered, focused, cohered the way a laser coheres light. It is a different kind of density relative to tannic density, somehow deeper in the wine than the tannins." But what about the interface between roots and soil? Do roots modify the pH of their immediate environment, causing the release of minerals? If water itself were enough to release significant quantities of minerals, then the vineyard soils wouldn't be stable over centuries, as they have been observed to be. However, the action of bacteria and other soil organisms can lead to decomposition of rocks, releasing their mineral constituents into the soil.

REDUCTION AS MINERALITY

Some people use "mineral" to describe aromas of wine; it's something they get on the nose. In this case, it could be that tasters are ascribing minerality to what is in reality the presence of certain volatile sulfur compounds in the wine, also known by the term "reduction." In its most raw state, reduction is caused by hydrogen sulfide, and smells like rotten eggs and sewers. This is rare in a finished wine, and wouldn't be classed as mineral. Far more common is the presence of complex sulfides and mercaptans (also known as thiols). These sulfur compounds, like hydrogen sulfide, are produced largely by yeasts during

fermentation, and are described in detail in Chapter 15. Their expression depends on their concentration and the context of the wine, but in some cases they can give a flinty or struck-match aroma that can be quite "mineral."

There is good reason to suggest that flintiness in white wines is a result of some low-level reduction. Great white Burgundies frequently show a little of this good reduction: a matchstick element to the nose is complexing. Some New World Chardonnay producers are now beginning to work out how to achieve this through winemaking practices. There's also a link here with terroir: some sites naturally have nutrient deficiencies that can stress the yeasts a little and cause them to produce more of these volatile sulfur compounds.

Scientists tend to prefer this second definition of minerality because they dispute the first—the more literal definition in which minerals in the soil end up flavoring the wine. This tends to make some believers in minerality act a little defensively when discussing the subject. "I fully understand that when I use the term it may have no scientific validity," says Jasper Morris, wine merchant and Burgundy expert. "I have been told by enough geologists that you shouldn't call Chablis flinty, because flint is not soluble in water and therefore it can't have a taste. But it is an image." Morris illustrates his point by comparing two Burgundy vintages. "I would use two vintages of white Burgundy to illustrate this point: 2007 and 2008. The acidity is higher in 2008, and when I taste those wines I feel that they are acid and not especially mineral. In 2007, the acidity is a little lower, but I find the wines distinctly mineral. By that, I mean that they have a fresh, zingy zest to them that in my mind this puts up the single word 'mineral'. In 2006 the wines were fat, rich, and round, and some of them have enough acidity to provide balance, but you don't feel 'mineral' when you taste the wines." He continues, "Also in Burgundy, in terms of the hillsides you expect minerality in those soils which have more active limestone, and you expect less in those soils which are more clay-based. For example, with any vineyard in any village which includes the words *charmes* in the name, you rarely get the concept of minerality." Morris is frustrated that the scientists seem to be criticizing minerality without offering an alternative explanation: "Scientific fundamentalists are denying that we get it right rather than offering any certainty in the other direction."

Paul Draper of Ridge, in California, is a winegrower who believes in minerality from the soil, even in the face of questioning from scientists. "Though I am well aware of what soil scientists say about minerals or other elements in the soil and the impossibility of them traveling through the vine and into the wine, the roots deep enough into those minerals are affected and the wine shows that effect," he states. "I think of minerality as a wet-stone quality in a wine. Our subsoils at Monte Bello are limestone and at times are at the surface or a meter below. In other places our backhoe pits find them several meters down. Perhaps 70% of our vine roots are deep in the limestone. I have seen minerality in some shales as well so I don't think the effect is necessarily limited to limestone. We see the most marked minerality (crushed rock, perhaps flint are other descriptors) is in our more eroded blocks where the limestone is closer to the surface. In the youngest blocks, where pits have shown considerable limestone, we don't see the minerality as yet but expect to when the roots are deeper."

Minerality remains enigmatic. As we begin to understand more about it, the picture seems multifactorial, with different mechanistic underpinnings for what wine tasters describe as 'mineral' in their tasting notes.

SPECIFIC DEFICIENCIES OF MINERALS

Famed vineyard soil scientist Claude Bourguignon points out that all the compounds produced in a living system are produced by enzymes, and many of these enzymes have metallic cofactors, such as magnesium in chlorophyll, or manganese and magnesium cofactors in enzymes that build up monoterpene molecules. Micronutrients such as these metallic cofactors all come from the soil. Bourguignon says that we still don't know all the enzyme pathways involved in the synthesis of aroma compounds or their precursors. His assertion is that hydroponically produced fruits have no taste, because without soil microbes, there are no micronutrients and thus no aroma, because enzymes need these micronutrients as cofactors.

Bourguignon may be onto something here, but in a slightly different way. Rather than great terroirs offering a surfeit of micronutrients, it may be that some offer slight but meaningful deficiencies, such that vine growth is not overly impeded, but the deficiency either causes a reduced enzyme activity in producing specific flavor compounds (or precursors), or results in musts that are slightly deficient in certain yeast nutrients, resulting in complexing characters in the wine—perhaps through a little reduction early on that then resolves into interesting complexity in the final wine. While plants wouldn't get far without chlorophyll, some of the flavor-active compounds or flavor precursors are nonessential secondary metabolites, so it is expected that there might be some variation in their levels in grape berries. This is all speculation, however. It would be great to have some more scientific studies on this topic.

THE TASTE OF TERROIR

"The vine may be cultivated advantageously in a great variety of soils … The sandy soil will, in general, produce a delicate wine, the calcareous soil a spiritous wine, the decomposed granite a brisk wine."
A Treatise on the Culture of the Vine, and the Art of Making Wine, James Busby, 1825

One of the frustrations of the topic of terroir is that while many people have described the physical characters of vineyard sites, there have been very few serious attempts to link soils with wine flavor. Many of us in the wine trade who taste regularly will have anecdotal experience of the way that wines from different soils taste different. For example, there's a dramatic difference between Touriga Nacional produced from the schist of the Douro Valley in Portugal and wines from the same variety grown in granite soils (such as the granite at Vale Meão in the Douro Superior, or with a cooler climate, the sandy granitic soils of the Dão region). A similar comparison can be made in the northern Rhône where Syrah is grown on both schist and granite. The granite soils produce lighter-bodied wines with more freshness and aromatic purity, tending to more floral notes. Wines from schist tend to be richer and more structured. I'm sure that

Mosel experts could probably blind-pick Riesling wines made from the different sorts of slate that form the majority of the vineyard soils here.

"If you taste the wines of Europe and start looking at the soils, these patterns start to emerge, so that one kind of soil will give you a consistent stylistic imprint on the wine," says Mike Weersing of Pyramid Valley Vineyards in New Zealand's north Canterbury region. He cites the differences between clay and limestone soils. Clay soils give flesh to Pinot Noir while limestone gives structure. Limestone alone can often be a little too thin, lean, and mean, so he's looking for both in the topsoil. "What do clay and limestone give Pinot?" asks Weersing. "They give it flesh, and they give it the girdling of the flesh, which gives it structure and length and energy. You can see this in Burgundy. If you look at the typical Burgundian slope, you have hard limestone at the top, then soft limestone, clayey limestone and then clay. And on the same slope, if you compare the wines, the wine from the hard limestone and soft limestone at the top (where there is not much clay) is very tight and vibrant, but doesn't have much roundness or richness. Once you move down into what the French call the kidney of the slope, where the clay and limestone mix, you get depth and richness, but you also have structure. Then, when you move off this into the clay Bourgogne soils, you can have wines that are pleasantly fat, but they don't age well."

"The French call this clay-lime complex *argilo-calcaire*," says Weersing, "and their simple observation (made over something like 1,200 years) is that Pinot Noir and Chardonnay, to be at their best, require the mix—neither only clay nor only lime at the surface, but a marriage of the two." He says that the bedrock isn't so important, although fractured limestone or chalk are perfect, because they release excess moisture, but retain sufficient water for the vine to drink evenly, without irrigation, throughout the growing season.

Terroir expert Pedro Parra thinks it's important to understand soils because this can then inform winemaking decisions. "What is important is to try to understand in which area we are located, if we want to relate terroir to wine," says Parra. "If we are on volcanic soils and we don't work well, then we end up with bitter tannins," he suggests, referring to red wines. "It is easy to get bitter tannins on volcanic soils. If you are on granite you risk going to dry tannins. If you are on schist, depending on the type of schist, you can go to dry tannins or bitter tannins. On calcareous soils, the risk is green tannins."

Parra states: "When you identify the lithology (the parental rock) you can understand the winemaking you need to do to make good wines." But, he adds, it is not always so simple. "In Apalta [Chile, one of the areas he consults in], there is a mix of volcanic and granitic soils, so you could have dry or bitter tannins."

CONCLUDING REMARKS

It is clear from this chapter that soils appear to be hugely important in shaping wine style. Climate determines whether you can grow a grape variety successfully in a particular location, but it is the soil type that seems vital in determining wine style and quality, given a suitable climate. This is currently an area of great interest to winegrowers, but there's still a lack of good research linking soils to vine growth, and onward to berry composition and then further on, completing the loop with the implications for winemaking and the sensory characteristics of the resulting wine. The wine world needs to focus a little more on the complex and interesting biology that takes place out of sight, under the ground, because this is critical to wine quality.

4 Precision viticulture

Welcome to the world of satellite imaging, yield monitors, global positioning systems, multispectral digital video, and state-of-the-art software. These may all jar with the traditional images of elderly *vignerons* carefully tending their vines, but they're some of the technological cornerstones of one the current hot topics in the wine world; precision viticulture ("PV" for short).

PV is a branch of precision agriculture, a relatively recent development in farming that was first proposed in the early 1990s. Its basis is that nature is uneven. Traditionally, farmers have ignored the natural variation in their fields, and have applied the same treatment across the whole area. To be fair, they have hadn't had access to tools that would let them manage their land in a more sophisticated manner. But over the last couple of decades affordable technologies have been developed that allow farmers to make accurate maps of this variation and then manage their fields accordingly. For example, some parts might need more fertilizer, others less. Or the soil properties might vary in such a way that certain parts will need irrigating while others won't. It sounds simple, and conceptually it is. The tricky part lies in the practicalities of making useful maps of this variation, and then devising ways of treating the various parts differently.

The need for dealing with natural vineyard variation is highlighted by the fact that, in vineyards, yields typically vary by eight- to tenfold from the least to most productive parts. Perhaps more detrimental to wine quality is the variation in grape ripening that this implies. If winegrowers can manage their vineyards to encourage homogeneous phenolic and sugar ripeness levels, then quality soars, but it only takes a small block of poor-quality grapes to bring down the overall quality level of a vineyard quite a lot. Despite these clear opportunities for PV, it's only very recently that precision techniques have been applied to vineyards. The pioneers of PV have been viticulturalists in Australia and California who have adopted similar principles, but in practical terms have taken somewhat different approaches.

In California, the emphasis has largely been on what is termed "remote sensing." This is where data collection is through gathering aerial images, either by means of satellites or, more commonly, flying an airplane over the vineyards. The pictures that are taken are "multispectral," meaning that they consist of overlapping images taken at different wavelengths, each giving different sorts of information.

In Australia, where many vineyards are mechanized, data collection is typically through the use of yield monitors, although multispectral remote sensing has been used quite extensively here, too. These are devices that are attached to conventional mechanical harvesters and which, when used with global positioning systems (GPS), enable yield maps to be generated. Each approach has its own benefits. One of the key developers of PV in Australia is Rob Bramley of the CSIRO. He explained why he has concentrated on using yield monitoring. "First you have to harvest the crop, so if you are doing this mechanically you may as well yield monitor, simply because it is far better to have some data than not," he explained. "Secondly, remotely sensed information will only ever give you relative information. It is also a fact that an image without ground-truthing is useless. However, I do think that remote sensing is useful, and we use it for mid-season monitoring and as a data layer to help understand the yield map."

Indeed, Bramley used remote sensing in a case study of PV at a vineyard owned by Vasse Felix, in the Margaret River region of western Australia. Crucially, this study showed that there is an economic benefit to be gained. In a demonstration of how straightforward it can be to use PV, Bramley and his colleagues took remotely sensed images of the vineyard and used these as a basis for targeted sampling of fruit just before harvest.

Yield map of a single vineyard
(4.3 hectares, Padthaway, S. Australia)

Highest-yielding bit:
lower quality grapes
– do something to
control vigour or
separate these
grapes out

Yield (t/ha)
■ < 8
■ 8 - 11
■ 11 - 14
■ 14 - 17
■ 17 - 20
■ 20 - 23
■ 23 - 26

Below An aerial image of vineyards in Stellenbosch, South Africa. It is an infrared image indicating the degree of vegetative growth in the vineyards. Picture courtesy of Warwick Estate

Above Variation in yield and elevation in a 4.3 hectare block of Shiraz grapes in Padthaway, South Australia. Interestingly, an aerial remotely sensed image of this same block acquired at *veraison* indentified the same characteristic zone of a high-yielding, high-vigor area at the center of the block. Winemaker assessment of the fruit classified the grapes from this central area as "C" grade, while the rest of the block was "B" grade. This suggests that differential harvest, and perhaps also differential management of this block, would be beneficial. Illustration courtesy of Dr Rob Bramley

On the basis of the results from this sampling, the grapes were then harvested into two separate bins based on their location. The fruit from one block was good enough for the Vasse Felix Cabernet Sauvignon and the rest went to the cheaper Vasse Felix "Classic Dry Red." Previously, all the grapes from this vineyard would have gone to the cheaper wine. Even taking into account the increased cost of the PV work and targeted harvesting, the result was a clear economic benefit.

Bramley has also done studies in vineyards in Coonawarra and Mildura. In these research projects, yield maps were constructed using harvesting machines fitted with yield monitors and a GPS device. He demonstrated that for both sites the patterns of variation were relatively consistent from year to year. This is good news for growers because it makes it possible for them to adopt targeted management strategies based on variation maps constructed over just a few vintages, cutting down costs significantly. These maps are constructed by means of some fairly complicated software known as geographical information systems (GIS), but the maps also have a predictive value. For example, consistently low-yielding areas can be expected to be low-yielding the following year, and so on.

In the USA, the leader of the PV field is Lee Johnson of NASA Ames Research Center in California. Johnson began using remote sensing of vineyards in 1993, well before PV was envisaged, as a means to track the spread of phylloxera. For this first project, NASA Ames and the Robert Mondavi Winery collaborated between them, and since this initial work Johnson has been continuing to work with Mondavi to develop PV technologies. Daniel Bosch, vineyard manager at Mondavi, has been involved with the project for a while now, and he has used the information gained from remote sensing to modify his approach in the vineyard. "We've changed our drip-irrigation system on the basis of this," he reports. "In some vineyards we now have two hoses." This allows separate blocks to be irrigated differently, depending on the vigor of the vines. Bosch also uses PV information to cultivate and harvest vineyard sections at different times. But perhaps the most elegant use of the technology is to mount a GPS device on a tractor, couple it to a computer with GIS software, and then selectively remove cover crops from areas of the vineyard that are low-vigor, leaving them in the high-vigor regions. This is done automatically as the tractor is driven through the vineyard, and can be targeted precisely, row by row.

Along with other California vineyards, Mondavi mostly use remote sensing as the basis of its PV. The early work was done with airplanes flying over the vineyard, but this is quite expensive at around $20,000 a flight, and to make it cost-effective growers need to band together to split costs. Now there are also commercial satellite operators offering a similar service, and it is likely that in the future growers will be able to pay for and access these sorts of images directly via the Internet.

But how do these colorful, complicated-looking pictures relate to what's happening on the ground? They are obviously indirect measurements of what is happening to the grapes, concentrating on the visible part of the grapevine, the leaf area. However, Johnson and his colleagues have spent a lot of effort correlating the imaging data with direct measurements of grape quality and yield. They've developed a complex model that has worked out the relationship between the remotely sensed images and physicalfactors affecting wine quality, and have identified vineyard canopy density, which they term leaf area index (LAI) as a key variable of interest. There's perhaps one advantage with this remote sensing over yield monitoring. "Remotely sensed data allows the grower to be proactive in terms of altering management practices in conditions observed during the growing season," Johnson explains. "Remote sensing can be used, for instance, to subdivide fields for harvest based on observed vine vigor, make irrigation decisions, support pruning decisions, and look for areas that may be more susceptible to mold." However, it needs to be borne in mind that the best time for acquiring these images is at *veraison*, relatively late in the vine-growth cycle, leaving only a short time for any proactive intervention. Is it expensive? "Commercial costs I've seen for remote sensing vary widely, but tend to run around $5 per acre." says Johnson.

The chief beneficiaries of PV are likely to be owners of larger vineyards, such as those found in the New World. It's hard to imagine a large take-up in an area such as Burgundy, where vineyards are already split into tiny plots and there have been many hundreds of years of experience of the performance of different sites. But one of PV's trump cards is that it allows a winegrower to gain an almost instant understanding of his or her vineyards that would have previously taken decades of patient observation to achieve. And even then, according to Mondavi's Bosch, the perceptions of experienced vineyard managers of where their vineyards vary are sometimes rather different to those revealed by aerial images or yield maps.

Aside from Australia and California, the main early adopter of PV is South Africa. In collaboration with consultant Dr. Phil Freese, who was involved in the early remote sensing work in California in his capacity as vineyard manager for Mondavi, Warwick, Thelema, and Rustenberg in the Stellenbosch region of the country are using PV.

Left A mechanical harvester in action in an Australian vineyard with a GPS device and yield monitor attached.
Picture courtesy of Dr Rob Bramley

Studying the terroir in Marlborough, New Zealand

As a part of a major research program on New Zealand Sauvignon Blanc, a subregional trial was carried out looking at five different sites in the Marlborough wine region, which is famous for its Sauvignons. The five sites were: Villa Maria (Fairhall); Squire (Rapaura); Booker (Brancott); Oyster Bay (Western Wairau); and Seaview (Awatere). Of these, the first four are on the Wairau Plain, with the fifth found in the cooler Awatere Valley.

Rob Agnew of the Marlborough Wine Research Center was one of the key players in this trial. "In 2004 when the program started we established five subregional vineyards. The basis for this was that Pernod Ricard (Montana, now Brancott) had made wines from three of these subregions (Seaview in Awatere, Brancott Valley, and Squire Vineyard in Rapaura), and had identified different flavors and aromas from these different subregions. The question was, was this a reflection of management, maturity of the grapes, or actual subregional differences?"

The Wairau Plain consists of deposits laid down by the River Wairau over thousands of years. The river is now confined to the northern side of the plain. The result of this river deposition is that there are large differences in soil depth and texture across the plain. The eastern side of the plain tends to have deeper silt-type soils, while the southern valleys have older soils. One of the main differences is in the depth of the topsoil before the river gravels are encountered. There is variation on quite a small scale, so that even within one row the soil type might be changing in texture, leading to very different wines.

These different sites were monitored through the different stages of development, and then the researchers picked grapes from each just before the commercial harvest, at 21.5 Brix. In one of the vineyards, the Squire Vineyard in Rapaura, a master's student, Tim Mills, monitored the soils extensively.

"His master's degree was done at the same time as a couple of other students, and one of those was looking at the above-ground effects," says Pernod Ricard's Andrew Naylor. "The researchers went across the vineyard block and did extensive trunk circumference measurements to map the vigor of the vines. Tim's holes identified that the vine vigor was closely correlated to soil depth—the depth of gravel and the amount of stones in the soil."

At this time, the only available way of measuring vigor was to look at trunk circumferences. A more recent technique, looking at the number of leaves on the vines, is now available, and is called plant-cell density (PCD). "Tim identified areas where he'd like to dig holes based on the vine vigor identified. Since then we have had a number of PCD images from that block, and they line up quite well with soil type as well. Tim was able to talk about water-holding capacity of the soil in relation to vine vigor," says Naylor.

Mills also did some nutrient work. "One of the things that was identified in that block is that the southern end was quite a deep silt. You could turn the irrigation off there for the season and it would probably survive. You can probably turn your fertilizer spreader off there as well," says Naylor.

The good thing about this sort of vineyard mapping is that it is pretty much independent of seasonal conditions. "Rob Bramley has been working with Mike Trought and has shown that if the data related to the soil type, then they have spatial stability," reports Naylor. "Sure, the level will go up and down with the season, but the relative differences will tend to stay the same. A lot of this variability is quite stable."

Agnew is convinced that soils are key to the flavor of Sauvignon Blanc, even if they haven't been implicated to any great extent so far. "I think in 10–15 years' time we might look back and think there is a lot more to the flavors and aromas that are coming from the soil," he says. "The Squire vineyard is a large vineyard and we have studied it intensely. Parts of it are a deep silt, 2 meters [6.5 feet] deep, and other parts have stones right at the surface. You can have this variation within 20–30 meters [65–100 feet] in the vineyard, because the river has wandered around across the Wairau Plain in the last few thousand years. Other vineyards, such as the Booker [Brancott] vineyard, have much older soils overlying clay."

The real outlier of the five is the Seaview vineyard in the Awatere. "This is much closer to the sea than the other four sites. It has a lower maximum temperature and a higher minimum temperature," reports Agnew. "This area traditionally grew processed crops, such as peas, and farmers considered it to be an early area, and this is probably because there is less frost risk, so they could plant earlier. With grapes, they are harvested a week to 10 days later, but this is not due to later budburst. It is the warmest site in winter, but the coolest site over the growing season."

He adds, "The Villa Maria vineyard is the coolest in the winter, but the warmest in the summer." Wind is also a factor: "The Awatere has 45–50% more wind run than the central Wairau Plain. At the airport, on the central Wairau Plain, it is calm 50% of the time; at Cape Campbell on the coast it is calm 3% of the time."

"We have compared the climate of the five sites. All the sites are cooler than the Blenheim weather station, which is the main regional one. The Brancott site (Booker block, right at the top of the Brancott Valley) is one I cannot get my head around with regard to accumulated heat and fruit maturity. It is on a much older soil type. It is a cooler site. The difference in GDD [growing degree days, a measure of cumulative temperature during the growing season] is only 150, out of a total of 1,370 GDD for Blenheim, so around all these five sites we are looking at a variation of 7%. The difference in the five sites is not great. For example, Central Otago only gets 75% of the GDDs of Blenheim. The differences in the subregional climate do not explain the large differences in flavor profile of the wines. For me, the climate is not telling the whole story. The Brancott vineyard is on a heavier soil type overlying clay, and here every single year the four cane vines have ripened a massive crop. Average Marlborough Sauvignon Blanc yield is about 11 or 12 tonnes per hectare (13 tons per 2.5 acres), but this site has yielded 10 kg [22 lb] per vine, which equates to 23 tonnes per hectare (25 tons per 2.5 acres), with some of the lowest accumulated GDDs."

Wines are made from each of the vineyard trial sites, and then sensory studies are performed on these. "We have eight four-cane plots in the vineyard. We get 200 kg [440 lb] of fruit so it is all amalgamated," says Agnew. "It is de-stemmed, crushed, pressed, cold-settled. Then it is decanted into stainless-steel kegs, each holding about 18 liters [4.75 gallons], and we have four kegs for each vineyard, with the fourth being a topper for the others. This results in three replicate wines in the winery from the field. These are then used for sensory work."

Agnew points out that while winemakers can often spot subregionality in the wines they make from fruit coming into the winery—after all, this was one of the motivations behind the subregional project—the sensory panels working on this study can only reliably

separate Awatere wines from those of the Wairau plain. "The sensory work showed that the Awatere is distinctively different from the others," says Agnew. "The other four vineyard sites show very little difference. This is a consistent story from season to season."

Interestingly, and somewhat counterintuitively, yield doesn't seem to have a huge effect on the wines that are made. "The crop-load effect is interesting," says Agnew. "There are no significant differences in the flavor and aroma characteristics of the wines despite the twofold differences in crop load."

"We have got a lot out of the subregional story, taking it through to wine and looking at the differences," he adds, "but personally I feel we need to look more at the soils, and the influence these have on the flavors and aromas. I don't think we understand this yet."

Several other South African estates are also trialing PV, along with others in Chile and New Zealand. Mike Ratcliffe of Warwick Estate explained some of the ways he was using PV to improve quality. "We've done some extreme viticulture in a couple of Cabernet blocks. At *veraison* [when the berries change color] individual bunches were marked as late-ripening [green berries], early-ripening [fully colored berries], and mixed bunches. The thinking behind this is that a few weeks after *veraison* is completed, the differing ripening stage indications are lost as all the berries were fully colored and identical. The different-colored markings were harvested individually at optimum ripeness. We then did a correlation of this information with our NDVI [normalized difference vegetation index, a remotely sensed measure of vine vigor] and found a statistically significant correlation." Ratcliffe goes on to explain that, although this work is labor-intensive, it reduces variation and improves quality significantly.

Another measure used by Warwick is leaf-water potential (LWP). This is a direct measure of how much water stress the vine is under. The viticultural goal is to stress the vine just enough, but not too much, at the right time, which enhances fruit quality significantly. "We cross-reference the LWP with the NDVI to detect trends and patterns in the vineyards," says Ratcliffe. "This is perhaps our most useful and practically implementable technology used at Warwick."

But if this all sounds rather high-tech, there are still lessons to be learned by winegrowers working in a more traditional, hands-on way with just small vineyards to worry about. "In principle, PV is for everybody," says the CSIRO's Rob Bramley, "simply because all vineyards are variable." Mondavi's Bosch suggests a low-tech implementation that would be in the reach of anyone: "Leaf-fall patterns are very similar to the patterns seen by remote sensing." There is therefore a window of about two weeks where a winemaker could go through their vineyard during leaf fall and map homogeneous blocks. The following growing season it would be easy enough to test ripeness levels a couple of weeks before harvest, and if these agreed with the maps then you could harvest the blocks in the appropriate order. PV in action!

Chilean terroir expert Pedro Parra recalls that when he worked for Don Melchor from 2001–2004, the only way they could map terroirs was by using NDVI to look at the canopy, and then using GPS to localize where there were big differences so he could dig pits and have a look at the soil. "When you work with NDVI on slopes, the shadow of the plants gives noise," says Parra, "and GPS is not accurate. To improve this we now work with DGPS, which gives just 20 cm [about 8 in] of error, and since 2003 we have had a new tool called electroconductivity, or 'resistivity'. This gives two readings at different depths, and you need to interpret this conductivity." Parra says that this sort of machine costs $40,000, and that in Chile three companies are doing this sort of mapping. "To map 1 hectare [2.5 acres] here costs about $60," says Parra, "but in France there are two companies and it costs 500 Euros a hectare [$650 per 2.5 acres] to do. This terroir analysis is not romantic, but people are starting to understand you can't vinify two terroirs in the same way."

While it's still relatively early days, it's clear that as the technology develops, PV in its various manifestations will become more widespread. Potentially, any data relevant to viticulture could be included in PV if the collection is cost-effective enough and it yields information that will result in targeted management that results in quality improvements. For example, if they were cheap enough, battery-operated data collectors could be spread around the vineyard in a targeted sampling effort, transmitting relevant information back to assist in management decisions. Another area of research concerns the use of ground-penetrating radar, a technique able to reveal subsurface characteristics of vineyards without the need for digging pits.

It seems that the tide has turned. While the work in the vineyard used to be seen as a dull, relatively unskilled prelude to the work of the celebrity winemaker, now it's the vineyard managers who are getting the attention. Will flying viticulturalists, brandishing their laptops and GPS devices, become as popular as flying winemakers were just a few years ago? I wouldn't be surprised.

5 Phylloxera and ungrafted wines

The global wine industry rests on a single, fairly fragile pillar. This pillar was put in place rather hastily just more than 120 years ago, when wine, as we know it, was almost extinguished forever by a tiny aphid, phylloxera. This tiny insect, with its complicated life cycle, caused a vine plague of epic proportions, which in the space of a few decades brought the world wine industry to its knees. Salvation came from an unlikely source—the same as the origin of the problem—native American grapevines. The pillar in question is the resistance of American vine rootstocks to phylloxera, and fortunately, it has proven amazingly durable.

The story of phylloxera is a remarkable one, recently retold for a general audience in Christy Campbell's award-winning book, and previously the subject of a more scholarly tome by George Ordish[1]. The focus of this chapter, however, is not the phylloxera story itself, but rather its legacy. The consequence of the phylloxera pandemic is that now almost all *Vitis vinifera* vines are not planted on their own roots. Instead, they are grafted onto resistant American rootstock. After a brief review of how phylloxera was beaten, I'll take a look at the nature of the grafting process, and discuss how grafting influences grape quality. Is grafting "natural?" Are wines from ungrafted vines different? And were the great wines made from pre-phylloxera vines better?

HOW PHYLLOXERA WAS BEATEN

The second half of the nineteenth century was not a particularly good time to be a winegrower in France. By 1850, a large proportion of the population had their livelihoods tied to the vine and its produce. According to Campbell, one-sixth of France's state revenues were generated by wine and one-third of the workforce derived a living from it. But within the space of a few decades the French wine industry was shaken to its roots by two natural plagues. One of these, the fungal disease oidium, was remedied fairly quickly by chemical means, but the other, phylloxera, caused devastation on an enormous scale and threatened to eradicate viticulture as we know it from the globe.

The grape varieties responsible for almost all the wines that we drink belong to a single species, *Vitis vinifera*. The hundreds of varieties in common use all belong to the same species, and the resulting lack of genetic diversity renders them vulnerable to attack by pest and disease.

The USA has a wide variety of native wild vines. Surprisingly, there is no historical record of these ever being cultivated or wine ever being made from them, despite their abundance. Eventually, European emigrants to the USA, anxious to produce wines like those of their homelands, turned to them out of necessity when their attempts to grow *Vitis vinifera* varieties had failed. The *vinifera* vines simply lacked innate protection against the mildew and phylloxera that were endemic to the USA, and which the American vine species had learned to exist alongside through co-evolution. But while the American varieties grew well and proved robust in the face of disease, their big flaw was that they produced fairly unpalatable wine, with a strong "foxy" taste.

In the Victorian craze for importing novel, exotic plant varieties, aided by the steamship, American vines began to be imported into France in the early decades of the nineteenth century, and by 1830 a couple of dozen varieties were growing in French nurseries. Nurserymen and *vignerons* keen to experiment disseminated these imported vines throughout Europe's wine regions. It turned out that they were carrying a deadly cargo of insect (phylloxera) and fungus (oidium).

1 Campbell C. *Phylloxera: how wine was saved for the world*. Harper Collins, London 2004
Ordish G. *The great wine blight*. Sidgwick & Jackson, London 1987

Above Phylloxera galls on the leaves of a rootstock variety, Ripestris du Lot. American vines can live with phylloxera because they have co-evolved with it. *Vitis vinifera*, however, cannot.

European vineyards, propagated vegetatively and thus vulnerable in their genetic uniformity, were hit first by the fungal disease oidium (powdery mildew). *Oidium tuckerii* (now known as *Uncinula necator*) was first discovered growing on a greenhouse vine in Kent, in southeastern England, in 1845. This previously unknown fungus almost certainly arrived from the USA on a botanical sample, and soon spread throughout Europe. Within a few years wine production in France had collapsed to less than one-third of its previous levels.

The response of *vignerons* was one of panic, but this time, science provided an answer fairly rapidly. French scientists discovered that dusting the vines with elemental sulfur provided protection against the fungus, and by 1858 the problem of oidium was in retreat. Some time following this, while France was in the throes of dealing with phylloxera, a second plant disease from America, downy mildew (caused by *Plasmopora viticola*) hit the vines in the 1880s. *Plasmopora* is usually referred to as a fungus, but recent taxonomic changes have it classified as an oomycete, a type of protist (a one-celled organism), known as a water mold. In this case, another chemical solution was found by serendipity, Bordeaux mixture, a combination of lime, copper sulfate, and water.

After the powdery mildew episode, French winegrowing then entered a short-lived golden age.

With the advent of the railroads, wine production became separated from consumption, such that northern industrial populations could have their thirst slaked by cheap southern wine. Between 1850 and 1880 the average consumption rose from 13 to 20 gallons per head per annum. But this prosperity boom for the *vignerons* proved to be brief.

Over a couple of decades a catastrophe unfolded in the vineyards of France that ended up putting global wine production in jeopardy, and ruined the livelihoods of enormous numbers of *vignerons*. The culprit was the root-feeding aphid, *Phylloxera vastatrix* (known today by scientists as *Daktylosphaira vitifoliae*). Monsieur Borty, a wine merchant in the town of Roquemaure in the Gard, had the misfortune of going down in history as inadvertently precipitating this natural disaster. In 1862, he received a case of vines from New York, which he planted in his small vineyard. Two years later, vines in the surrounding area mysteriously began to wither and die. The malady spread. By 1868, the whole of the southern Rhône was infected and phylloxera had become endemic throughout the Languedoc. Within the space of a decade, it had moved through France and gone global, reaching Portugal's Douro by 1872, Spain and Germany two years later, Australia in 1875, and Italy by 1879. The last French wine region to be affected was Champagne, in 1890.

The response to phylloxera was, again, one of panic. At first, the outward signs—the leaves of affected vines turning yellow and falling prematurely, followed by death—gave no clue as to the specific nature of the disease. When affected vines were extracted from the ground, their roots were found to be rotten and crumbling, but with no obvious pest present. It was down to Professor Jules-Émile Planchon, a botanist appointed by the government commission established to investigate the outbreak, to identify the culprit. He dug up roots of a healthy vine and observed clumps of minute wingless insects happily gorging themselves. Clever parasites don't kill their hosts, and in their homeland of the USA phylloxera and grapevines had co-evolved such that they existed alongside each other. However, this cozy relationship hadn't developed with *Vitis vinifera,* and the arrangement was a hopelessly imbalanced one: the parasitism by phylloxera was too harsh, resulting in eventual death to the vine.

Phylloxera has a complicated life cycle, which was only worked out after the plague had broken, and is beyond the scope of this piece. Like many aphids, phylloxera is parthenogenetic, which means it doesn't need sex to reproduce. The root-growing form settles on suitable roots and punctures them with its mouthparts. It injects saliva, and this causes the root cells to grow into a structure known as a gall, which increases the supply of nutrients and provides some protection. Then it lays eggs, which hatch, move along the roots, climb the trunk, before being blown into the air. If they find a suitable feeding site they'll carry on proliferating, and populations can increase rapidly. In some regions there are asexual forms of phylloxera that can develop on the leaves. These also induce gall formation, and in this case the protective structure that encloses the feeding phylloxera extends below the leaf and opens to the upper leaf surface to allow crawlers out.

There are three potential mechanisms for vine damage by phylloxera: removal of photosynthates; physical disruption of the roots; and secondary fungal infections of damaged roots. It is unlikely that the severe, usually fatal vine damage is caused by the first mechanism. Much more likely is that the vine damage occurs through secondary infections by fungal pathogens.

Faced with ruin of the entire wine industry, the French government initially offered a cash prize of 30,000 francs for a remedy, later upgraded to 300,000 francs as the scale of the disaster increased. Initial attempts to halt the spread of phylloxera involved uprooting and burning affected vines, but these failed. Remedy after potential remedy was suggested and tried, ranging from the almost plausible to the downright bizarre. Some of the proposed treatments sounded surprisingly modern, such as biological control (finding a natural enemy of phylloxera) and planting more attractive hosts for the aphid between the grapevines. They didn't work, though.

Vignerons growing vines on the plains discovered that, where it was possible, seasonal flooding was reasonably effective in killing this subterranean pest. Vineyards in sandy soils were also observed to be immune, but transplanting vine production to sandy areas was hardly a practical solution. The chemists searched rather randomly for a magic bullet. At the time, it was known that carbon bisulfide was an effective insecticide, despite the fact that it was highly volatile and explosive when mixed with air. But how could growers get it to where it was needed, deep in the soil? An elaborate array of devices for injecting this rather dubious chemical into the root zones were invented, and before long this was seen as the best bet for heading off the pest. The problem with carbon bisulfide was threefold: it was a relatively expensive treatment, and beyond the reach of many *vignerons*; it would need repeating, year after year; and it wasn't terribly effective, mitigating the worst symptoms of the malaise without being curative. As one commentator at the time noted, keeping the vines going by constant treatment "inevitably reduced them to the condition a sick man kept alive by constantly stuffing him with medicine."

Another controversial solution was proposed. American vine species, which harbored the invader in the first place, had natural protection against phylloxera. Where they had been planted they thrived, while all around them was a scene of devastation. Yes, the wine they produced tasted fairly bad, but better odd wine than no wine at all, argued the faction, which became known as the *américainistes*. Tastings of wines made using the American vines were set up, but the depressing conclusion was that these foxy wines really weren't good enough.

Then, someone had a brilliant idea. In 1869, a Monsieur Gaston Bazille suggested grafting *vinifera* varieties onto American rootstock. With the benefit of hindsight, this seems a brilliant solution, but at the time there were many unknowns. Chief among these were first, how long would the graft last? Second, would the rootstock impart some of that American vine "foxiness" to the wine, altering the qualities of the grapes of the *vinifera* by this unnatural union of scion and stock? And third, how resistant would the various rootstocks be? It was already known that some American vines were more resistant to phylloxera than others. A period of intensive testing then ensued.

One of the first documented applications of this grafting was in 1874, when Henri Bouschet displayed an Aramon (*Vitis vinifera*) vine grafted onto American rootstock at a Congrès Viticole in Montpellier. In a short time it became clear that, somewhat counterintuitively, the wine made from *vinifera* vines grafted onto American rootstocks retained the full character of the *vinifera* variety while benefiting from the phylloxera resistance of the American roots. It completed a rather neat circle. The plague had come from America, but so had the salvation.

Still, though, the battle between the two factions of the chemists (who advocated carbon bisulfide) and the *américainistes* (who supported the use of resistant American vines or their rootstocks) continued. The turning point that shifted things in favor of the latter school was the International Phylloxera Congress in Bordeaux in 1881, where discussions reached the conclusion that *vinifera* varieties grafted onto American rootstock maintained their characteristics.

The idea spread. The technique of grafting was easy enough to learn that just about anyone could attempt it. The supply of American vines was more of a problem, because many wine regions began to prohibit their importation in order to prevent the remaining phylloxera-free vineyards from succumbing to the pest. However, not everyone was won over by this radical idea of grafting. Many resisted replanting, with the attendant three-year loss of production, and clung tenaciously to their chemical treatments. In the end, good sense prevailed and the grafters won the day. The lengthy work of replanting France's vineyards began. It was not a straightforward process, and some of the grander estates, reluctant to pull out their vines kept them going with insecticide treatments as long as they could. The choice of appropriate grafting material was also complicated by the fact that it took a while to find American vine species that were well-adapted to the chalky soils that predominated in some of France's key regions.

One thing that replanting did facilitate is a change in the viticultural landscape, with some sites being abandoned and others being replanted with new varieties. In addition, the choice of rootstock, with its effect on graft parameters, such as vigor, became an additional variable, and a new item in the viticultural toolbox.

THE GRAFTING PROCESS: HOW IT WORKS

Grafting is an ancient practice that takes advantage of the fact that plants don't have an immune system, and thus different varieties or even species can unite and grow as one. It is recorded as far back as the 323 B.C.E. by Theoprastos for apple trees, and by Cato in *De Agri Cultura* for grapes in the second century B.C.E.

It is easy to see why, when the proposal to graft all *vinifera* varieties onto American rootstocks, growers would have objected. Why fuse their precious vines with foreign vines that only ever make bad wine? There is also the widespread emotional—and dare I say "spiritual"—significance accorded to the soil that the vine is rooted in. Traditional notions of terroir have frequently encompassed a belief that the vine roots are extracting something special from the vineyards they are growing in, and that this involves the extraction of some component that confers a sense of place on the wine. In effect, the concept is widely (if unconsciously) held that there is a conversation going on between the earth and the vine, mediated by the roots, which then determines the character of the wine. According to the viewpoint of biodynamics, this sort of interaction involves unquantified (and probably unquantifiable) life forces. The fact that vines are all grafted must seem to the biodynamic practitioner as some sort of admission of defeat, for if the vine were truly in harmony with the life forces of the vineyard it would be able to repel the damaging attentions of hungry phylloxera, and war would be averted.

Above Newly grafted vines growing in a nursery.

Physically, grafting involves the union of two plants, the scion and the stock, with the critical marriage being of a thin layer of cells known as the vascular cambium. This runs as a cylindrical layer of cells close to the outer circumference of the stem or trunk of the plant, producing phloem on the exterior side and xylem on the interior. Phloem is the vascular tissue that conducts sugars and nutrients, and xylem is the tissue that conducts water and dissolved minerals, as well as providing structural support.

Grafting takes advantage of the wound-healing response in plants. When an incision is made in a plant stem, the response is the growth of de-differentiated cells (cells that have reverted to an unspecialized state). Thus when branches or trunks that have already initiated secondary growth are damaged, they respond by producing callus, a tissue of undifferentiated cells produced by proliferation near the wound site. This same callus production happens when the scion and the stock are joined together. This callus provides the tissue through which vascular continuity is restored, and signals from the vascular tissue of both scion and stock presumably influence the undifferentiated callus cells to become cambial cells. From these, the new vascular tissue, which is seamless between scion and stock, develops.[2]

Grafting a *vinifera* vine onto an American vine stock is a relatively simple process. Incisions are made in both the rootstock and scion such that they marry together closely, giving some physical resilience to the join. The critical feature is to have the two cambial layers in apposition to each other. Typically, the cleft made in the rootstock is designed to allow the maximum area of cambial contact. Most grafting is done with cuttings (bench grafting), but it is also possible to graft in the field. And as well as joining American rootstock with *vinifera* scion, grafting is commonly used to transplant one *vinifera* variety to another in the field. In any graft situation, both scion and stock retain their own, separate genetic identity, and the graft merely facilitates vascular transportation. Mexican grafting teams who can field-graft enormous numbers of vines during the growing season, with staggering efficiency, are in high demand worldwide. The advantage for the grower is they can switch the identity of their vineyards without the three-year loss of production that replanting would incur.

ARE WINES FROM UNGRAFTED VINES BETTER?

So we reach the central question of this piece. Have vines that have been altered by the grafting union lost something intrinsic to their identity? Were the wines made from ungrafted vines, pre-phylloxera, somehow better? "When I first met wine people, it was the great debate," recalls Hugh Johnson, author and wine expert. "There was still lots of evidence of how good wines had been before phylloxera, but I don't remember a single conclusive demonstration that they had become less good since, or that if they had, it was blamable on the louse." As the number of surviving pre-phylloxera bottles has diminished, the pre- and post-phylloxera debate has become less voluble. But there are still people in the trade with enough experience of these pre-phylloxera bottles to have formed an opinion. Renowned wine critic and writer Michael Broadbent is one. Are the pre-phylloxera wines better? "Who knows?" he replies. "The quality was undoubtedly high from 1844 until 1878; one has read about them and one has tasted them." A wine writer and leading authority on wine Serena Sutcliffe is another who has experienced plenty of these old bottles. Are they special? "Yes, they are different. They are more intense; they have total, concentrated heart," says Sutcliffe. "They also have the most incredible scent and long, lingering finish. But you have to mix this in with the fact that yields were so much lower then: so how much is due to that, how much to the original vines? I suspect it is a combination."

Hugh Johnson also picks up on this theme. He agrees that pre-phylloxera wines may have been different, but this is not necessarily because of grafting onto American rootstocks. "Of course, it was not one event," maintains Johnson. "The oidium before it and the mildew at the same time made owners more proactive in their vineyards than they had ever been. They manured like mad to make their vines healthy and strong. They sulfured the vines, used carbon sulfate

2 Esau K. *Anatomy of seed plants*, John Wiley, New York 1977

on the soil, invented Bordeaux mixture. Yields went up dramatically. They started chaptalizing, tried to control fermentation temperatures, started bottling much earlier, in fact, changed so many things in a short period that it's amazing that some of the 'pre-phylloxera' wines made during this time were as good as they were."

Sutcliffe adds that she doubts that the top wines from top vintages that are made now will last 80–100 years. "But, how many people today require their wines to do that? Virtually no one. We are in the age of instant gratification, so it is perhaps irrelevant." Sutcliffe also adds a potential explanation for the longevity of the old wines. "A fascinating fact is that the vast majority of these pre-phylloxera gems (the real ones, not the endlessly cloned Glamis stuff which pops up everywhere!) are very low in alcohol, often 10% or even under, as with some old Ausone from the nineteenth century: absolute proof that alcohol in itself does not assure longevity—perhaps the contrary. So beware all those 14% numbers out there! It seems to eat up the wine over time."

As well as experience with pre-phylloxera wines, there are two further lines of argument that might give some insight as to whether ungrafted wines are somehow better or not. The first is the fact that around the world there are some significant areas where vinifera vines are still grown on their own roots. Perhaps most notable is southern Australia, which phylloxera hasn't reached yet, and which is kept clean by a strictly enforced quarantine. The Barossa in particular has a number of old vine vineyards, planted on

their own roots. These Barossa old-vine vineyards make some extremely fine wines, albeit in a rather different style to the Old World classic regions, which renders a comparison rather difficult. Most Chilean vines are also ungrafted, but Chile has so far struggled to make world-class wines, so these data points are also of limited use. The Mosel still has many old, ungrafted vines, with phylloxera struggling to grow in the impoverished slate soils of the region. The wines made from these old Riesling vines can be remarkable.

Useful evidence comes from particular isolated ungrafted vineyards still being cultivated in the middle of otherwise grafted wine regions. One example of an ungrafted vineyard where wines are made that can be compared with those from grafted vines on the same estate is Quinta do Noval's Nacional, a six-acre vineyard in the Cima Corgo of the Douro. Phylloxera has never affected this vineyard, which was first planted in 1925. "There are some very old vines in the parcel, but I would say the average age is about 40 years. When a vine dies we replant, but without grafting to any American rootstock," says Christian Seely, who took over running the quinta on behalf of AXA in 1996. Each year, the grapes are picked, vinified, and matured separately. On average, just 250 cases are produced annually. "Although we don't declare it every year, the Quinta do Noval Nacional is always extraordinary," says Seely. "The remarkable thing about the wine is that although we vinify it in exactly the same way as the other wines from Quinta do Noval, it is always very different from any of the other lots of wine

Left An ungrafted old vine, around 100 years old, planted in the Erdener Prälat vineyard in Germany's Mosel region. Phylloxera can't flourish in these soils.

from the estate. It marches to quite a different drum than the rest of the vineyard. Sometimes it can produce a vintage Port that is among the greatest wines of the world when the rest of the vineyard is having a late-bottled vintage- (LBV) quality year (as in 1996); at others it is not even of vintage quality when the rest of the vineyard is making great vintage Port (as in 1995), but it is always different to the rest of Noval. Even in years, such as the 1997, for example, when I believe the Quinta do Noval Vintage and the Nacional to be equivalent in quality, the wines have an entirely different character." Even here, though, we can't say that it is the fact that the Nacional vines are ungrafted that makes them produce different wines. There may well be something different about the soil of the Nacional vineyard that neuters the threat of phylloxera. This difference could reasonably be expected to influence the nature of the wine. As Seely puts it: "Nacional is a supreme example of the importance of terroir."

The second line of argument is a theoretical one, using science to answer the question of how the grafting of *vinifera* varieties onto American rootstocks is likely to affect wine-grape quality. These theoretical considerations are more informative.

How does the rootstock affect wine-grape quality? It's clear that rootstocks can influence the growth pattern of the scion quite considerably. Apples are instructive here. In the early 1900s scientists at the East Malling Research Institute in Kent, in the UK, did a lot of work on rootstocks for apple trees. There are around 20 well-known rootstocks for apples, and these determine how the apple tree (the scion) will grow. With one rootstock, a certain variety of apple will produce a tall tree. With another, the same variety will produce a dwarf plant just 5 feet high. Not only does the rootstock provide a supply of water and mineral nutrients for the plant, it also communicates with the scion by means of hormonal signals. That the roots of grapevines signal to the aerial part of the plant is well-known, particularly from work done recently on partial deficit irrigation and precision root drying. The roots are able to signal information about soil-water status to the aerial parts of the plant by means of the stress hormone abscisic

acid, which is produced in response to deficit. This results in the leaves closing their stomata to avoid losing water before they are themselves suffering any water stress. It is likely that the rootstock will be signaling all manner of information by means of plant hormones, and there will be communication the other way also, from shoots to roots. While the scion and rootstock have subtly different genetic makeup, the rootstock choice is likely to have physiological consequences for the *vinifera* scion and will, to a degree, shape its growth pattern. This interplay is effectively a new viticultural tool, provided that viticulturists understand enough about it to manipulate it effectively. This will in turn have an impact on grape, and thus wine, quality. But there is no reason why this should be a negative impact, because in many instances it could be positive. Science has no evidence to support the claim that grafted vines produce grapes that are necessarily inferior to those produced by the same variety grown on its own roots.

CONCLUDING REMARKS

I think the fact that so many expert commentators over the ages have argued that there was something special about the pre-phylloxera wines is strongly suggestive that they were very good wines indeed, and perhaps even better than the wines that followed, made from grafted vines. There are numerous confounding factors here, not least that old-vine vineyards were being uprooted at more or less the same time, and being planted with new vines. While no one knows for sure why it is the case, young and adolescent vines simply don't perform as well as their older siblings. It could be this dip in quality following widespread replanting that caused commentators to note a qualitative deficit post-phylloxera, which may then have been wrongly attributed to grafting. But it could also be that other factors involved in the replanting are responsible for this deficit. From theoretical considerations, it seems that pointing the finger at the grafting onto American vines is misguided, and changes in winemaking, viticultural "advances," and stylistic changes have produced wines that may impress more when young, but lack the longevity and purity of the very old wines made before the phylloxera plague hit.

6 *Lutte raisonée*, IPM, and sustainable winegrowing

Science's contribution to viticulture is often cast in a rather negative light. If I ask you to think about science applied in the vineyard, you might well conjure up images of chemicals being sprayed with abandon, obliterating all life but that of the vines. Or, your mental picture might be one of space-age trellising and irrigation technologies forcing vines to pump out heroic yields of grapes that are then turned into mega-gallons of soulless industrial wine. Or, perhaps you imagine geeks in lab coats genetically engineering a new generation of supervines. But this is a misrepresentation. It is science that brings us an understanding of the true complexity of natural systems. The insights from the science of ecology are teaching us how to work with the checks and balances of nature, and encouraging a new, rational, limited-input, environmentally sound means of vineyard management that offers a third way between the ideologically driven approach of biodynamics and conventional chemical-based agricultural systems.

This new viticultural approach is known in France as *lutte raisonée*, literally the "reasoned struggle." This is based on a scientific concept known as integrated pest management (IPM), and borrows some of the insights from organic farming but without the straitjacket of strict regulation that typifies organics and biodynamics. One of the strengths of this new viticulture is that it reconciles the needs of growers with those of the environment in, if you'll excuse the cliché, a win–win scenario.

NATURE'S STRUGGLE

Imagine a picture of a meadow on a midsummer's afternoon. The sun is shining, the air is buzzing with insects, and there's a lovely contrast between the various shades of lush, green vegetation. On the surface things look peaceful enough, but in reality this portrait of natural harmony is actually one of a continued fight for existence. Each organism is in a struggle to grow, survive, and reproduce in the face of hostile, ever-present competition. Plants compete with each other for light, water, and nutrient resources, at the same time balancing the need to open their stomata (pores in their leaves) to allow gas exchange while restricting the loss of precious water. These plants also represent a useful food source for a number of herbivorous insects and mammals that, in turn, have their own predators to worry about. And if that's not enough, there are a host of fungal, bacterial, and viral diseases all waiting to colonize a willing plant recipient. Indeed, a lot of plant morphology and chemistry has been shaped by evolution to make plants unpalatable to herbivores and resistant to microbial attack. It's quite a challenge just surviving.

With all these organismal interactions that occur in the environment, a series of checks and balances has evolved, some of them quite sophisticated. Here's an example. When some plants are munched on by certain insect herbivores, they release volatile compounds called semiochemicals. These can be detected at even tiny concentrations by predators of these herbivores. Remarkably, the chemicals released by the plant can give information about the particular organism that's feeding on them, thus specifically alerting the appropriate predator. Effectively, the plant's SOS call has become a "Dinner's ready!" signal to the predators. This sort of clever balancing act helps ensure the continued survival of the plant even in the face of attack, through an evolved cooperation with other organisms. It is these types of complex interactions that IPM attempts to understand and then harness.

SHIFTING THINKING

The emergence of IPM reflects changing attitudes toward pest control. To illustrate this, let's take a larger-perspective view of agriculture and, more specifically, viticulture. Since the dawn of agriculture, humans have had problems with pests and diseases. Growing just one crop in a field—monoculture—is asking for trouble. It is shifting the odds firmly in favor of the pest or disease species, because these unwanted guests have adapted to reproduce fast and explode in numbers once they find a suitable habitat (this may well be your field or vineyard). But if we are going to farm effectively, then growing crops in near or absolute monoculture is a prerequisite. However, because natural pest reduction is less likely to occur, the larger the extent of monoculture, the higher the risk of losing a substantial portion of the crop to some pest. And as agriculture has developed, the extent of monoculture has developed, such that ever-larger areas are covered with just one crop. In the pursuit of increased yields, weeds have been eradicated by pre-emergence herbicides, and hedges and patches of scrub or woodland have been removed.

In the "natural" situation of an ecosystem, checks and balances develop. Biodiversity ensures that there are not only pests, but also predators of those pests present. Any system that is hopelessly imbalanced is unsustainable, so it is selected against. This means that the ecological systems that have survived are, by definition, sustainable, unless they experience some externally applied change, such as climate change or someone planting a large vineyard in the middle of them.

Quite naturally, farmers don't like to lose a portion, or all, of their crop to pests and disease. Traditionally the response has been to turn to chemists to provide magic bullets to eradicate these problems. This approach is flawed, however, because development of resistance on the part of the pest or disease is almost inevitable given enough time. Apply a strong selective pressure such as a pesticide to a rapid-life-cycle insect, and you are challenging natural selection to come up with a resistance mechanism. Just one mutant that has a degree of resistance will be strongly selected for, and will in a short space of time be able to restore the population to damaging levels. Natural selection almost always wins. And bear this in mind: if you have knocked out the pest population, you'll likely have also knocked out the enemies of these pests. Then, when resistance arises to your chemical of choice, the situation will be worse than before. You'll be without a chemical bullet, and your pest will be without its enemy. You'll sit there looking at your devastated crop, and long for the days when you only lost 10% of it.

This is the kind of scenario that led to the generation of IPM, a new, smarter paradigm for countering pests that began to be developed in the 1970s. IPM is a thoroughly scientific, whole-system approach (call it "holistic" if you will), that has largely replaced the rather naïve agrochemical-dependent way of doing farming, and it is increasing in significance.

The five bedrocks of IPM are knowledge, monitoring, anticipation, economic thresholds, and timing. IPM rests on a thorough scientific knowledge of the biology of pest, weed, and disease organisms, in the context of the larger ecosystem. Practitioners use this knowledge to monitor populations of potential problem organisms and then anticipate when they will reach damaging levels. This is where the concept of economic thresholds steps in. Rather than try to eradicate all pests, farmers need to decide what population levels they can tolerate economically. Finally, timing is a crucial concept that is key to effective implementation of IPM. Chemical inputs

Far left and left Downy mildew on a vine leaf at the end of the growing season. It becomes evident as "oil spots" on the leaf surface, with the growth of the mildew seen on the underside of the leaf.

may still be needed, but through careful timing these are greatly reduced. Another benefit of IPM is that because it is a multipronged strategy and doesn't just rely on the outmoded notion of a chemical bullet, resistance in the pest, weed, or disease population is much less likely to arise.

In essence, IPM is about reconciling rather disparate aims. Farmers want to reduce crop losses while at the same time reducing environmental degradation and avoiding pest-resistance buildup. Farmers using IPM are making choices based on a broad-perspective outlook that takes into account the whole ecosystem, rather than just a part of it.

THE IPM TOOL SET

The strength of IPM is that it offers many potential solutions to agricultural problems. Many of these are still rather experimental while others are tried and tested. They include the following.

BIOLOGICAL CONTROL

Biological control is one of the foundations of IPM. It's conceptually quite simple. If you have a pest problem, introduce the natural enemies of this pest—whether they are predators or diseases—and let them control the problem. In practice it is not quite as simple. In order for the natural enemies to complete their life cycle and establish themselves in a vineyard, there has to be a large enough population of pests, and unless growers want to be reintroducing populations of the natural enemies each time treatment is required, then they must tolerate a degree of pest loss from the continued (but manageable) presence of a sustaining population of pests. They might also have to introduce refuge areas of non-

crop-plants in and around their fields to sustain diverse populations of insects throughout the year, some of which will be beneficial. And biological control that isn't properly thought out can go horribly wrong. One of the best examples of failed biological control is the cane toad, *Bufo marinus*. Many years ago, sugarcane growers in Queensland, Australia, had a problem with cane beetles, and some clever academic identified the cane toad as a natural pest of these beetles. So, in 1935, the cane toad was introduced as a control, and adapted well to life in Australia. However, it quickly became a real problem. Not only did it decide not to go after the cane beetles it was intended to control, but it has also proved poisonous to local animals that have tried to eat it. A milky secretion from a gland on the back of its head can prove fatal to any cat, dog, or dingo that decides to try eating a cane toad. Even today, cane toads are spreading unchecked through northeastern Australia because they have no natural predators. The recommended humane way of eliminating them is to place them in a plastic bag and put this in a freezer, but locals are known to indulge in the rather more brutal art of cane-toad bashing. In an ironic twist on the tale, Australian government researchers are now developing biological control agents of the cane toad. Admittedly, this is a rather extreme example of badly thought-out biological control, but it illustrates the potential dangers when this control method is attempted with insufficient knowledge of the ecological systems involved.

NATURAL ENEMIES OR "BENEFICIALS"

Many IPM strategies rely on the identification of natural enemies of pest arthropods. These are also known as "beneficials." Natural enemies might be predators which eat the problem species, or parasitoids, which are insect parasites. An example of the latter is the parasitic wasp. It might lay eggs in a problem caterpillar. The eggs produce larvae that then grow inside the caterpillar, using it as a food source and killing it in the process. So, in the vineyard, planting flowering buckwheat (*Fagopyrum esculentum*) attracts the parasitic wasp *Dolichogenidea*, which is a common parasitoid of leafroller larvae, a serious pest of grapevines.

BIOPESTICIDES

These are pesticides that use specific microbes as the active agents. They aren't used as widely as they could be, largely because of a lack of knowledge about them coupled with an absence of commercial drive. One example is that of *Trichoderma harzianum*, a fungal enemy of another fungus, *Botrytis cinerea*, which causes bunch-rot on grapes, and *Ampelomyces quisqualis*, an antagonist of powdery mildew. Some are already being used in vineyards.

"REFUGES" OR "ECOLOGICAL COMPENSATION AREAS"

It's all very well introducing parasitoids or predators of pests into your vineyard, but they will need somewhere to live, and they might not find your vines an ideal home. Added to this, clean-cultivated vineyards are barren places during the dormant season, providing nowhere for overwintering insects to hide. This is where ecological compensation zones come in handy. These are patches of ground given over to specific patterns of vegetation, such as scrubland, woods, or hedges that can act as refuge areas for natural enemies of problem species. This sort of biodiversity can offset some of the negative effects of monoculture. It is likely that the efficacy of these compensation areas will be enhanced by the use of cover crops or allowing some vegetation to grow between vine rows. Ecological compensation areas are now being trialed in some vineyard regions in France (*see* below). These compensation zones are not a universal cure for all vineyard problems, however. While they are intellectually and emotionally appealing, their use needs to be carefully thought through because there is a risk that growing

certain types of vegetation near vineyards could encourage the presence of insect species that actually turn out to be a problem, either directly as pests themselves, or indirectly by acting as transmission agents of viral or bacterial diseases.

COVER-CROPPING

The ground between vine rows is normally kept clear of weeds by pre-emergence herbicide applications or heavy tillage. Cover-cropping is the practice of growing plants between rows instead of leaving them clear, and offers several potential benefits. These include the promotion of soil life, preventing erosion, and enhanced populations of beneficial insects. There exist a range of plants that are potentially suitable for this purpose. It is now common in some wine regions to grow a cover crop over winter when the threat of erosion is likely to be highest, and then this crop is tilled into the soil in spring. But newer strategies involve growing cover crops throughout the growing season when they can act as refuges for beneficials. Studies have shown that these summer cover crops support season-long high populations of predators that can provide biological control for some vineyard pest species. However, cover-cropping has one potential drawback. In drier, nonirrigated vineyards the cover crop can compete with the grapevines for scarce water resources.

SEMIOCHEMICAL STRATEGIES

Chemical signaling is vital to most insects. They use their acutely sensitive olfactory and pheromone systems to detect food sources and

Above A cover crop growing between rows in a vineyard. This is the Calvert vineyard in New Zealand's Central Otago region.

to find mates. Where the specific sign chemical (semiochemical) strategies the insects use are known, they can be exploited. For example, mating-disruption strategies involve the use of insect sex pheromones to confuse and thus disrupt the mating behavior of target species. Experimental work is focusing on ever more sophisticated semiochemical interventions. For example, some insects use their sense of smell to dictate egg-laying behavior. They will choose not to lay eggs on those plants that have already been laid on, and they can smell the difference. If the semiochemical involved can be identified, then it can be used to mark crop plants to discourage the pest insect from laying eggs. Push-pull strategies can also be employed. Plants can be introduced into crop zones that smell repellent to pests, while adjacent target zones of noncrop plants can be planted to pull in pest species, thus diverting pests from the crop.

WEATHER MONITORING

This can help reduce the number of chemical inputs by predicting when certain pests or

diseases are likely to be a problem. Any spray programs can be scheduled intelligently, applied only when they are really needed. This climatic monitoring is inexpensive to implement, and is likely actually to save money because sprays and the labor required to administer them are costly.

LUTTE RAISONÉE/IPM IN ACTION

To get a better idea of how these techniques might look like in action, I spoke to a researcher involved in trying to implement ecological compensation areas (known as *Zones Ecologiques Réservoirs* in France) in French vineyards. Marteen Van Helden works for l'École Nationale Ingénieurs des Travaux Agricole (ENITA) (National School of Agricultural Engineers) in Bordeaux, and his research concerns developing the science behind ecological compensation areas so they can be used as an IPM tool in vineyards. Currently he has set up a number small vineyard plots with experimental hedges, and he is monitoring the population dynamics of insect species on the grapevines and in the hedges. "Viticulture is particularly interesting for IPM," explains Van Helden, "because there is very little risk of increasing the pressures of diseases or pests on the vines: this is not the same with other crops." The danger with many crops is that planting ecological compensation areas increases the habitats for natural enemies, but it may also end up pulling in a larger population of pest species, too. Vineyard insect pests are usually a problem late in the summer. The idea is that in vineyards in early summer there will be a buildup of natural enemies on the hedgerows, which eat pest species there and then move to the grapevines later. "Our experiments have been ongoing for five years now," says Van Helden. "We don't have solid results, but a lot of farmers are interested." He explains that this is partly because winegrowing is a matter of image, and producers like the idea of an attractive countryside for wine tourism. IPM also fits right into the "natural," "wholesome" image that most wineries would like associated with their products. Van Helden is taking advantage of this enthusiasm

Left A weather station in a vineyard. Data from the vineyard can be transmitted wirelessly, and this information can then be used to make informed decisions about crop protection, reducing the need to spray by calendar.

to see if he can create larger sites to try out IPM rather than just his small experimental plots.

As well as his work in Bordeaux, Van Helden is involved in a project that will see ecological compensation areas being trialed across a whole appellation, Saumur-Champigny in the Loire. "We want to see whether we can recreate functional biodiversity in an existing situation," says Van Helden. "We want to see what we can adapt; we don't want to redo the landscape entirely." It will be an interesting project because it will help to explore which sorts of landscape elements are most significant for encouraging viable natural enemy populations. It is also important that this experiment is taking place at the scale of a whole appellation. In ecology, scale is quite important, because lots of little isolated plots aren't as useful as a few larger plots, perhaps linked by features such as hedges. Small plots suffer from what is termed "island" effects. If you have a certain area of a particular habitat, it will better support a larger diversity of species if it is all one plot than if it is broken into several different islands, even if the total area is the same in both cases.

Hedges are a vital component in this type of project, because they act as refuges and are also

1 Vines require chemical inputs to protect them from disease. They are one of the most intensively sprayed crop plants.

2 A cover crop planted between the rows in Neudorf Vineyards, Nelson, New Zealand. These can improve soil structure and fertility, increase soil life, prevent erosion, and encourage beneficial insects.

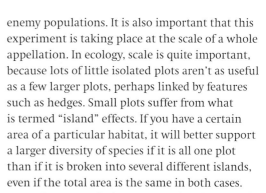

3 Spray residue can be seen on these grapes and leaves in Mendoza, Argentina.

4 A vine suffering from grapevine leaf roll virus in South Africa's Stellenbosch region. It's a major problem, and not just in South Africa. The virus is spread by mealy bugs (an insect), which need to be controlled if uninfected vineyards aren't to be affected.

5 Botrytis, a fungal disease, is affecting part of this bunch of almost-ripe grapes.

linking elements and corridors, but alone, they might not be enough. Some natural enemies prefer larger natural sites, such as patches of scrub or woodland. The hedges can act as "roads" directed toward the vineyards. In the vineyard, small landscape elements such as undergrowth provide important refuge areas. In the Saumur-Champigny experiment, Van Helden will be advising growers what sorts of plants might be useful for ecological compensation areas and vineyard undergrowth. Farmers might find space for these around their plots, or plant hedges in or at the boundaries of their vineyards. He estimates that the cost of planting a hedge is around 4–5 euros ($5–6) per yard, with some soil preparation, mulching, and a year or two of follow-up also involved. There are possibilities for funding this through the local Chamber of Agriculture or local government. He hopes that a few enthusiastic farmers will start the ball rolling and then others will be encouraged to get involved. On a larger scale, the two organizations planting most hedges in France are the railroads and highways. They can be encouraged to collaborate with farmers in these sorts of projects. In some ways, generating successful biodiversity is not just a scientific venture, but also a social and cultural one, and it needs to be approached this way.

CHAMPAGNE

The Comité Interprofessionnel du Vin de Champagne (CIVC) provides technical support and advice to growers in the region. It claims to be the first wine region in the world to run environmentally friendly programs at the scale of the region. IPM in Champagne goes under the name of *viticulture raisonée*, and since 2001 it has issued a practical guide for growers. I've seen the 2003 version, and it's a 200-page full-color guide to viticulture with an emphasis on reducing chemical inputs. It's an impressive document, with lots of attractive color pictures. The technical services of the CIVC have actually been working on *viticulture raisonée* for a decade; they have found they don't need to enforce it because growers are embracing it readily. They have weather stations throughout the region and provide growers with alerts and advice. This sort of project is extremely encouraging, mainly because with

Above Sustainable Winegrowing New Zealand certifies a large proportion of the country's vineyards. As a result, there have been impressive reductions in agrichemical inputs to vineyards in New Zealand.

its high take-up and official sanction, it is likely to have a widespread effect on entire vineyard regions, and is relevant to all types of producers, from boutique wineries to industrial giants.

SUSTAINABLE WINEGROWING

While organics and biodynamics have attracted a lot of attention for their green credentials, there's a growing interest in a third way: sustainable winegrowing. Sustainability certification programs are becoming increasingly popular, because they rely on IPM techniques and offer a scientifically rational way of reducing chemical inputs into vineyards. For example, Sustainable Winegrowing New Zealand has a virtually industry-wide take-up and, as a result, the agricultural inputs used in New Zealand vineyards have been dramatically reduced. A paper by Walker and colleagues on pesticide use reduction in New Zealand showed that, over the period 1999–2007, the wine sector reduced its total agrichemical loading of insecticides by 72% and fungicides by 62%. Many countries have similar certified sustainable schemes for winegrowers. While critics argue that some of these schemes don't go far enough, the relatively low barrier to entry results in a high take-up, and gets growers thinking about issues of sustainability, which has to be a good thing.

Headaches for winegrowers: the key diseases and pests

Viruses These are a major problem in viticulture. Perhaps the most common is leafroll virus, common throughout vineyards worldwide, and seen as a downward rolling of the leaf late in the growing season. It doesn't kill the vine, but substantially delays ripening, reducing wine quality. Some spread is by humans employing poor nursery practices, taking cuttings from infected vines. It is also widely spread by a vector, the vine mealybug. The only way to counter leafroll virus is to plant virus-free vines. However, if mealybugs are not controlled, leafroll virus can rapidly reinfect newly planted vineyards. Other viruses also infect grapevines, and insects and nematodes, as well as poor cultivation practices, transmit some. Virus diseases are frequently overlooked because they rarely kill vines, instead having an effect on grape quality, for instance, by delaying ripening. It is likely that many of the clonal differences seen in different grape varieties are not genetic, but instead reflect different levels of viral infection.

Fungi Fungal diseases are one of the major problems for winegrowers. *Vitis vinifera* lacks innate resistance to downy and powdery mildew, fungal diseases imported to Europe from the USA in the nineteenth century (although downy mildew is technically not a fungus, but an oomycete). The result? Growers have no choice but to spray, either with traditional chemicals such as sulfur and Bordeaux mixture, or with more modern systemic fungicides. *Botrytis cinerea* is a fungal disease that causes gray rot of grape bunches, although when it infects certain already-ripe grapes it can become the beneficial "noble" rot that is a crucial component of certain styles of sweet wine. There is currently limited biological control against these fungi in the form of biopesticides, which are not yet widely used. In the absence of biological control, IPM can still help target chemical treatments more precisely. Knowledge of the biology of these fungal diseases has facilitated the development of risk-assessment indices, based on weather conditions. By calculating the index score, a winegrower can know how much and how often to spray for effective control. Particularly in Europe, trunk diseases, such as esca and eutypa, have recently emerged as a big problem for which there is currently no cure.

Arthropods Arthropods—the taxonomic group that includes insects and spiders—can be major pests in vineyards. In Europe, the key problems are: the grape berry moths *Eupoecilia ambiguella* and *Lobesia botrana*; leafhoppers, which act as vectors of viruses and bacteria; and spider mites. In the USA, Pierce's disease, spread by a leafhopper, is a matter of urgent current concern. This isn't forgetting phylloxera, which almost destroyed viticulture worldwide in the latter half of the nineteenth century (see Chapter 5). As yet, though, there is no effective biological or even chemical control strategy against this aphid pest. "Insect-eating nematodes will kill phylloxera," says entomologist Jeffrey Granett, "but because the phylloxera are so small the nematodes cannot complete their life cycle."

Grape berry moths were already present before vines were widely grown, but adopted grapevines when they became a monoculture in certain regions. Their larvae feed on grape flowers and cause loss of yield, and then later in the season they attack berries, predisposing them to invasion by botrytis. Models based on weather conditions and the results of population sampling by sex-pheromone traps can predict the likelihood of their attack. Thus spraying can be withheld until the risk is deemed high. In addition, mating-disruption strategies based on the use of synthetic sex hormones similar to those emitted by females can be employed.

Leafhoppers are a significant problem in European regions, especially so for organic growers, where control methods are extremely limited. Perhaps the most serious problem is *Scaphoideus titanus*, which is able to transmit the mycoplasma disease flavescence dorée, with disastrous consequences. Like phylloxera, this is an American introduction to European viticulture, coming over some time after World War II. The only control is insecticidal sprays, and in some French regions that are affected, spraying is obligatory for all growers. The hunt is on for effective biological control agents.

Pierce's disease (*Xylella fastidosa*), a bacterial disease spread by the glassy-winged sharpshooter, a leafhopper that feeds on vines, is a serious problem in parts of California, and the subject of much current anxiety. Its range is expanding northward toward premium viticultural regions. *Xylella* blocks the vessels of affected vines, killing them. There is no effective chemical or biological control, and the disease is so serious that viticulture has to be abandoned in affected areas (this is the reason no grapes are grown in Florida, for example).

7 Biodynamics

Many readers will be surprised to see a chapter on biodynamics in a book on wine science. The language of biodynamics, with its references to the alignment of the planets, undefined life forces, and the use of bizarrely fashioned preparations, seems totally at odds with a rational, scientific worldview. As a consequence, most scientifically literate people have dismissed biodynamics altogether. Alternatively, they have regarded biodynamic practices as affectations and explained any benefits merely in terms of increased attention to vineyard management. My goal in this chapter is to try to integrate the insights and practices of biodynamics with a scientific understanding of viticulture. Why would I bother in the first place? Simply because so many of my favorite winegrowers are working biodynamically. You could drink very well indeed, if all you were to consume from this point onward were wines made from grapes grown biodynamically.

I'll begin by trying to capture the essence of biodynamic viticulture, and answer some key questions. How does biodynamics differ from conventional and organic agriculture? Does it actually work, and if so, how? I'll also address which, if any, aspects of biodynamics can be explained scientifically.

Back in 1997, the sales team and directors of Corney & Barrow, an independent British wine retailer, visited Domaine Leflaive in Burgundy. Anne-Claude Leflaive poured them two wines, blind, and asked them which they liked best. Twelve out of the thirteen preferred the same wine. What was the difference? Well, both were her 1996 Puligny Montrachet 1er Cru Clavoillon, and so, technically, the same wine. But the wines were made from adjacent plots of vines, one organic, and the other farmed with biodynamics, an alternative system of agriculture that represents the focus of this chapter. This latter wine was the one that the Corney & Barrow team had singled out almost unanimously as the favorite. The following vintage, Domaine Leflaive went fully biodynamic.

Anecdotal observations like these don't constitute hard scientific data, but they are common enough—and come from people making serious enough wines—to merit proper attention. Indeed, the roll call of biodynamic producers forms a star-studded list, and it is growing steadily.

"Whether it is biodynamics or soils, people are always accusing me of being faith-based," says Mike Weersing, of Pyramid Valley, in New Zealand's north Canterbury region. "It is not faith so much as a long, cumulative inability to continue to disbelieve. It is empiricism. The more you see it, the harder it becomes to doubt." That's how, as a scientist, I feel about biodynamics.

WHAT IS BIODYNAMICS?

It is helpful to think of biodynamics not primarily as an agricultural system, but rather as an altered philosophy or worldview that then impacts on the practice of agriculture in various ways. In other words, to farm biodynamically, first you have to think biodynamically.

It has its roots in a series of lectures delivered by Austrian philosopher-scientist Rudolf Steiner in 1924. Steiner's life mission was to bridge the gap between the material and spiritual worlds through the philosophical method. (Already, I can feel the scientists' hackles rising with the use of these terms.) To this end, he created the "spiritual science" of anthroposophy, which he used as the basis of the Waldorf School system that persists to this day.

It was late in Steiner's life that he turned to agriculture. His eight lectures, entitled *Spiritual Foundations for the Renewal of Agriculture*, were delivered just a year before his death, but they nevertheless remain as the foundation of biodynamic farming.

THE FARM AS A LIVING SYSTEM

Key to biodynamics is considering the farm in its entirety as a living system, and seeing it in the context of the wider pattern of lunar and cosmic rhythms. In this holistic view, the soil is seen not simply as a substrate for plant growth, but as an organism in its own right. The idea of using synthetic fertilizers or pesticides is thus an anathema to biodynamic practitioners. Instead, they use a series of special preparations to enhance the life of the soil, which are applied at appropriate times in keeping with the rhythms of nature. And disease is seen not as a problem to be tackled head-on, but rather as a symptom of a deeper malaise within the farm "organism." So, if you correct the problem in the system, the disease will right itself. Where biodynamics differs significantly in practice from organic farming is in the use of these preparations and the timing of their application. In other ways the techniques employed are quite similar.

Just from this outline, we can see why it is difficult to discuss the theoretical basis of biodynamics from a scientific viewpoint. Chief among these is the postulation of undefined "life forces." These aren't specified in any concrete way, and we have no means of measuring them. While the concept of lunar rhythms is one that we can frame in scientific forms, the cosmic rhythms that come from the alignment of the planets are similarly undefined. It's hard to see what physical, measurable effect the planets and their movement could have on life on Earth.

AN AUDIENCE WITH NICOLAS JOLY

What better way to try to catch the flavor of the underlying philosophy of biodynamics than to attend a Nicolas Joly seminar? Joly, who owns Coulée de la Serrant in the Savennières region of France's Loire Valley, is probably the most celebrated and widely quoted proponent of biodynamic viticulture. "I was trained to be a banker, but I turned out to be a winegrower," he says. Joly had a career in finance, with spells working in the UK and USA. When he returned to his family's estate in the Loire in 1977, he decided that he wanted to make wines that expressed the "spot" of Coulée de Serrant. Early on in his tenure, an official from the chamber of agriculture visited

Left Nicolas Joly, from the Loire, one of the most evangelical of all biodynamic winegrowers. His estate is Coulée de Serrant in Savennières.

him. "They told me that my mother had been running the estate well, but in an old-fashioned way, and it was now time for some modernity. I was told that if I started using weedkillers, I'd save 14,000 francs." Joly took this advice, but soon regretted it. "Within two years I realized that the color of the soil was changing; insects like ladybugs were no longer there; all the partridge had gone." Then fate intervened. Joly read a book on biodynamics. "I wasn't attracted to the green movement, but this book fascinated me, and I had the crazy idea of trying to practice this concept."

Joly's prime emphasis is on living forces, and the correct timing of viticultural interventions. "The soil has to be alive. Organic manure is from different animals. Each animal produces very different manure. Some animals are dominated by heat, like a horse. If you force a cow against its will it will go down—the earth forces dominate. Wild boar and pigs feed on roots, so their manure will work on the roots. All these different fertilities are essential."

He continues, "Spring is good for us. For a vine, spring is the victory of sun forces over earth forces. In autumn, the law of death comes into force: the law of gravitation comes into force and leaves begin falling. Look how tired we get in the evening. On the first day of spring the days are a bit longer than the nights. The sun attraction is stronger than gravitation.

"The vine is one of the few fruit trees strictly linked to the season. The vine is dominated by the earth forces. It goes downward so it has immense strength in its roots and only goes up a little bit. It couldn't flower in the spring like the cherry or the apple. The more a plant leaves its gravitational forces, the more it can develop its flowers.

"The vine is waiting for sun to land on earth. This is what happens at the summer solstice. It withholds its flowering process for the time when the sun lands on earth. The summer solstice is a very important day for a vine. If you taste wines where they flower too early, they have a very good short mouth but a bad second mouthful. The vines flowering closest to the solstice produce the best wines."

So what is the difference between biodynamics and organics? "In *biodynamie* we are connecting the vine to the frequencies it needs. Like tuning a radio, we are tuning the plant to the frequencies that bring it life. Organics permits nature to do its job; *biodynamie* permits it to do its job more. It is very simple."

What does Joly make of inorganic fertilizers? "Fertilizer is a salt. It takes more water to compensate salt. You are forcing growth through water: the plant has to overdrink, so it grows, and carries on growing after the solstice. The process of growth ends up conflicting with the plant's act of retiring to seed and fruit. The result of this is rot, so you need to counter this with lots of chemicals."

And disease? "Disease is a process of constrictive forces and contractive forces. Disease itself doesn't exist. The living agents that bring diseases are just doing their duty. There is no point in fighting hundreds of new diseases."

Joly's biodynamic philosophy extends to winemaking, too. "The more you help the vine to do its job, by means of a live soil, proper vine selection, and avoiding poisonous treatments, the more harmony there is. If the wine catches this harmony well you have nothing to do in the cellar: potentially it is all there." He chooses to use natural yeast, rather than inoculating with yeast cultures: "Re-yeasting is absurd. Natural yeast is marked by all the subtleties of the year. If you have been dumb enough to kill your yeast, you have lost something from that year."

Very quickly, I realized that Joly is taking an approach to agriculture that is at odds with my training as a scientist. He is using an altogether different way of describing natural processes—a "picture" language that jars alarmingly with the Western rationalistic worldview. This is more the language of religion than that of scientifically based viticulture. Yet at the same time I have immense respect for the vision of viticulture he is expounding. It has a life and vitality of its own, which exposes some of the intellectual and environmental bankruptcy of chemical-dependent conventional viticultural regimes. Above all, he is making profound, interesting wines.

BIODYNAMICS-LITE?

Perhaps the easiest way to illustrate the differences between biodynamic and conventional viticulture is by asking the following question. If I were a winegrower, what would I have to do differently to become a certified biodynamic producer? Is there a minimum set of criteria I would have to meet?

I asked Nicolas Joly what shifting to biodynamics would entail. "First you move to organics," he explained, "and if you are confident, then putting in *biodynamie* would require as little as six days' extra work in a year for a 15-hectare [37-acre] *domaine*. The problem is moving to the new understanding of nature. Recreating a model takes a bit of time." Dominique Lafon comes up with a similar figure, estimating that it requires 100 hours more work a year on 14 hectares [36 acres] of vineyards to implement *biodynamie*.

Anne Mendenhall of Demeter USA (the American arm of Demeter, the leading international certification body for biodynamic agriculture) was more specific about what would be needed to obtain biodynamic certification. "The full use of biodynamic methods would be required for two years. That is, you'd need to use the two field-spray materials, BD 500 and BD 501, and compost made with the other six BD preparations." She explained that these preparations can be purchased ready to spray and, because of the small quantities involved, are not expensive. Would I have to keep animals on the farm? "Not absolutely, but it is highly recommended that some livestock be integrated. Chickens running in the vines during the growing season and sheep grazing during the winter have been successful. They are there more to provide the astral component of the farm." And while most biodynamic practitioners would consider the correct timing of interventions to be crucial to their success, this is not a requirement of the Demeter certification. Mendenhall states that, "No one has been decertified for improper timing in the USA."

Above Biodynamic winegrower James Millton, from Gisborne in New Zealand, is preparing BD501, ground quartz (silica), for spraying on the vines.

Above Compost is a vital part of organic and biodynamic farming.

Above This is BD500, a horn manure preparation. It has been buried in the ground, in cow horns, and has just been dug up.

Interestingly, despite the antipathy of biodynamics and organics to any chemical treatments, *vignerons* applying these techniques still have to rely on a chemical solution to treat the problem of fungal disease. Thus I would need to use a copper-based treatment, such as Bordeaux mixture, in conjunction with wettable or powdered sulfur in order to ward off mildew and rot. Dominique Lafon mentions that in Europe a limit on the use of copper is being proposed, with a figure of 4 kg of copper per hectare [about 3.5 lb per acre]. "This is possible in Burgundy if you are careful," he says.

I asked biodynamic consultant Jacques Mell, who is based in Reims, why this concession is allowed. "In vineyards there is no crop rotation. Vines stay in the same soil year after year, so they are living on their own excrement. They become feeble because there is no reviving of the soil, and this weakens them. They are in a state of weakness where they are liable to attack."

MEET JACQUES MELL

A common way for a *vigneron* to make this transition to biodynamics is by hiring a consultant, and Jacques Mell may well be the first flying biodynamic consultant in Europe. At any one time he consults for around 25 growers, and in addition to his contracts in France, he currently has three clients based in Italy, although he can't give me their names because of confidentiality clauses.

Trained as a lawyer, Mell discovered organic agriculture in 1967 through his involvement in beekeeping, and then biodynamics ten years later. He formed his own consultancy in 1989. At this time there were only six winegrowers who practiced biodynamics in France; now he estimates that there are more than 100 (Demeter alone currently have 56 certified vignerons on its books). Mell deals with general agriculture as well as winegrowing, but is currently seeing a higher take-up among *vignerons*. To hire Mell's services would cost some €1,500 (nearly $2,000) a year, which seems reasonable. "My aim is to make it affordable," Mell explains. "It is not just something for the rich farmers."

One of Mell's clients is Francis Boulard of Champagne Raymond Boulard. Boulard is not yet fully biodynamic, but he has been curious enough to experiment with part of his production (in 2003, around 2.5 acres) to see what difference biodynamics makes. Two years into this trial, he has noted consistent improvements in the plot he has farmed this way, and he plans to continue with it. Many growers seem to convert this way. They try biodynamics a little, like what they see, and gradually adopt the system more widely.

Boulard tells me that he is one of the growers participating in a five-year trial that was initiated in spring 2002 by the CIVC. They are systematically comparing three different viticultural regimes: organic, *lutte raisonée* (an integrated approach with limited, selectively targeted chemical inputs, see Chapter 6), and biodynamics. According to Boulard, the CIVC is taking samples of soils and grapes, and then comparing finished wines. It will be fascinating to see the results of this experiment, but the CIVC won't comment on it until all the results are in.

CERTIFYING BODIES

One of the complexities of producing a list of biodynamic producers is that, similarly to organized religions, there are many flavors or streams of biodynamics. Added to this, many growers take a pick-and-choose attitude, implementing some aspects of mainstream *biodynamie*, but omitting others. For winegrowers in Europe, there are two main certifying bodies, Demeter and Biodyvin. Demeter is the larger of the two and covers all forms of agriculture, with a rigorous set of hoops for growers to jump through. Biodyvin is a newer organization set up especially for winegrowers, with its own rules; Olivier Humbrecht is the current president. Some practitioners prefer not to belong to any certifying body. One such unofficial biodynamicist is Dominique Lafon. "For me, *biodynamie* is a tool," he told me. He regards Biodyvin with some skepticism, as a "machine to promote biodynamics in a commercial way."

BRINGING TOGETHER BIODYNAMICS AND SCIENCE

By now, if you've been reading attentively, you will probably have some idea about the nature of biodynamics. You'll probably also be surprised that I'd want to make any attempt to bring biodynamics and science together: surely they are using such a different language and stem from such different fundamental worldviews that there simply isn't a way for scientists to begin to enter a dialogue with biodynamic practitioners. But I think there is, if we are prepared to strip biodynamics down into its component parts and then consider or test the efficacy of its various aspects.

Michel Chapoutier, in France's Rhône valley, began farming biodynamically in 1991. All 620 acres of Chapoutier's Rhône vineyards are now farmed this way, making him the largest biodynamic winegrower in France by some distance. Unlike many practitioners, he thinks that understanding the science behind biodynamics is important. "Biodynamic culture has an interesting future if we have an open attitude to fundamental science." Chapoutier suggests that unless the observations of the effects of biodynamics are underpinned by a theoretical science understanding, biodynamics is in danger of becoming a sect. To this end, he is keen to uncover the scientific explanations behind the various treatments. "Steiner had the genius of finding a great idea," he explains, "but he is considered so highly that people think he got everything right, even the details. People like Steiner are good with big ideas, but not so good with the details."

Certainly, a scientific underpinning to biodynamics would aid its wider acceptance by people currently deterred by its rather esoteric, cultish image. Many practitioners of biodynamics would probably see this as undesirable. To them, conventional science only offers a limited perspective on the natural world. However, scientific respectability could improve the take-up of biodynamics dramatically.

Rigorous research on biodynamics faces a number of obstacles, though. First, because biodynamics sees the whole farm as a single "organism," the idea of separate, adjacent plots being farmed by different methods, in a trial-type scenario, doesn't really fit. A second difficulty is persuading research-funding agencies to pay for these studies. Professor John Reganold, a scientist at the University of Washington

Above Michel Chapoutier, from France's northern Rhône, is one of the best-known proponents of biodynamics.

(Pullman), one of the leading authorities on organic agriculture, told me that because his research proposals have contained the word "biodynamics," funding agencies vetoed some of his research proposals. "Many scientists won't even look at biodynamics," he reports.

Despite these problems, proper studies have been carried out, and generally they seem to suggest that biodynamics (or at least part of the system) really does work. In 1993, Reganold and colleagues compared the performance of biodynamic and conventional farms in New Zealand, a report published in leading scientific journal *Science*. They found that the biodynamic farms had significantly higher soil quality, with more organic matter content and microbial activity. Then, in 1995 Reganold published a review of the different studies that have examined biodynamics and have met basic standards for scientific credibility. The conclusion was that biodynamic systems had better soil quality, lower crop yields, and equal or greater net returns per acre than their conventional counterparts. But what could the mechanism be? A tantalizing clue is offered by some experiments carried out by a graduate student of Reganold's, Lynne Carpenter-Bloggs, on the effects of biodynamic preparations on compost development. In an experimental setting, biodynamically treated composts showed higher temperatures, faster maturation, and more nitrate than composts that had received a placebo inoculation. Reganold is clearly impressed: "Of all the farm systems that I've seen, biodynamics is probably the most holistic."

In May 2002, the results of a 21-year study comparing organic and biodynamic farming with conventional agriculture were published, also in the respected journal *Science*. A group of Swiss researchers, led by Paul Mäder of the Research Institute of Organic Agriculture, showed that while biodynamic farming resulted in slightly lower yields, it outperformed conventional and organic systems in almost every other case. The biodynamic plots showed higher biodiversity and greater numbers of soil microbes, and more efficient resource utilization by this microbial community.

"This appears to be the sort of detailed, long-baseline work we are after," stated Douglass Smith and Jesús Barquín, in an article addressing the credibility of biodynamics in *The World of Fine Wine* (Issue 12, 2006). "But buried in the supporting material, only available online, is the methodology behind the study. There we find that the biodynamic and organic farms began with composts prepared differently. Certain chemicals were added to the organic fields that were not added to the biodynamic ones. And these were only the main differences. What were the others? We aren't told." Smith and Barquin conclude that the results from Mäder's experiment may simply be due to other factors such as the differences in the original composts or in the chemical additions to the organic plots.

More recently, Reganold published the results of a long-term, replicated study on a 12-acre Merlot vineyard near Ukiah, California. Beginning in 1996, the vineyard was split into eight management blocks, and these blocks were randomized to either biodynamic or organic farming. Thus the goal of this study was to see whether biodynamics had any efficacy beyond conventional organics. All management practices were the same except for the additions of the biodynamic preparations. No differences were seen in soil quality over the first six years, and no change was seen in a range of other measures, including nutrient analysis of leaf tissues, cluster weight, berry size, and yield per vine. However, there were some differences. Ratios of yield to pruning weight in years 2001–2003 indicated that the biodynamic vines had ideal balance but the control vines were slightly overcropped. Biodynamically treated wine grapes had significantly higher tannins in 2002, and higher (but not significantly so) tannins, phenols, and anthocyanins in 2003. Not a dramatic improvement over organics, by any means, although this is just one study in one vineyard area. The results might have been different in a region where there was a different set of environmental pressures in the vineyard.

Aside from these publications in the mainstream scientific literature, several practitioners of biodynamics have also tried to test its effectiveness in a scientific way, although the experiments have frequently lacked the rigor to be convincing. The question remains: if these studies have been carried out carefully enough, why weren't they submitted to proper journals to undergo the peer-review process? This would have added greatly to their credibility.

One phenomenon that seems to support the efficacy of biodynamics, and which hasn't yet been explored scientifically is that of "rescued" vineyards. Of course, these accounts are anecdotal and are delivered by advocates, but there is little reason to suppose that they are lying. Anne-Claude Leflaive of Domaine Leflaive reports that their plot in Bienvenues-Bâtard-Montrachet was rescued in such a way. In 1990, the then 30-year-old vines were in bad health, and they were advised to replant. The leaves were chlorotic and the wood was small; the vines had been yielding badly. The new team of Pierre Morey and Anne-Claude decided to do an experiment on these "lost" vines. They stopped using herbicides, opened up the soil and employed the biodynamic preparations. "We were the first to be astonished by the response of the vines to the new treatment," she recalls. "Now these vines are the oldest of the *domaine*, over 50 years old."

In the Napa Valley, leafroll virus is a problem that has led many vineyards to be replanted before they are 20 years old. Ivo Jeramaz of Grgich Hills recalls how a 50-year-old vineyard in Yountville that now yields his best Cabernet Sauvignon came within a whisker of being yanked out of the ground. "We had very low yields and the vineyard couldn't ripen properly," recalls Jeramaz. "We'd be lucky to get to 22 Brix and the grapes were pink." Instead of replanting, Jeramaz cultivated the vineyard with biodynamic principles. The results were that the heavily virused vineyard suddenly sprang back to life. "After three years the vineyard had rebounded dramatically, and it now makes our most expensive Cabernet. There are fewer red leaves and the vineyard wants to go to 30 Brix." Barry Wiss of Trinchero Napa Valley reports a similar experience in his 23-acre Chicken Ranch vineyard in Rutherford. The Cabernet Sauvignon vines in this vineyard had a bad leafroll virus problem, and within a few years of farming with biodynamics this was cured. This sort of reversal could easily be examined scientifically.

So, we can reach the tentative conclusion that biodynamics seems to work when tested scientifically, albeit not as dramatically as many would claim. But this is a qualified endorsement. We still don't know exactly which elements of the biodynamic system are contributing the efficacy, and because the biodynamics works to a degree, this doesn't lend validity to the mechanistic assumptions of practitioners.

From a scientific perspective, I would suggest that some elements associated with biodynamics, such as the use of specially prepared composts, are much more likely to have benefit than others. Composting could increase microbial diversity, and some of the foliar sprays could have a scientifically explainable effect—they are likely to possess biological activity. I doubt that "dynamization" of preparations has any benefit, and like many scientists I'm skeptical about the use of serial homeopathic dilutions, which would likely only serve to reduce efficacy. Practices that tend to go along with biodynamics, such as allowing limited weed growth, could have benefits in terms of creating competition to keep vegetative growth in check, or acting as refuge areas for beneficial predators of vine pests.

There's likely also a large placebo element. As winegrowers adopt biodynamics, they are entering into a philosophical system that acts as a framework to help them maintain a careful approach in the vineyard.

Despite the odd and seemingly antiscience nature of some of its practices, biodynamics merits further scientific study. By and large, winegrowers operating within this rather unusual philosophical framework are making interesting, wines filled with personality—something the world desperately needs more of. And the limited scientific studies that have so far addressed biodynamics have come down in its favor. But it's an open question as to exactly how biodynamics has its effects, and by extension it is therefore unclear which elements of its theory need to be adopted by *vignerons* in order for them to accrue its benefit. This is something that could be tested. The difficulty remains that much of biodynamic practice is so esoteric and has such a pseudoscience ring to it that mainstream scientists are afraid to be associated with it and would find it very difficult to get funding to do the work that is needed.

Table 1 Selected list of biodynamic producers

France	*Bordeaux*	Pontet Canet	*Loire*	Cousin Leduc
	Burgundy	Comtes Lafon		Roches Neuves
		Domaine de la Romanée Conti		Huet
		Domaine Leflaive		François Chidaine
		Leroy		Coulée de Serranr
		Domaine Trapet	*Rhône*	Chapoutier
	Alsace	Josmeyer		Marcoux
		Marcel Deiss		Montirius
		Zind Humbrecht	*Champagne*	Léclapart
		Marc Kreydenweiss		Fleury
		Pierre Frick		Larmandier Bernier
		Bott Geyl		
		Dirler Cadé		
Austria		Meinklang		Nikolaihof
		Sepp Moser		
Italy		La Raia		Foradori
		Nuova Cappellatta		
Portugal		Aphros		
Chile		Antiyal		Emiliana
		Sena		Matetic
USA		Bergström		Brick House
		Littorai		Grgich Hills
		Beaux Freres		Frey
		Ceago		
Australia		Robinvale		Cullen
		Gemtree		Cape Jaffa
		Ngeringa		Paxton
		Pertaringa		Walter Clappis
		Castagna		Jasper Hill
New Zealand		Burn Cottage		Seresin
		Felton Road		Millton
		Quartz Reef		Rippon
South Africa		Waterkloof		Reyneke

Table 2 The different biodynamic preparations

Preparation[1]	Contents	Mode of application
500	Cow manure fermented in a cow horn, which is then buried and overwinters in the soil	Sprayed on the soil typically at a rate of 85 oz per acre in 9 gallons of water
501	Ground quartz (silica) mixed with rainwater and packed in a cow's horn, buried in spring, and then dug up in the fall	Sprayed on the crop plants
502	Flowerheads of yarrow fermented in a stag's bladder	Applied to compost along with preparations 503–507. Together these control the breakdown of the manures and compost, helping to make trace elements more available to the plant
503	Flower heads of chamomile fermented in the soil	Applied to compost
504	Stinging nettle tea	Applied to compost. Nettle tea is also sometimes sprayed on weak or low-vigor vines
505	Oak bark fermented in the skull of a domestic animal	Applied to compost
506	Flowerheads of dandelion fermented in cow mesentery	Applied to compost
507	Juice from valerian flowers	Applied to compost
508	Tea prepared from horsetail plant (*Equisetum*)	Used as a spray to counter fungal diseases

[1]All these preparations are activated or energized by a special stirring process known as "dynamization."

8 Partial root drying and regulated deficit irrigation

Partial root drying (PRD) is a brilliant idea, and a good example of a concept developed in the laboratory and then applied to real-world agriculture. While the credit for the hard work developing PRD in the field must go to Australian viticulturalists, the theory underpinning this practice is largely British. The UK hasn't made many significant contributions to viticulture, but the concept of PRD—now one of the buzz subjects in wine science—was devised in lab experiments carried out at the University of Lancaster, in England, in the late 1980s. Initially, this work had nothing to do with grapevines. According to Lancaster's Dr. Mark Bacon, this research was "purely for the academic pursuit of understanding how plants communicate information regarding the soil water status from the roots to the shoots." The chief focus of these studies was the plant hormone abscisic acid (ABA), and we'll need to understand a bit about ABA if we're going to have enough context to appreciate how PRD works.

ABA

ABA is one of a group of chemicals known as plant hormones or plant-growth regulators. The core members of this group are auxin (first discovered in the 1920s), cytokinin, gibberellin, ethylene, and ABA. Other molecules, such as brassinosteroids, have made recent bids to be admitted to the hormone club, but we'll ignore those for now. These hormones are responsible for coordinating plant growth by acting as signaling molecules. Sent from one part of the plant to another, they give each cell instructions on how to behave and grow. An obvious example of this signaling is what happens if you prune a rosebush, fruit tree, or vine. By cutting off the growing tip, the loss of the hormone signaling from this tip causes dormant lateral shoots to become active, and they grow out. Until fairly recently, researchers used to ascribe one or more functions to each hormone, but now it is becoming clear that hormone signaling pathways aren't arranged in a linear fashion but in a complex network of interactions, dubbed as hormone "crosstalk." This makes it complex to unravel the precise roles of each, but recent progress in molecular genetics is proving extremely helpful in spurring this field on.

ABA is actually a bit of a negative hormone. It usually appears when things are going wrong. In particular, when plants are stressed, the first thing they do is make ABA. One of the most threatening stresses affecting plants is drought. During conditions of water stress, the roots synthesize ABA—the first part of the plant to experience the drought—and send it to the shoots and leaves. This alerts the aerial parts of the plant to the fact that hard times are on the way, and they stop growing and close the small pores called stomata in the leaves. Stomata are important because while they allow in the gases needed for photosynthesis, they also leak out precious water vapor. As a result, the plant makes a calculation: if things are too hot and dry, it will simply close stomata and stop growing, because the risk of water loss then outweighs any benefit from continuing to grow. Even though vines are adapted to warm climates, in very hot summers they'll stop growth at the warmest times of day, as occurred widely in the European summer heat wave of 2003, retarding development. The French term for this heat-induced shut down is *blockage*. Paradoxically, therefore, periods of extreme high temperatures during the growing season can actually slow grapevine development, because photosynthesis has to stop.

Researchers were eager to show that it is the roots that are producing these drought signals, and not the shoot itself responding to water stress, and also that ABA actually is the chief hormone involved. They did this in two ways.

First, they showed that shoots stop growing before the development of decreased water potentials in the aerial parts of the plant. Thus a long-distance signal, rather than reduced shoot-water availability, is mediating this response. Second, they used what is known as a split-pot system. In this elegant work, the root system of a single plant was divided and put into two separate pots. The scientists showed that by watering one pot and letting the other dry out, shoot growth was stunted even though the plant—in this case a young apple tree—actually had enough water. This simple manipulation showed that the plant had been tricked into thinking it was stressed. Removal of the roots in contact with the dry soil also removed the growth inhibition, confirming that this signal was indeed coming from the water-stressed root system. Further research has nailed ABA as the chief culprit in this root-to-shoot signaling, although it doesn't act alone. "Cytokinins, ethylene, and ABA have all been shown to play a role," says Bacon.

Although this split-pot system was initially an experimental tool for proving a theory, it didn't take long for others to see its potential for commercial usefulness. However, I know that there's a gulf between clever ideas or inventions and things that are useful or feasible in practice. With PRD, this translation between the lab and the field is where the real work is being done.

While Brits devised the theory underlying PRD, it took an Australian to apply it to grapevines. The Australian in question was Brian Loveys, who coined the term PRD and put in a mammoth amount of development work to make this idea practical in the vineyard. The potential benefits are clear. Grapevines, like many other plants, will put their efforts into making shoots and when conditions are benign. Give a grapevine plenty of water and nutrients, and it will produce luxuriant foliage, but relatively poor-quality grapes. Stress, on the other hand, encourages vines to invest in the future and concentrate their energies on fruit growth. The problem is, stress the vine too much and, although fruit growth is the preferred option, the quality of the fruit will be compromised, and the vine might even die. Good vineyards—those producing the highest-quality fruit—tend to be those that allow the vine just enough water at the right time in the growing season. Grapevine vigor is the enemy of wine quality.

French researcher Gérard Seguin carried out an influential study along these lines in the 1980s. He conducted a survey of the properties of the soils in the Bordeaux region and concluded that it was the drainage properties of the soil affecting the availability of water that mattered most (this is covered more extensively in chapter 3). The physical properties of the soils, regulating the water supply to the vine, are all-important in determining wine quality. Seguin maintained that the best terroirs are those where the soils are free-draining, with the water table high enough to ensure a regular supply of water to the vine roots, which then recedes a good deal on *veraison* (when the berries change color) so that vegetative growth stops and the vine concentrates its energies on fruit-ripening. Of course, if you are not blessed with such a terroir, then there's not much you can do about it. But what you can do is to try to implement viticultural interventions that aim to make your terroir as favorable as possible. And if you are irrigating, then you have some degree of control over water availability,

Far left In Mendoza, Argentina, many vineyards are still flood-irrigated, with a network of irrigation channels. Meltwater from the Andes is readily available and the vineyards are pretty flat.

Left A young vine in Portugal's Douro Valley being watered by drip irrigation. Once this vine has established its root system it will no longer need irrigation.

which, according to Seguin, is probably the key vineyard factor influencing wine quality.

REGULATED DEFICIT IRRIGATION

One technique that has attempted to replicate this change in the grapevine's water supply is called regulated deficit irrigation (RDI). It sounds simple enough in theory: just reduce or cut irrigation at the right point and you get an increase in fruit quality. It's an artificial attempt to replicate the ideal conditions identified by Seguin for high-quality grape production. In practice, it's more complex. You need to know exactly how much water deficit the vines are experiencing, or things can go wrong. The benefit of this technique is that it is relatively simple to apply, and can be used in any irrigated vineyard without the need for expensive modifications. Experience with RDI has shown that it works best if the water supply is cut just after flowering and berry set, but then restored at *veraison*. This seems a little at odds with the results of Seguin's study of Bordeaux terroirs; perhaps this has something to do with the climate differences between Atlantic-influenced Bordeaux and hot areas where irrigation is typically practiced.

RDI has been particularly successful in Australia. According to Dr. Michael McCarthy, who has been one of the pioneers of the technique, "RDI is probably the most rapidly adopted bit of new management that we have seen for a long time, starting with no industry awareness of it in the early 1990s to something like 50% adoption now in irrigated vineyards in Australia."

PARTIAL ROOT DRYING

PRD is more complicated than RDI but from a scientific perspective it is perhaps more elegant, and has the potential to be more effective. Physical separation of a vines root system isn't feasible in the vineyard, so the split-root system is created by use of a dual-drip system that irrigates either side of the vine. The irrigation regime is then switched from one side to the other at intervals of 7–14 days. This allows the roots system to dry out enough for it to signal to the shoot and leaf system that there's some water stress, but not enough for damage to occur. This also ensures repeated bursts of signaling from the root that help keep canopy growth restricted. The watered roots on one of the sides maintain an adequate supply to the vine so it can still function, but because of the root-signaling, vigor is reduced, water use is decreased and, potentially, the grape quality is enhanced.

The chief benefit of PRD is the reduced water use. "Most users of PRD would do so because of the documented improvement in water-use efficiency," says Brian Loveys of Australia's CSIRO. This is a critical issue because in vineyard areas where irrigation is practiced, water is scarce, and is likely to prove a limiting factor for viticulture in hot climates. This is a situation likely to get worse in the future with increasing competition for water resources and a predicted rise in global average temperatures.

Whether or not PRD actually creates better wines is still a little controversial. "The question of wine quality is much more difficult," agrees Loveys. "Grape quality has so many different components, some of which are certainly altered by deficit irrigation methods." He has done some research showing that PRD changes the types of anthocyanins in red grapes, and that some flavor compounds, including β-damascenone, β-ionone, and trimethyl dihydronapthalene, are also influenced positively. Experimental wines made by his group from PRD-irrigated vines also show improved flavor characteristics.

Moving away from grapes, experiments by the University of Lancaster group on tomatoes have demonstrated measurable positive effects of PRD methods. In these studies PRD caused a significant reduction in vegetative mass of the tomato plants, but not the fruit. The tomatoes were smaller, but 21% more concentrated as determined by total soluble solid measurements. A commercial taste-test panel expressed a marginal preference for the PRD-grown fruit. The most impressive statistic was that the water-use efficiency (WUE) was increased by 93% during the ripening period.

MECHANISMS

So how does PRD work? Its physiological basis is still being worked out. As mentioned, ABA seems to have a key role by traveling from the roots to the shoots and telling them to stop growing. But why does ABA reduce shoot growth and not fruit growth? "PRD may have some effect on fruit growth, but in wine grapes, even though they may be smaller, the overall effect on yield is minimal—or not apparent," explains Mark Bacon. "There is some feeling that fruit may be relatively isolated from the vegetative part of the plant, in respect of xylem connections and function." For those whose plant physiology is rusty, the xylem is part of the conduction system of the plant, transmitting water and various solutes from the roots to the shoots. The other component of this system is the phloem, which conducts nutrients such as the sugars produced by photosynthesis. "After *veraison*, all water to the fruit is provided by the phloem, not the xylem," continues Bacon. "As signals like ABA travel in the xylem, the hypothesis goes that signal transport to fruit is much reduced." He adds that this will require a lot more work to unravel in detail, but the concept of chemical isolation of the fruit—that the root signal doesn't actually reach the fruit—is potentially a good explanation for why the grapes carry on developing normally, even though shoot growth has largely stopped.

A second mechanism for improved fruit quality is that PRD may alter the way that a plant distributes its nutrient resources. One of the important physiological processes in plants is how carbohydrates, produced by photosynthesis in the leaves, are partitioned to the various parts, a topic known as source-sink relationships. When there is a steady supply of water, the actively growing shoots act as what is called a "sink," taking more of the carbohydrates than the fruit. There is some evidence that the reduced growth of the shoot system means that its sink strength is reduced, such that more of the resources are then allocated to the fruit. In tomatoes, developing side shoots are the main competing sink for carbohydrates, and if their growth is curtailed then resources are likely to be redirected toward the fruit. That's why when you grow tomatoes it's a good idea to pinch the side shoots out. The same probably applies for grape vines. The shoot stops growing and its strength as a sink is reduced relative to the still-developing grapes. This is also the proposed mechanism for the effectiveness of RDI.

Of course, PRD is only possible where irrigation is practiced. You can't do it in areas that have significant growing-season rainfall, or where irrigation is against the rules, so this excludes many of the classical European winegrowing regions. However, the theory behind it has been applied by Bart Arnst, viticulturalist at Seresin Estate, in unirrigated vineyards in New Zealand's Marlborough region. He has begun a system of alternate mowing between vine rows, such that one half of a vine's roots will face competition from the cover crop at a time. It's an interesting implementation of the PRD theory.

The requirement for a dual-irrigation system is one further obstacle toward take-up of PRD. "PRD is proving complex to implement on large vineyards because of the complexity of, in effect, running an extra irrigation system," reports Michael McCarthy. He is hoping that by using subsurface drip irrigation for RDI he'll be able to achieve the same water savings as PRD but without the complexity of the dual-irrigation system. "I now prefer the terminology "strategic irrigation management" to encompass both PRD and RDI," he says.

It's early days, but how far have these sorts of technologies spread? In Australia, "most of the major wine companies would have tested or implemented PRD and/or RDI," says Loveys. "We are also aware of fairly significant use of PRD in California, Spain, and South America." From what has been shown so far, it does seem that these deficit irrigation strategies are an important new tool in viticulture. However, there are dissenting voices. "You should be aware that there is a school of thought, coming mostly from the USA, that there are no benefits of PRD over and above those achievable through other techniques that reduce irrigation amounts relative to what has become industry practice," reports Loveys. "This is not surprising, as all these techniques rely on the stimulation of the same natural stress-response mechanisms of the plant. We believe that PRD offers a safer and more reliable way of achieving the desired result of improved water use efficiency."

Of course, it may well be that PRD will yield improved grape quality as well as improved water use efficiency. "In any event," Lovey concludes, "what the debate has done is show that it is often possible to grow a commercial crop of grapes with considerably less water than would have been considered necessary a decade ago." It's not the most glamorous of conclusions, but perhaps—considering the growing urgency of efficient water use—it's the most important one.

9 Pruning, trellis systems, and canopy management

The grapevine is a something of a freeloader. It has adopted a growth strategy where it can't be bothered to support itself and relies on others instead. In nature, plants are in competition for two sets of resources: those from the earth and those from the sky. In most environments the latter struggle is the key one, the struggle for enough light to drive photosynthesis and hence food production. Trees often win this battle by getting their leaves 30 feet or more from the ground, but to do this they have to spend years slowly building a woody trunk with enough mechanical strength to support this elevated growth habit. Grapevines, like other climbers, have seized the opportunity to make the most of this third-party effort. They realized that by climbing they could save themselves the trouble of developing a self-supporting stem. Not having to develop supporting girth permits rapid growth, so grapevines are experts of growing up other plants until they break through to uncompeted-for light on the outside of the canopy of their hosts.

The growth habit of the vine is finely tuned for this lifestyle. Shoot structure is simple, with each node having the capacity to produce tendrils or flower buds opposite each leaf. Gripping to the host plant via these tendrils, the shoots grow rapidly toward the light, seeking the gaps in the canopy. Where the vine breaks through to sunlight, tendrils are discarded in favor of flowers, resulting in fruit production. At the other end of the vine, the roots are capable of growing deeply, eking out water and mineral resources in competition with the pre-existing root system of the host plant.

The science of viticulture attempts to manipulate vines to get them to produce good yields of high-quality fruit. It takes into account the natural growth habit of the vine and adjusts it to suit the context of the vineyard. Because vines are climbers, this usually involves some means of support for the freeloading grapevine to grown on. It's an area where good controlled scientific experiments are rare, in part because of the immense difficulty in doing them. The way vineyards look today is a reflection of tradition, trial and error, guesswork, specific environmental constraints, and convenience. As a result, vineyards across the world's winegrowing regions differ markedly in their appearance. This chapter sets out to discuss some of the scientific considerations that shape the way that vines are pruned, trellised, and managed.

VITICULTURAL GOALS

The holy grail of viticulture is to get high yields of high-quality grapes with minimum cost in terms of vineyard labor and inputs. Almost invariably, a compromise is involved: sometimes yield must be sacrificed for quality; sometimes it's the other way around. Good viticulture also takes into account the economic objectives of the wine that's going to be made from the crop. Vineyards are therefore managed to produce grapes of appropriate quality and at the right yield, for the right cost.

If you are starting a vineyard from scratch there are several key choices that need to be made. These include choosing the appropriate variety, rootstock, vine spacing, trellis method, and making decisions regarding irrigation, pruning, and canopy management. Making the right choice is important, because the vines will have a productive lifespan of 20 years or more. This sort of timescale makes innovation based on experimentation tricky, and so in areas where vineyards are already established, people frequently copy the vineyard style of their neighbors. Often viticulturalists have to work within the confines of vineyards that have already been planted, in which case there is limited room for modification. In the following sections I'll give a brief overview of some of the scientific principles behind trellis systems, pruning, and canopy-management decisions, and how these all affect wine quality.

Current viticultural thinking is that the control of vine vigor and fruit-zone light exposure is the key to successful vineyard management. Vigor is an important element of viticulture. If the vine is growing actively all through the season, developing a huge, dense canopy, then the actively growing shoots will represent a powerful sink for the vine's resources that will inhibit the sugar accumulation, which normally occurs during the fruit-ripening phase. The immediate result is delayed fruit development and lowered quality, but also of importance is the effect of the profuse canopy in shading the inside-growing shoots. This is because light is of crucial importance for bud fertility. In the fertile regions of the shoots (next year's canes) grapevines have uncommitted bud primordias that can form either tendrils or flowers, and because the development of these buds takes two seasons, it is the light that the fruiting canes received in the preceding season that determines their fertility in the current year. This is understandable, and makes sense for a grapevine in the wild, where it will be growing on trees because it only wants to make fruit where the shoots poke out through the host canopy into the light. In addition, shaded grapes may maintain high levels of vegetal-tasting methoxypryazines, which are undesirable in most wines and are dissipated through light exposure. In support of these ideas, precision viticulture studies have emphasized that in many wine regions, the parts of vineyards that produce the highest-quality grapes are those with the lowest vigor (see Chapter 4). Excessively shaded canopies have another drawback in that they increase the risk of disease. This is because of the limited air circulation and the extended drying-out time after the canopy is wetted.

Another key to successful viticulture is getting grape flavor development (known as phenolic or physiological maturity) to coincide with sugar maturity. In warmer regions, the risk is that by the time the grapes have reached flavor maturity, the sugar levels are very high, resulting in overly alcoholic wines. This is currently a major problem in many New World wine regions. In cooler regions, flavor maturity often occurs at much lower sugar levels and the challenge can be getting the grapes ripe enough in terms of sugar before the grape loses its photosynthetic capability or the fall rains set in.

With trellis systems and canopy management there is no one-size-fits-all solution. Viticultural methods have to be adapted to local conditions, and while the general principles remain the same, factors such as soil fertility, climate, water availability, grape variety, and skill and

Some common viticulture terms

Basal leaf removal Basal leaves are removed to expose the fruiting zone, allowing access to light and encouraging air circulation, which prevents disease.
Cane The stem of a grapevine that is one season old and has become woody, and which can either be cut back to spurs (up to four buds) or canes (typically six to fifteen buds) for the following season's growth. Canes are also sometimes called "rods."
Cordon The woody framework of the vine extending from the top of the trunk. A cordon-trained vine has a trunk terminating in one or more cordons, which are then spur pruned.
Head training Where the head of the trunk is pruned to either spurs (gobelet system) or canes (such as the Guyot system, which was named after its inventor).

Hedging Also known as shoot tipping, this involves cutting back excessive growth at the top and sides of the canopy midway through the growing season. The aim is to leave enough leaves to ripen the fruit, while preventing excess growth that will lead to shading and competition for resources with the fruit. A balanced vine will typically have two fruit clusters and 15 leaf nodes on each shoot.
Shoot Green growth arising from a bud.
Spur A short cane that has been cut back to between one and four buds to provide the following season's shoots.
Trunk The main, permanent vertical growth of the vine, which supports the canes or cordons. Grows in girth only.

availability of vineyard workers should be taken into account in the consideration of the most suitable management choices.

PRUNING

Pruning is an intervention that aims to improve vine fertility, encourage optimum canopy development, and regulate crop load in line with the quality objectives of the grower. Vine pruning can seem a little complex for those unfamiliar with it, so here is my attempt to present a digestible introduction. There are two different styles of pruning—both widely used— known as cane pruning and spur pruning.

Cane pruning involves selecting one or two (rarely more) shoots from the previous season's growth and cutting them back to between, say, six and 15 buds. These then form the basis for the following year's growth when they are tied down horizontally. A renewal spur is also left for generating new canes. Typically, with cane-pruned vines the only permanent vine growth is a vertical trunk. The actual practice of cane pruning is more challenging than spur pruning and is usually employed with varieties that have low fruitfulness in basal buds, and where the vineyard workers are up to the task. Results can be very good. It's ideal for cooler climates and certain varieties where basal buds have low fruitfulness.

Spur pruning involves cutting the previous season's growth back fairly drastically to just a few (up to five, but more normally two or three) buds. These will be borne on a more substantial permanent vine structure, usually consisting of a trunk plus horizontal cordons. Spur pruning is technically much simpler and requires less skill on behalf of vineyard workers. It can be part mechanized by using mechanical pre-pruning (sometimes called "bushing") that may then be tidied up by hand.

Minimal or mechanical pruning is a relatively recent development. This involves no real pruning, just cutting the vine's growth back rather crudely using mechanical means. It makes a mess of the vineyard, but proponents claim that after a couple of years the vine gets into balance and produces good yields, with small bunches of grapes all over the canopy, rather than clustered in a fruiting zone. This

is therefore only compatible with mechanical harvesting. It is used in situations where manual vineyard labor isn't practical or economically feasible, and only really works in warm climates. A variation on the theme is mechanical cutting back of vine growth, which is then tidied up by vineyard workers manually. This is evidently only going to work for spur-pruned vines.

TRELLISES AND CANOPY MANAGEMENT

There exists a confusing array of different styles of trellis, each with their own advantages in specific situations. Trellis systems are a vital part of the canopy-management tool set.

Canopy-management techniques are aimed at achieving optimum leaf and fruit exposure to sun, while reducing the risk of disease and pushing the quality-to-yield ratio as far as possible. Open canopies help prevent disease in two ways. They allow better spray penetration and also better air circulation, with faster drying-out times. Canopy-management strategies aim to get the vine into some sort of balance and they have been particularly successful in situations of high vine vigor, often caused by fertile soils and irrigation. The modern canopy-management techniques that involve, for example, split-canopy trellis systems, such as the Smart-Dyson, are not, however, of much value in low-vigor sites, such as the major vineyard areas in the classic Old World regions.

Richard Smart's work has been particularly influential in this area. He dubs the traditional canopy-adjustment techniques of trimming, shoot thinning, and leaf removal in the fruit zone as "band aid viticulture," because they are interventions that have to be reapplied annually. His solution is to alter the trellis technique to increase canopy surface area and decrease canopy density—a once-only intervention. In particular, the use of high vertical-shoot positioning (VSP) systems and divided canopies (such as the Smart-Dyson and Scott-Henry trellis systems) have been effective means of getting highly vigorous vines into balance. The basis of this work is to manage the vine vigor, achieving optimum leaf-to-fruit ratio. Smart considers this to be as important a consideration as the traditional vineyard currency of yield.

1 Standard vertical shoot positioned (VSP) trellis.

2 A permanent cordon, with short spurs, on a single wire trellis.

3 A single-wire trellis late in the growing season showing the "umbrella sprawl" of the shoots. This works well in warm climates where the fruit is partially shaded.

4 A split-canopy system with shoots growing down as well as up. This is often used in high vigor vineyards.

5 A bush-vine in Stellenbosch, South Africa, just before harvest. You can see the way that the grapes are partly, but not fully shaded by the leaves, which is ideal in a hot, sunny climate. Red grapes enjoy dappled light.

6 The Wehlener Sonnenuhr vineyard in Germany's Mosel region. Here the vines are trellised using the pendelbogen system, with two canes tied up to a single pole in a heart shape.

7 In the northern Rhône vines are grown as bush-vines, but each is supported by a single pole, as illustrated here in Hermitage.

8 Interventionist viticulture at Dry River, Martinborough, New Zealand. This is a split-canopy system with complete leaf removal in the fruit zone, and an "Extenday" white horticultural fabric under the rows that reflects sunlight to the bunches. In this cool climate, it helps the Pinot Noir and Syrah grown here achieve appropriate ripeness.

9 A single Guyot pruning system in Sancerre, Loire, France.

I asked viticultural consultant David Booth, who worked in Portugal (and, sadly, died prematurely in 2012 at the age of 47), about his views on canopy management. "I think it is probably one of the most important tools we have," he responded. "But I do have a very broad definition, much more than just putting up technically advanced trellis systems. My definition encompasses a range of vineyard-management practices, including winter pruning, shoot thinning, shoot positioning, leaf thinning, and hedging. The skilled viticulturalist should be able to look at the soil profile (before planting) at any particular site and anticipate future vine vigor. Then they can make a series of decisions about trellis system, spacing, and rootstock. A high VSP is probably the easiest to manage and I reckon this should be the natural first choice. The more advanced trellis systems are for when the anticipated vine vigor is so high that you have doubts about your ability to accommodate the growth within the VSP, or several years after planting you realize you have blown it and underestimated vine vigor and need to modify the existing trellis. My first choice for divided-canopy systems is Smart-Dyson, since it is easy to manage, does not require wide rows and is easy to machine harvest." Booth adds, "A key factor to think about in canopy-light environment is not just sunlight striking bunches, but also leaves shading other bunches, which has negative implications for wine quality, for reasons that are not understood but probably have a lot to do with potassium balance."

There are two dogmas in viticulture that are worth addressing here. One is that reduced yield equals higher grape quality (and vice versa); the other that old vines produce better wines. The scientific bases behind both of these assertions are not entirely clear. Evidently, it is possible to lower grape quality by means of excessive yields. And in classic Old World regions, which are typically low-vigor sites, reducing yields by pruning canes short does have the effect of raising quality, to a degree. But in higher vigor, irrigated vineyards in warm regions, pruning short will not result in a vine that is in balance, and no quality gain is likely to be seen. In such high-vigor situations, moving to a split-canopy system often has the effect of bringing the vine into balance,

raising yields, and improving quality at the same time. The second dogma is that old vines produce better vines. It is repeated so often, the suspicion is that there must be some truth in it.

If it is indeed the case, what is the scientific explanation? One suggestion is that it has to do with the amount of overwintering perennial wood on the vine. Some researchers have noted that training systems with more perennial wood, and thus more carbohydrate storage area during the dormant period, produce better wines. Older vines tend to have more perennial wood and this could be to their advantage. However, a more likely explanation is the one offered to me by David Booth. "Young vines are harder to manage because they are less buffered against any environmental stresses. But as you know, they can give great quality, probably because they are naturally low vigor (small root system) and have good leaf and bunch exposure. Old vines are also naturally low in vigor, due to wood disease and exhaustion of nutrients. I reckon the problem is more in the middle years, especially in high-vigor soils with inadequate trellising systems— then you get the classic shading problems."

I asked him whether it is possible to increase yield and maintain or even improve quality by viticultural interventions, such as canopy management. "Sure, but only really in the case that I have just mentioned of the middle-aged vigorous vine on the high-capacity site," replied Booth. "These sites are more common than you might think. Look for the small yellowing leaves in the inside of the canopy, as an indication of leaf shading. A full-on trellis conversion may be the solution, but I tend to work first with competitive cover crops, nutrition, irrigation management, leaf pulling, and shoot thinning. As you might have figured out by now, there is no silver bullet, just a raft of tools that often need to be used in conjunction."

1 A four-cane pruning system in the Brancott Vineyard, Marlborough, New Zealand. This is Sauvignon Blanc, and the Brancott Vineyard can produce high yields of good-quality Sauvignon, which is why there are four canes rather than two.

2 A pergola system in Mendoza, Argentina. This can produce very high yields.

3 A pergola system in Rías Baixas, Spain, just after pruning. The vines are suspended off the ground with marble pillars. The yields here will be high, as you can see from the number of buds left on the canes.

4 A spur-pruned vineyard with a permanent cordon. This is in the Lisboa region of Portugal.

5 The Slyvoz pruning system shown here is designed for high yields, with lots of short canes bent downwards. This is pictured in Marlborough, New Zealand, where it is sometimes used for sparkling base wines, where high yields of only just ripe grapes are needed.

6 The Lyre split-canopy training system, here seen in Austria. It is complicated, but can work well, giving good sunlight exposure and air-flow around the canopy.

Examples of different training systems

Gobelet An old, and probably the simplest, training technique. Here the spurs are arranged around the head of a trunk or short arms coming from the top of the trunk. This is only really used in warm, dry climates in low-vigor situations. It doesn't need a supporting trellis, but shading of fruiting zones can be a problem (although it could be argued that the dappled light the low-vigor canopy provides is ideal). The style is popular in the Mediterranean regions. In the New World, this is known as the bush-vine; in Italy it is called the *alberello*.

Guyot One of the most popular cane-pruned systems, with a single or double cane layered horizontally from the head of a trunk. One or two renewal spurs are also left. Simple and effective, it is particularly suited to classic Old World, low-vigor vineyards.

Cordon de Royat Simple spur-pruned system, usually with a unilateral cordon spreading from a low trunk. Variations on the theme include a double cordon. It is a simple and effective system.

VSP (vertical shoot positioning) Widely adopted system where the shoots are trained vertically upward in summer, held in place by foliage wires. Leads to relatively tall canopies, and is suitable for mechanization. VSP is good option for most sites.

Scott Henry A split-canopy trellis system where the shoots are separated and divided into upward and downward growing systems, held in place by foliage wires. It is useful for high-vigor situations and is suitable for both spur and cane pruning. The advantage is lower disease pressure, improved grape quality, and higher yields. Looks like a wall of vines in practice, growing from the ground to about 6 feet in height.

Smart-Dyson Developed by John Dyson and Richard Smart, this is a variant of Scott Henry trellis, with curtains trained up and down from just one cordon. Popularized by the influential work of Smart, the world's foremost flying vitculturalist.

Lyre A split-canopy system, this has two canopies facing each other forming a "V" shape from a single cordon. It is complicated, but can give good results.

Dopplebogen The "double bow" system for Riesling vines common in the Mosel-Saar-Ruwer region of Germany, where each vine is singly staked and two canes are bent around into a bow shape. Single-staked vines are also found in France's northern Rhône. This sort of system is adapted for steep slopes where any other sort of trellis system would be impractical.

Éventail French term for "fan," this is a cordon system with a number of arms arising from a short trunk, each bearing a short cane. This is popular in Chablis and also used in Champagne.

Geneva double curtain A rather complicated split-canopy system, with cordons grown high on two parallel horizontal trellis wires, with the shoots bending down. A variation on this theme is the lyre system that is relatively common in Austria, but here the split canopies have upward-growing shoots, and angle outward slightly. Both are hindered by the fact that they require wider rows and complicated trellis systems.

Sylvoz A high-yielding system sometimes used for making sparkling base wines, with many downward-pointing "hanging" canes.

Tendone The Italian term for the arbor or pergola system common in parts of Italy (e.g. Veneto), Portugal (Vinho Verde), Argentina, and Chile. The vines are trained high off the ground on a series of wooden frames. They look attractive, and yields can be heroic, but fruit shading is a problem and quality suffers as a result. They are hard to work, too. Of all trellis systems, this replicates most closely the growth of the vine in the wild.

Single-wire umbrella sprawl Commonly found in Australia, the shoots sprawl in an umbrella fashion from a single wire. It looks untidy, but the fruit is shaded in dappled light, preventing sunburn in an otherwise risky situation.

Section 2
In the Winery

10 Oxygen management and wine quality

Oxygen and wine is a big topic, and covers a great deal more ground than simply the topic of oxidation. Many years ago, no less an authority than Louis Pasteur described it as the "enemy of wine," but a more modern understanding of wine chemistry has shown that this is far from the truth. Yes, there are times when exposure to oxygen can be a bad thing, but there are other times when wine quality can suffer because of a lack of exposure to oxygen. A more correct view would be that oxygen is an important tool to shape wine style. Further, a good understanding of when, and how much, oxygen should be allowed contact with the developing wine is vital to the winemaking process.

We'll begin, though, by dealing with oxygen's negative effects. Of all wine faults, oxidation is the easiest for students of wine to experience. It's hard to teach someone about wine "faults," such as reduction, *Brettanomyces*, or even cork taint, unless you have examples on hand, and that's not always easy. If you want to experience oxidation, simply pour a glass from a bottle of wine, reseal the bottle, and a week later you'll have an oxidized wine. But as with the other faults—with the noble exception of cork taint—there's some gray area with oxidation. First of all, some wine styles involve deliberate exposure to oxygen to induce a degree of oxidation. The best example would be certain types of Sherry, tawny Ports, and Madeira, where long aging in barrels that aren't fully filled to the top results in a kind of oxidation that is part of the wine style. Second, old wines are often greatly appreciated, and part of long aging in bottle involves a degree of oxidation. It's not always clear-cut, because one taster's old claret aged to perfection may be considered over the hill and oxidized by another. In fact, it may be simplistic to think of oxidation as simply a wine fault. Even where there is no evidence of oxidation, there may be a significant quality loss to the wine through inappropriate exposure to oxygen.

Wine is a complex liquid consisting of a multitude of different chemicals, many of which are created by yeasts during the fermentation process. As with any mixture of chemical entities, they will rearrange themselves into the most favorable energetic state. This is the principle of entropy. What this means, in simple terms, is that the various molecules will swap tiny charged particles called electrons, and the nature of this swapping is determined by the reactivity of different components and what is known as the redox state of the wine. In any chemical reaction between two partners, one entity gains electrons (i.e. is reduced), and the other loses them (is oxidized). These processes occur in tandem, and it follows that you don't always need to have oxygen present for oxidation to occur. The redox (reduction–oxidation) potential of a particular wine determines what sort of state of its component molecules is most energetically favorable. That is, the reactions that take place in wine will be affected by the redox potential. This potential will shift according to how much oxygen the wine is exposed to. It follows that it will be lowest in bottled wine and highest in barrels after replenishing and stirring. It will also be lower in wines sealed with closures with lower oxygen transmission (such as tin-lined screw caps) than it is in those with higher gas transmission (such as most synthetic corks). Exposure to air results in the redox potential rising because oxygen will become dissolved in the wine and this will fuel oxidation reactions.

So what sorts of reactions happen when wine is exposed to oxygen? This is where it gets a bit complicated. Oxygen, which is often assumed to be highly reactive, doesn't react directly with wine, but requires an oxidizing agent to

Above A red-wine barrel cellar at Thelema, Stellenbosch, South Africa. Despite their incovenience and cost, barrels are still used worldwide because of the way that they allow just a little bit of oxygen exposure during winemaking, in addition to their flavor impact.

become truly reactive. "If we take the aroma compounds that we have in the wine and we put them in contact with oxygen in water, most of them will just stay as they are," explains wine scientist Maurizio Ugliano, previously of the Australian Wine Research Institute, now working for synthetic closure company Nomacorc. This is a remarkable statement, but it chimes with experience. There's something distinctive about wine that renders it susceptible to oxidation, and means that it rapidly loses its fruitiness in a way that, for example, fruit juice or soft drinks don't.

So how is it that oxygen becomes reactive? "What happens is that the oxygen interacts with a lot of things that have nothing to do with the aroma fraction, particularly phenolics, and through that first interaction you will generate species that will then drive a lot of chemical reactions in wine," explains Ugliano. "You have a first interaction between oxygen and phenolics, and the outcome of this reaction will be the reactive species that will then be able to affect aroma precursors and aroma compounds, and push the wine toward oxidation."

Professor Roger Boulton of the University of California-Davis adds to this: "As a wine is exposed to oxygen, the key initial reaction is the oxidation of monomeric phenols with a special reactive group to form hydrogen peroxide," he explains. Hydrogen peroxide is very reactive, and then goes on to interact with other wine components. "The peroxide can be consumed by a number of other reactions, either being quenched by tannins

and other phenols (dominant in red wines, much less in whites) or forming acetaldehyde by reaction with ethanol," says Boulton.

Acetaldehyde is an important molecule in the oxidation of wine. Also known as ethanal, it's the oxidation product of alcohol, and has an aroma described to be like fresh-cut apples, and it tastes nutty and apple-ish. Sherry and Madeira, made in a deliberately oxidized style, have high levels of acetaldehyde, which gives wine a flat texture in the mouth. Indeed, one description of oxidized whites is "sherried."

Transition-metal ions are also a vital part of the process. "You need metal ions to start the reaction between phenolics and oxygen," says Ugliano. "In the absence of metal ions, such as iron or copper, this would not happen. What we have observed when we have tried to see whether the concentration of metals was one of the drivers for oxidation, was that the concentrations of metals that we seem to have today in the wine industry are not making any difference. If we add metals to different levels, staying in the range that is reported from studies, we don't observe a difference in oxidation." So it seems that the metal ions are necessary but they are not the limiting factor in oxidation in the concentrations seen in modern wines. Without them, there wouldn't be oxidation. But we can't eliminate them from wine to the level required to stop oxidation.

Phenolic compounds are necessary to generate reactive oxygen species that then go and do the damage, in both red and white wines. "There have been a lot of studies done recently in France, and we did one in Chile looking at catechin and epicatechin, the so-called flavonoids," says Ugliano. "They interact with oxidation through this mechanism that involves iron and copper, and they form quinones. Quinones are very powerful reactive agents. For example, they can react with amino acids and form aldehydes. Today, a lot of what we observe as oxidation, especially in white wines, is not the contribution of aldehydes, but the loss of the fruity aromas that are due to thiols. In Sauvignon Blanc, oxygen plays a big role in the natural decay of the passionfruit aromas. We think that a lot of the oxidation concerns that exist in the wine industry today should be more directed to managing this loss of fruitiness, rather than simply looking at oxidative faults. Wines end up flat and not expressing the characters that you'd expect."

Sulfur dioxide (SO_2) is almost universally used in winemaking to protect wine from microbial spoilage and oxidation. But SO_2 is not an antioxidant, and it doesn't directly protect the wine from oxidation. SO_2 doesn't actually react with oxygen. Instead, it reacts with the products of the first stages of oxidation. "Sulfur dioxide will be recycling the quinones back to the original, nonreactive phenolics," says Ugliano. "But they will consume sulfur dioxide, which will also bind directly to these reactive species, blocking them in a state where they will not be reactive any more. Sulfur dioxide doesn't really react with oxygen. It is a very slow rate of reaction."

So when SO_2 is binding up the products of the initial oxidation step, does it capture all of them, or do some get away and cause a little oxidation—some collateral damage? "It is difficult to answer the question because there haven't been many studies on this," replies Ugliano. "A lot has to do with how much is around. If you have enough sulfur dioxide, you can in theory catch all the reactive species, but this would be a transient scenario because in this reaction sulfur dioxide will be consumed. Relatively quickly you would get to the point where there is not enough anymore. So there are going to be some that escape the blocking capacity of sulfur dioxide and just do other things. The interesting part is that one of the other things that these reactive oxygen species can do is to react with other phenolics. So, the phenolic component itself has a buffering capacity in the wine to prevent the onset of the negative oxidative mechanism, which could be aroma oxidation. There is also the role of other phenolic parts in blocking these reactive phenolics. We don't understand this very well yet, because characterizing the phenolic fraction of a wine is not so simple. So, we are not really capable of saying that a wine that has this particular family of phenolics in very high concentrations will be capable of blocking the presence of the reactive oxidative phenolics."

Red wines have a higher capacity to absorb oxygen without showing signs of oxidation. This is because the extended skin contact that red winemaking typically employs means that more phenolics are present, and these are able to act as "buffers." White wines tend to lack this sort of buffering capacity, and so need to be protected from air to a greater extent.

So let's look at SO_2 and its role in protecting wine by reacting with the products of the initial stages of oxidation. If you look at the equations, for every mg/liter of oxygen consumed, 4 mg/liter of free SO_2 will disappear if the only outcome of this oxygen consumption is reaction with SO_2. So if this ratio of 4:1 is observed, then there is no oxidation of wine components. If the ratio is below 4, then other oxidation mechanisms are occurring, such as the formation of aldehyde or the loss of aromatic thiols. Andrew Waterhouse, an expert on oxygen in wine, (working at the University of California-Davis) has compared the amount of oxygen consumed with the amount of free SO_2 consumed, and in one white wine found a ratio of 1, suggesting that SO_2 is doing a poor job in this wine. Unsurprisingly, a sensory panel reported that this wine tasted oxidized. Waterhouse and colleagues found that in lees-aged white wines the SO_2 seems to work better. Following this up they noticed that lees-aged wines have less pyruvic acid in them, and that pyruvate binds to free SO_2, enough to make it unreactive, but not as tightly as acetaldehyde would bind. Pyruvate and some other wine components are able to bind free SO_2 so that it isn't available to work in a protective way, but the binding is a fast equilibrium, meaning that in a titration experiment of the type used to measure free SO_2 levels, the binding is released and the read-out for free SO_2 is higher than it is actually present at in the wine. Waterhouse points out that this is why people are typically worried about white wines when free SO_2 drops below 10mg/liter, because in these situations in the wine there may be no free SO_2 available. During barrel aging with lees present, the yeasts and bacteria break down the ketones and aldehydes, reducing them to the alcohol forms. Thus the wine is better protected with the same level of free SO_2.

Waterhouse proposes that the ratio of free SO_2 consumption to total consumed oxygen could be a useful measure of the oxidation resistance of a wine. Those with a ratio closer to 4 are wines that are much less susceptible to oxidation, and therefore need less protection. But this is a complicated measurement to make. His conclusion is that we really need a better way of measuring free SO_2 that measures what is actually present in the wine, unlike the current measurement system which doesn't recognize transiently bound SO_2.

OXYGEN MANAGEMENT

During winemaking, exposure of the developing wine to oxygen is crucial at certain stages of the process, and getting this right is one of the keys of successful winemaking. The problem is that there's a lot of guesswork involved. Traditional winemaking practices have serendipitously controlled oxygen exposure by using oak barrels and processes such as racking, often to good effect; sometimes not. One of the successes of modern winemaking has been the use of stainless-steel tanks and practices that protect the must and evolving wine from oxygen throughout the winemaking process—this is known as "reductive" winemaking, and it has been central to the development of fruit-driven wine styles. Some winemakers will, however, deliberately allow their white-wine musts to have oxygen exposure, resulting in the oxidation of many of the phenolic compounds present (known as oxidative juice handling). This must goes a dark color, but from then on the wine is handled reductively. The resulting white wine is actually longer-lived and more resistant to oxidation.

MACRO, MICRO, AND NANO-OXIDATION

During winemaking, oxygen isn't always an enemy of wine. Traditional red winemaking usually contains stages where as much oxygen as possible is introduced into the wine during fermentation, either through punching down the cap of skins, which is in contact with air, or pumping wine from the bottom of the tank over the surface of the skins. Some fermentation tanks now even have jets fitted for pumping oxygen directly into the wine. The yeasts consume oxygen, and the fact that they are producing carbon dioxide means that there's little risk to oxygen-sensitive wine components at this stage. As fermentation slows down and stops, the time window for this macro-oxygenation closes, and from this point onward the wine will need to be protected from oxygen to varying degrees, depending on the stylistic goals of the winemaker.

From here on, it is not just the cumulative dose of oxygen that matters, but also the rate of delivery. Traditional winemaking involves the use of oak barrels, which allow exposure of the wine to small amounts of oxygen over a long period. The level of oxygen the wine is exposed to during barrel-aging will depend on a number of factors including the size of the barrel (smaller barrels allow a larger volume to surface area in contact with the oak); the thickness of the staves; whether or not the barrels are replenished—keeping them full with a minimal ullage (headspace) protects the wine; how long the wine is in barrel; and, of course, whether the barrel is racked. Racking is the process of moving wine from one barrel to another, or moving the wine from barrel to tank, and this can be accompanied by quite a lot of oxygen pick-up. Indeed, some racking stages deliberately involve exposure to oxygen, although there are ways of moving wine under the protection of inert gas. Regular wine movements, such as pumping wine from one tank to another

Above Pumping over a tank of red wine to help keep the cap of grapes skins wet, oxygenate the wine, and aid color extraction.

Above Splash-racking a fermenting red wine, in order to introduce oxygen.

Above Racking barrels in a large winery in Rioja. The wine in a barrel is transferred (racked) aeratively into another clean barrel.

can also involve oxygen pick-up, although it is possible to move wine with minimal uptake. There is also the deliberate exposure of wine to oxygen called micro-oxygenation, which is discussed later in this chapter. Small doses of oxygen can be helpful during *élevage* (the wonderful French term describing the process by which the wine is "brought up" or "raised" in the cellar), particularly for red wines, where they help modify the structure of the wine in complex ways.

Then, as the wine is bottled, oxygen management becomes crucial. At this stage, large doses of oxygen are almost always going to be harmful. Care must be taken moving the wine and during the bottling procedure itself. However, after bottling there is good evidence that a limited amount of oxygen transmission (OTR) by the closure can be of benefit. Certainly, the actual level of closure OTR will have an affect on the wine. The question is, how much OTR is appropriate for which wines? This has been dubbed "nano-oxygenation," in contrast with macro- and micro-oxygenation.

OXYGEN IN WINE RESEARCH INITIATIVE

This is where I need to introduce an interesting research initiative, begun in 2007 by synthetic closure company Nomacorc, with the goal of studying the impact of oxygen in winemaking and after bottling. Nomacorc's vested interest in sponsoring oxygen research is clear. It is currently offering closures with a range of different OTR levels, and so would see itself well placed if winemakers were to start choosing closures offering a specific level of OTR to match their wine styles.

Nomacorc has partnered with respected wine-science institutions across four continents, each of which has been looking at particular grape varieties. One of the keys to this research is the development of a quick-and-easy noninvasive method of oxygen measurement. To this end Nomacorc has worked with German company PreSens® to develop luminescence technology for measuring oxygen concentrations throughout the winemaking process and even after bottling. Nomacorc is now distributing this technology under the brand name NomaSense, and the oxygen-analyzer equipment package it sells

consists of a PreSens® Fibox™ 3 LCD Trace oxygen analyzer, together with a range of reusable sensor spots. The way this works is that a sensor spot is attached inside a bottle, a light is directed onto the probe and a readout taken. Alternatively, if the technology is to be used invasively, a dipping probe can be introduced into a tank or barrel. The great advantage of the sensor-spot technology is that the dissolved oxygen in a bottle of wine, for example, can be followed in time without the need to open the bottle. The reusable NomaSense® oxygen sensor spots are precalibrated and withstand normal winery cleaning practices. They come in two different sensitivities, PSt3 and PSt6, which cover all possible oxygen ranges in the winery.

With this sort of technology, it is possible for wineries to carry out thorough oxygen audits. This is something that is much needed because very few wineries have any idea how much oxygen they are introducing into their wine through different stages of the winemaking process.

Controlling oxygen in the winery is important if the concept of matching closures to wine type is ever to become a reality. Currently, the differences in oxygen pick-up at bottling, for example, are frequently so large that they render differences in closure OTR insignificant. The noise in the system from poor oxygen management obliterates the nuances of wine-style fine-tuning that closures are capable of. Nomacorc's Dr. Stéphane Vidal explains the concept of oxygen management in wines as consisting of three phases. First, there is macro-oxygenation, which is the relatively large dose of oxygen that wine experiences during primary fermentation. Then there is micro-oxygenation, the smaller exposure to oxygen occurring during barrel aging, or aging in tank with the deliberate introduction of small doses of oxygen. Finally, there is nano-oxygenation, which is the very limited exposure to oxygen that occurs post-bottling through closure OTR. He points out that it makes little sense to alter this sequence suddenly by exposing wine at bottling to a high dose of oxygen (macro-oxygenation again), which is what commonly happens with poorly managed bottling conditions.

There were two tiers to the Nomacorc-sponsored project. The first objective was to gain a greater understanding of ways to control oxygen during winemaking in terms of quality control. Nomacorc's Malcolm Thompson describes this as the "low-hanging fruit". In other words, a relatively easy way to improve the consistency of wine development. "We have identified significant improvements in how to manage oxygen, particularly at bottling," he claims. As an example of this work, Rainer Jung and his team at Geisenheim have been studying the effects of different bottling parameters on Riesling wines, looking at the effect of factors such as headspace composition, fill height, and oxygen pick-up during bottling. Using NomaSense®, Jung has shown that the headspace represents a significant source of oxygen contributing to wine evolution. The project evaluated the evolution of a Riesling wine under different bottling and post-bottling conditions. 375 ml (about 12.5 fl oz) bottles equipped with sensor spots were filled using two headspace volumes, each containing three different concentrations of oxygen, emulating conditions typically encountered during actual bottling runs. Dissolved oxygen in the tank was 0.3 ppm, and after bottling the dissolved oxygen levels ranged from 0.9–1.3 ppm (0.3–0.5 mg/bottle). Headspace volume was either 5–6 ml or 17–19 ml, and two or three different oxygen concentrations were used in each case. Two closures were employed, a Nomacorc and a screw cap, with the former being stored in either 21% oxygen (air) or zero oxygen. The levels of total pack oxygen (TPO) measured during bottling ranged from 0.2 mg per bottle to a high of 6.0 mg, which is a dramatic range.

TPO measurements showed that the wines started with very different levels, but that all oxygen was absorbed by 300 days. As well as the oxygen present at bottling, over the 300 days of the trial the Nomacorc contributed 2.5 mg of oxygen per bottle, while the Nomacorc in an atmosphere of zero oxygen contributed just over 1 mg of oxygen per bottle, which must have come from the body of the closure itself. The screw cap contributed around 0.3 mg/bottle. The TPO level correlated highly with decline in free SO_2 levels and also the change in color of the wine. "Our results show that headspace oxygen, which has largely been ignored by the industry, is a critical factor impacting wine development and, more specifically, a wine's oxidation resistance

influencing shelf-life performance," reports Jung.

Let's try to make some sense of these figures. Australian wine scientist Richard Gibson (who runs the Scorpex consultancy service) has provided some calculations of the likely impact of oxygen pick-up at bottling on wine longevity. He begins with the observation that 1 mg of oxygen reacts with 4 mg of SO_2. (Of course, as we discussed earlier, in reality the 1:4 ratio is seldom seen in wine; it is usually more like 1:2.5.) A headspace of 5.95 ml consisting of air will contain 1.24 ml oxygen, equivalent to 1.78 mg, which in turn equates to 2.37 mg/l of oxygen once it is dissolved in the wine. This can react with 9.5 mg/l of SO_2. The amount of oxygen entering the bottle through the closure can be calculated from the closure oxygen-transfer rate. A rate of 0.01 cc/day equates to 0.019 mg/l of oxygen entering the bottle each day, which can react with 27.7 mg/l of SO_2 over a year. From these figures, the expected shelf life of a wine can be calculated. For example, a wine with 35 mg/ml free SO_2 at filling, 2 mg/l of dissolved oxygen, 0.5 ml of oxygen in the headspace, and a closure OTR of 0.008 cc/day, will have a shelf life of 217 days. These sorts of calculations make it clear that both the TPO at bottling, and also the closure OTR levels are vital factors in determining how the wine will develop over time. From Jung's experiments, the significance of the difference between 0.2 and 6 mg TPO in terms of shelf life are readily apparent.

In fact, it could be argued that until wineries get a handle on oxygen pick-up at bottling, it is premature to begin thinking in terms of the second objective of Nomacorc's research project, described as "winemaker intention." The idea is to use knowledge of oxygen's effects on wine to put winemakers in a position where they can integrate closure design into winemaking.

WINEMAKER INTENTION

The Australian Wine Research Institute (AWRI) conducted an influential closure trial, beginning in 1999 and following the same wine under 14 different closures for several years. It was this work that first raised the possibility of using post-bottling oxygen management as an active winemaking tool. "An important outcome of the AWRI's research on wine closures is the recognition that when a wine is bottled under different

Nomacorc Oxygen in Wines Research Program

Institution	Lead researcher	Project	Varieties studied
University of California, Davis, USA	Dr. Andrew Waterhouse	Mechanisms governing oxidation	Chardonnay, Cabernet Sauvignon
INRA Montpellier, France	Dr. Véronique Cheynier	Oxygen influence on evolution of polyphenolics	Grenache
Geisenheim Institute, Germany	Dr. Rainer Jung	Influence of bottling conditions	Riesling
Pontifical Catholic University of Chile, Santiago, Chile	Dr. Eduardo Agosin	Influence of oxygen on aroma development	Carmenère
AWRI, Adelaide, Australia	Dr. Elizabeth Waters	Influence of oxygen ingress on reduction and oxidation	Sauvignon Blanc, Shiraz
DLR Rheinpfalz, Germany	Professor Uli Fischer	Influence of winemaking and oxygen exposure on sensory and chemical composition of Pinot Noir, before and after bottling	Pinot Noir
Centro Ricerca e Innovazione (CRI), Italy	Dr. Fulvio Mattivi,	The responsiveness of different wine varieties to oxygen	Various
University of Zaragoza, Spain	Professor Vicente Ferreira	Factors responsible for wine aroma associated with oxidation	Various

closures, different wines begin to be created from that point onward," state the scientists involved in the trial. "Other workers have apparently expanded this concept to other bottling variables such as the filling height, the concentration of free SO_2 at bottling, and the mixture of gases in the headspace of bottles post-filling. The ability to link such variables to wine development after it has been bottled creates the possibility of reliably predicting, and therefore optimizing, wine development in bottle." The notion is that if we were able to gather enough information about how different wine styles respond to oxygen post-bottling, and couple this with our knowledge of closure OTR, we could then offer winemakers the information they need to match closure type to wine style. In this scenario, closure choice becomes an active winemaking decision.

The data from the Nomacorc-sponsored research project has been used to construct a piece of software called the Nomacorc closure selector. This allows winemakers to answer various questions about their wine, and then have the appropriate closure from the Nomacorc Select Series range, which comes in an assortment of OTR options, recommended to them. However, not everyone is comfortable with the idea of using closure OTR to shape wine style. "Intentional oxidative maturation in the bottle is a high-risk strategy that may benefit a small number of wines for a short time," says Richard Gibson. "Inevitably, it will lead to inconsistency, loss of quality, and consumer dissatisfaction in many other wines. Relying on oxygen transmission through the closure to carry out or complete oxidative maturation of wine involves serious risks, as it is not known how long after bottling the wine will be consumed."

Gibson's view is that the appropriate time for oxidative maturation, if it is desirable, is in the cellar. While he thinks that small amounts of oxygen may be needed for avoiding the risk of reductive problems, aging in the bottle is basically an anaerobic process. "The concept we are talking about here [completing oxidative maturation in bottle] also means that wines can't be consumed too early," he adds. "Wines released to market may be astringent and undeveloped. You'd have a short window between optimal development and the wine being oxidized."

The results from the Nomacorc-sponsored studies have also begun to fill significant gaps in our knowledge about oxygen in wine. In particular, there have been some interesting results coming from the laboratory of Dr. Véronique Cheynier at INRA Montpellier (France). Cheynier's laboratory has been studying the influence of oxygen on the evolution of polyphenolics in red wines made from Grenache. Polyphenolic compounds are important in red wine, and include the anthocyanins (which account for color) and tannins. To cut a very long story short, the anthocyanins and tannins change form throughout the winemaking process, combining with each other and other molecules. Exactly how they do this is very important for red-wine quality, and oxygen plays a major role.

For this work, a matrix of 16 different Grenache wines was created, comparing extraction techniques, winemaking processes, and closure OTR. In the first instance, two extraction techniques were used: traditional maceration and flash release. Flash release (also known as *flash détente*) involves the rapid heating of grapes to 200°F for 6 minutes by steam and then cooling them down rapidly in a vacuum. This increases the extraction of polyphenolic compounds from the skins. Then, after fermentation each wine was further subdivided into two batches, one of which was subjected to micro-oxygenation and the other of which wasn't. These four different wines were then each bottled in 375 ml (about 12.5 fl oz) bottles in order to amplify the effects of oxygen post bottling, and were sealed with synthetic (Nomacorc) closures. Each of the four wines was subdivided further into four batches with varying OTR conditions, thus completing the matrix of 16 different experimental wines. The first batch was sealed with Nomacorc Light closures and stored in air (21% oxygen); the three remaining batches were sealed with Nomacorc Classic closures and stored in air (21% oxygen) or in stainless steel drums at 4% oxygen or 0% oxygen. All wines were kept at a temperature of 73°F. Calculated OTR rates for these four scenarios were 11.9, 8, 1.9, and 0.8 μg O_2 per bottle per day, respectively.

Using a technique called high-performance liquid chromatography (HPLC) the researchers were able to analyze wines for a wide range of polyphenolic compounds. They also did analysis for free and total sulfur dioxide, and looked at the color of the wines using spectrophotometry. The wines were studied 10 months after bottling to look at the effect of post-bottling oxygen exposure. The researchers used a technique called principal component analysis (PCA) to compare the color and phenolic composition data from the 16 different wines. PCA is a statistical technique used to find patterns in a mass of data. It pulls out from the data the factors that account for the variance in the variables under study—here the color of the wine and the phenolic composition. This showed that the most important factors in explaining the differences in these wines at 10 months are (1) the OTR, and (2) whether it was a flash-release wine or traditional-maceration wine. Of 20 color and polyphenolic parameters measured, OTR had a significant effect on 18, extraction technique affected 14, and micro-oxygenation affected 11.

What's the message here? After some time in the bottle, closure OTR appears to have more influence on red wine polyphenolics (and hence color, structure, and mouthfeel) than standard winemaking practices do. But this is just a chemical analysis, and what really matters is linking these changes detected by chemical analysis to changes in how the wines are perceived by consumers. For this reason, Cheynier's laboratory extended this study to sensory analysis.

In a second study, the same 16 wines were subjected to sensory analysis by a panel of 18 trained judges. The panel selected 12 attributes to describe these samples and then rated the wines blind for each, both at bottling (to assess the winemaking differences) and 10 months after bottling. They found that OTR affected significantly eight of the 12 assessed attributes (while five attributes were affected significantly by winemaking at bottling, only one of these remained significant 10 months after bottling). Wines stored under high OTR had more color intensity and appeared more orange, and had differences in odor (higher in "red fruits," "caramel"; lower in "vegetable" and "animal"). These significant differences are interesting. What would be even more interesting would be to present a range of wines aged under different OTR to untrained consumers, to see which they prefer.

Another interesting set of research findings is coming from work conducted by the AWRI looking at Sauvignon Blanc. Managing oxygen is particularly important for this variety because it presents particular challenges, owing to the fact that closely related volatile sulfur compounds could be both olfactory defects and also desirable aromatics. There are three important sulfur-containing compounds that contribute to the aroma of Sauvignon Blanc: 3-mercaptohexan-ol (3MH), 3-mercaptohexyl acetate (3MHA), and 4-mercapto-4-methyl-2-pentanone (4MMP), contributing aromas of grapefruit, boxwood, and passionfruit. These polyfunctional thiols (the name of the chemical class of these compounds) are susceptible to oxidation, and so the ideal closure for Sauvignon Blanc wines would seem to be the lowest OTR closure possible, which is currently a screw cap with a tin/Saranex™ (barrier film) liner. However, results from the AWRI's studies have also shown that using such a low OTR closure can increase the risk of reductive aromas from the shifts in sulfur-compound chemistry that can occur in the low-redox potential environment of a wine sealed this way.

This creates a paradoxical situation, where winemakers who are using screw caps with tin/Saranex™ liners play it safe by using prebottling copper fining to remove the sulfur compounds responsible for reductive off-flavors. The problem is that these sulfur compounds are very close in reactivity to the good thiols (3MH, 3MHA, and 4MMP), so this copper fining can negatively affect varietal aroma.

The AWRI has shown that copper additions at bottling can reduce the concentration of 3MH in Sauvignon Blanc wines. Interestingly, this effect was only observed at higher concentrations of SO_2 (60 mg/l) and not at lower concentrations (30 mg/l). But there was an even more interesting observation: after eight months of storage, even in the wines bottled with 30 mg/l of SO_2, a decrease in 3MH, 3MHA, and 4MMP was seen in copper-fined wines that in most cases was larger than the decrease associated with the use of higher OTR closure. This indicates that for Sauvignon Blanc a winemaker is better off using a higher OTR closure than using a tin/ Saranex™ screw cap in conjunction with copper fining. The authors point out that because copper is a powerful promoter of oxidation (a catalyst), excessive copper fining can accelerate oxidation reactions, resulting in a higher risk of premature oxidation.

A study from leading cork-company Amorim's Paulo Lopes has also looked at the effect of oxygen dissolved at bottling and transmitted through the closure on the composition and properties of a Bordeaux Sauvignon Blanc. This study involved both chemical and sensory analysis, and followed the wine for two years after bottling. Using a colorimetric technique involving an oxygen-sensitive dye, Lopes and colleagues compared the OTRs of a range of different closures. Lowest was screw cap with a tin/ Saranex™ liner, and highest was a synthetic cork (Nomacorc classic). Microagglomerate cork (Neutrocork), agglomerate cork, screw cap with a Saranex™ liner, natural cork, and colmated cork were in between these two extremes, in that order. In addition, wine was bottled in a sealed glass ampoule, with no OTR.

The chemical analysis showed that the levels of "good" thiols analyzed for (3MH and 4MMP) were highest in the low OTR closures and lowest in the higher OTR closures; the exception here was the screw cap with the Saranex™ liner. Here, the level was lower than might be expected from the OTR, leading the authors to suggest that the Saranex™ was scalping the thiols. The analysis for sulfide (H_2S) showed much higher levels in the ampoule and screw cap with tin/Saranex™ liner, consistent with the observation that very low OTR closures carry with them an enhanced risk of reduction.

In the sensory analysis after 24 months in the bottle, the two extremes of OTR were not successful for this wine. The Nomacorc-sealed wines showed evidence of oxidation, which masked the fruity characters. The ampoule and screw cap with the tin/Saranex™ liner showed reduction issues, which, again, masked the fruity characters. There has been some criticism of the wine chosen for this study, which may not have been the optimal choice. The intermediate OTR closures performed best in terms of fruit expression, but the agglomerate-sealed wines were ruined by contamination by TCA (2,4,6-trichloroanisole)—analysis showed these levels to be 1–3 mg/l, which masked the fruity characters of the wine. The authors concluded that, "An oxygen-sensitive variety such as Sauvignon Blanc benefits from some low-oxygen exposure after bottling, at the levels provided by cork stoppers. These wines retained high enough amounts of varietal thiols to maintain the typical box tree and tropical fruit aroma of Sauvignon Blanc but, at the same time, kept the deleterious sulfides at very low levels."

TEMPERATURE DURING TRANSPORT AND STORAGE

One important topic that doesn't get as much coverage as it deserves is the effect of environmental stresses in the supply chain. After bottling, there is usually a delay before the wine reaches the consumer. This delay may involve transportation over long distances, storage in a warehouse, shorter journeys from warehouse to retailer or restaurant, and then time spent on a retail shelf. During this period, wines may be exposed to swings in temperature, high temperatures and even light, all of which could cause loss of quality.

A recent study from the University of California-Davis looked at the combined effects of storage temperature and packaging on a Cabernet Sauvignon over a six-month period. This involved glass bottles sealed with a synthetic cork, a natural cork, and a screw cap, as well as two types of bag-in-box, kept at 50, 68, and 104°F. At the highest temperature, all the wines showed oxidized characters, and in general the storage temperature made more difference to the wine than the packaging. Thirty sensory attributes differed significantly among the storage temperatures, and 17 differed significantly among the packaging variables. Together with other related studies, this indicates that exposure to variable—and, in particular high—temperatures in the supply chain has a significant impact on wine quality, largely through acceleration of oxidation reactions. Unfortunately, evidence from data loggers suggests that wines are frequently abused in this way, especially when they are shipped long distances.

MICRO-OXYGENATION

One hi-tech manipulation that is currently the focus of much attention is micro-oxygenation (or microbullage, as it is sometimes known). The principle behind it is quite simple—it's a winemaking technique for adding very low levels of oxygen to a developing wine over an extended period. Small "micro-bubbles" of oxygen are fed through a special ceramic device placed at the bottom of the tank. The tank needs to be very tall to do this effectively, and the flow rate can be carefully controlled such that the oxygen dissolves into the wine completely before it reaches the top. The idea is that it allows winemakers to simulate the slow-controlled oxidation that occurs during barrel-aging for wines that are kept in stainless-steel tanks. A Madiran winemaker, Patrick Ducournau, developed the apparatus involved in the early 1990s. Producers in this region in France's southwest experienced problems when they started putting their Tannat-based wines into stainless-steel tanks. Tannat is a red grape that has a tendency to be tough and tannic, and without the benefit of oak aging to soften it, can be quite uncompromising. Ducournau first employed micro-oxygenation commercially in 1991, and set up a company, Oenodev, which now offers this technology worldwide.

It's a near-miraculous technique, if its proponents are to be believed. Among other things, micro-oxygenation is supposed to build optimum structure, reduce herbaceous or vegetal characters, provide color stability, stabilize reductive qualities, and increase the suppleness

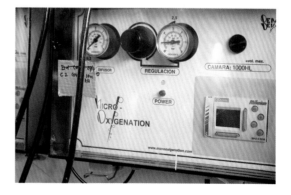

Above A micro-oxygenation controller unit in a Chilean winery.

or roundedness of the wine. Robert Paul, of Wine Network Consulting in Australia, a company providing micro-oxygenation facilities to around 30 wineries, claims that, "Treated wines can gain a more savory, structured palate." Randall Grahm, of Bonny Doon in California, is an enthusiastic supporter of micro-oxygenation. He asserts that, "Microbullage, if practiced appropriately, is the most useful tool for the mastery of *élevage*."

However, there are conflicting ideas about what micro-oxygenation actually achieves. A popular notion is that it makes red wines drink well earlier. This is largely based on the huge commercial success of the branded wines from Australian company Rosemount, which was in part attributed to the use of micro-oxygenation permitting the reds to come to market in the same year as production. Robert Paul says, "The technique is very successful, but the hardest part is convincing people that it is not about softening wines to make them drinkable earlier." Apparently, it's something of a paradox. According to Paul, "The treated wines are more apparently rounded. I would argue that this is because they are better wines fundamentally—they have better balance, stability, and structure and are cleaner with more apparent fruit."

On a trip to Portugal a couple of years ago I mentioned to winemaker David Baverstock that I was researching micro-oxygenation. He suddenly became quite animated; he's clearly excited by this technology. "We're doing quite a bit of micro-ox in the Alentejo, particularly with Monte Velho" (a commercial label of Herdade de Esperão, where he is winemaker). Baverstock claims that it gets rid of a lot of the "green tannins" and softens the wines up. He tends to use this technique with the more commercial wines. "There's always some fruit that is never quite as good as the top stuff, perhaps from younger vineyards or from less well-controlled growers." He uses micro-oxygenation in conjunction with different sorts of oak chips, matched to the type of the wine.

There seem to be two common motivations for the use of microx, as it is nicknamed. First, it is a useful remedial technique for removing unwanted green characters or sulfides from red wines. Second, there's a cost saving. Instead of using expensive barrels for mid-range wines, it's possible to produce the same effects with micro-oxygenation in combination with barrel staves or oak chips in tank. The micro-oxygenation adds the structure to the wine, while the oak chips or staves add the wood flavor.

But for all the scientific claims made by its proponents, micro-oxygenation is still something of a black art, based on trial and error rather than precise knowledge of the underlying mechanisms. Roger Boulton expresses some reservations in this regard. "The chemical effects are likely to be several, but there are no independent scientific measurements of the changes; mostly supplier claims and selected satisfied testimonials." He continues, "There is little indication the proponents of this treatment have developed a strong understanding of the changes occurring and there are both short-term and longer-term results that cannot be predicted. In terms of science, treatments that are shrouded in secrecy, selective results, and 'proprietary' methods will have little scientific acceptance until they can be independently reproduced and validated. There is little evidence that "micro-oxidation" has come close to a scientific method and it seems that some people prefer it that way. Please note that in terms of oxygen consumed and acetaldehyde produced it is multiples of what most wines ever see during barrel-aging, even when racked frequently, so the term 'micro-oxidation' is something of a misnomer." Ken Fugelsang of Fresno State University, in California, agrees that a lot of scientific uncertainty surrounds the technique. "More people are using micro-oxygenation, but we still clearly don't understand the full potential impact on wine," says Fugelsang. "It produces a more aged structure in young wines, but what impact does this have as wine goes into bottle?" Not all red wines adapt well to this technique. "It is highly dependent on the natural structure of the wine before the process starts," he explains. "It is potentially dangerous for a light wine: it could oxidize it rather than build up structure"

Who is using microx? Clark Smith suggests that its use is rather widespread, with Chile leading the field. "Chile has perhaps only 120 wineries, but some 80 of these use micro-oxygenation, or at least own the equipment," says Smith. In California, he says that all the enormous Central Valley producers use it, and of the ultrapremium North Coast wineries, perhaps one-third are at least experimenting with it. He estimates that around 5% of producers are using the techniques in France and Australia.

A MORE IN-DEPTH VIEW

Smith outlined his explanations for the chemical changes that micro-oxygenation brings about. He refers to this technique as a tool for "integrated tannin management," which involves the use of oxygen for building tannin structure in red wines, and then harmonizing these wines during aging. "It's like making a tannin soufflé," he explained, the implication being that it is a process of some complexity that's difficult to do right, but worth it when it works. Smith explains that to understand the use of introduced oxygen in wine properly, it is necessary to look at its impact at various stages in a wine's evolution. The first exposure, hyper-oxygenation, is done prior to fermentation where it is needed. It is the opposite of micro-oxygenation because it causes tannins to drop out of the must, thus decreasing the polyphenol content of the wine. The second type of exposure is macro-oxygenation, which can be used during fermentation for the purpose of boosting yeast health. "During fermentation it is the yeasts that take up almost all the oxygen, not the phenolics," explained Smith. "There is growing evidence that oxygen applied at the right time can help fermentation proceed." Reasonably large amounts can be added at this time without too much risk. Then, immediately after alcoholic fermentation is an important time to begin the third exposure, micro-oxygenation, for the purpose of building structure. At this stage the goal is to encourage the phenolic compounds present to polymerize. The added oxygen is thought to oxidize ethanol to acetaldehyde, which then encourages this polymerization process. Relatively high levels of oxygen can be safely added at this stage because the excess aldehyde produced can be consumed during malolactic fermentation. This also has the effect of fixing color, as the anthocyanins complex with tannins. It is important that SO_2 isn't added prior to micro-oxygenation because this inhibits the structurization effects. The structurization

process can continue during and after malolactic fermentation, but to a much lesser extent, and far smaller levels of oxygen can be added during these phases. The next stage is harmonization, when the wine begins to settle down, develop aromatic complexity, shed any vegetal characteristics, and generally grow up a little. It's the wine's teenage years; it has gone through its growth spurt and now it needs to become civilized and responsible. This can take place in tank or barrel. This is a critical stage in the process, and the only way to assess the progress of the wine is by tasting, paying particular attention to the characteristics of the tannins. This is tricky, and requires some experience. Too much oxygen exposure and the tannins will turn hard and drying; too little, and reduction characters might appear. As Smith points out, "We don't have instrumental measures for running micro-oxygenation."

Although micro-oxygenation is quite a new technique, the enthusiasm with which winemakers have adopted it suggests that there must be something to it, even though the exact details of the underlying science aren't clear at this stage. Personally, I like the concept of integrated tannin management, beginning in the vineyard and carrying through into the winery. As soon as the science catches up with the technological push, micro-oxygenation looks set to become an established, mainstream technique to assist in red-wine tannin management throughout the winemaking world. I'll conclude with Randall Grahm's ringing endorsement. "I am continually amazed at how misunderstood this technique is, even by some otherwise extremely competent members of the trade. It is an extremely useful tool to enable a winemaker to master long *cuvaisons*, so as not to extract excessive bitterness in the wine, to soften tannins and preserve color. Microbullage, if practiced appropriately, is the most useful tool for the mastery of *élevage*. It is an extremely powerful lens that enables winemakers to observe where in the life cycle of development their wine might be and when is the moment just to put the wine into bottle. I would liken microbullage to the advent of temperature control in fermentation, perhaps a technology that was considered 'unnatural' at one point, but is now largely considered indispensable."

Examples of levels of oxygen transmission rate (OTR) by various closures

Closure	OTR (cc O_2/closure per day)[1]	Comment
Screw cap (tin/saran liner)	0.0003–0.0007	Data from Paulo Lopes' indigo carmine dye method
Screw cap (tin/saran liner)	0.00002–below detection limit	Jim Peck's data obtained by MOCON in 36-month American Vineyard Foundation (AVF) Study
Screw cap (tin/saran liner)	0.0002–0.0008	AWRI measurements Godden et al 2004
Screw cap (Saranex™ liner)	0.001–0.0006 in 36-month	Jim Peck's measurements by MOCON AVF study
Technical corks (not specified, likely TwinTop or microagglomerate)	0.0001–0.0006 dye method	Data from Paulo Lopes' indigo carmine
Microagglomerate technical cork	0.0001–0.0006 wet method	Jim Peck's MOCON measurements, 2010,
DIAM 10 (very low OTR)	0.0007	Data from DIAM
DIAM (low OTR)	0.0015	DIAM 2, 3, and 5 are available in low or medium OTR, data from DIAM
DIAM (medium OTR)	0.0035	OTR, data from DIAM. DIAM 2, 3, and 5 are available in low or medium
Natural cork 45 mm (1.75 in) super select 47 mm (1.85 in) flor 54 mm (2 in) flor	0.001–0.0002 Below detection limit–0.001 Below detection limit–0.0009	Jim Peck's measurements by MOCON using his inverted, wet-cork method
Natural cork	0.0017–0.0061, initial period, then 0.0001–0.0023 after 12 months	Data from Paulo Lopes' indigo-carmine dye method
Nomacorc Select 100	Over first 3 months: 0.0029 Over first year: 0.0023 After 1 year: 0.0021	Data from Nomacorc using NomaSense
Nomacorc Select 700	Over first 3 months: 0.0134 Over first year: 0.0065 After 1 year: 0.0040	Data from Nomacorc using NomaSense
Extruded synthetic 38 mm (1.5 in)	0.0025–0.0030	Jim Peck's MOCON measurements 2012
Extruded synthetic 44 mm (1.75 in)	0.0019–0.0026	Jim Peck's MOCON measurements 2012
Vino-Lok	0.0001–0.0002	Measured by PreSens (the same technology as NomaSense)

[1] Synthetic closures and screw caps typically show a tight standard deviation, whereas natural cork can vary significantly. The variation in OTR shown for the same closure type has to do with many factors, including methodology and the measurement technique used. Also, most closures show an initial higher rate of OTR for the first few months, as the oxygen in the body of the closure (if any) is released (outgassing), and then a steady, lower rate of OTR after this. MOCON is one of the measurement techniques used to measure OTR, as is PreSens/Nomasense (a luminescence technology).

11 Red-winemaking techniques: whole-cluster ferments and carbonic maceration

It's time for a quick detour to examine the science behind some red-winemaking techniques, specifically carbonic maceration and whole-cluster (or whole-bunch, or with stems) fermentation. The two are related in that the former is a subset of the latter, but is usually used for very different purposes.

The world of fine wine is big and diverse, and it's dangerous to make generalizations. But if we take a wide-angle look, winemaking trends seem to move in cycles. It seems we are currently in a phase where elegance and complexity are being pursued by winegrowers at the expense of power and strength. Visit wine regions worldwide and you'll find very few young winemakers aiming to produce bigger wines—certainly not at the high end. They tend to prize elegance, freshness, and definition above all else. The monster 100-point wines of the 1990s and 2000s are increasingly looking like yesterday's wines.

It's perhaps for this reason that there is increasing interest in winemaking techniques that foster this elegance and complexity. For example, there's a marked shift away from small new oak as the primary vessel of *élevage*, with renewed interest in concrete and larger, more neutral oak. Wild ferments used to be a novelty; now they almost seem normal. While many are suspicious of the natural-wine movement, even those people outside of it have begun to work more naturally in the cellar because they believe this is likely to result in wines that express their sense of place better. And there's increasing discussion of and experimentation with the topic of this article, that is, the use of stems in red winemaking.

There is, of course, nothing new about making wine with the stems included, which is usually referred to as whole-bunch or whole-cluster fermentation. Through history, wines would have been made from intact bunches that were then either pressed immediately to yield juice for white-wine fermentation, or macerated during the fermentation process for red-wine production. The only way to remove the stems prior to red-wine fermentation would have been manually picking off each berry. This is time-consuming and therefore very expensive, although there is one famous Bordeaux estate that practices it for its first wine, Château Pape-Clément, and in Burgundy Domaine de la Vougerie does this on a smaller scale for its *grand cru*. And in Chile, Lapostolle's Clos Apalta is also made from grapes that are destemmed by hand. But the development of the crusher destemmer allowed winegrowers a quick, economical way of separating the stems from the berries, and the vast majority of red wines are now made from grapes that are first destemmed, either by such a machine or, increasingly, in the vineyard by the machine harvester.

Some anatomy. When a bunch of grapes is picked, it consists of the grapes, plus some other material that holds the cluster together. The main axis of the cluster is the rachis, and the pedicel attaches the berries to this. The part that attaches the cluster to the vine, and which is cut through to release the bunch at picking, is called the peduncle. Together, this material, which we are referring to with the broad term "stem," consists of about 2–5% of the weight of the cluster. Depending on the region and that year's climate conditions, stems can vary a lot in their appearance. This is because they start out as green, photosynthetic material, and then undergo a process of lignification, which is the transition from green, fleshy plant to woody plant, achieved through the deposition of

Far left These grapes have just been hand-harvested. In many cases they would be destemmed before fermentation, but in a whole-bunch or whole-cluster ferment they are not: the grapes are fermented in the presence of the stems.

Left Inside a destemmer. The rubber teeth revolve, and the grapes are separated from the stems as gently as possible.

lignin in the spaces in the cell walls between the cellulose fibers. So having stems in the fermentation can mean very different things in terms of wine outcome, depending on the degree of lignification of these stems.

WHO DOES WHOLE BUNCH?

Burgundy is the region most associated with whole-cluster fermentation, and by association Pinot Noir is the grape variety most closely linked with this technique. In part, this could be because Pinot Noir as a variety lacks acylated anthocyanins, which are a form of pigment. This explains why Pinot Noir is usually lighter in color than other red wines, but in addition, anthocyanins also have important interactions with tannins in wine, and form pigmented polymers, which are important in wine structure and color. Some of the wood tannins leached from the stems could be making up for this shortfall in Pinot Noir. But traditionalists in the northern Rhône with Syrah have also used it. Increasingly, New World producers working with Pinot Noir have been exploring the use of whole clusters, and it is also beginning to catch on with Syrah producers who are looking for elegance.

The Burgundian domaines most famously associated with whole bunch are Romanée-Conti, Dujac, and Leroy. "Clearly, in Burgundy at the moment there is a tendency to move toward stems," says British wine writer and Burgundy expert Jasper Morris. "I can see two main reasons for this," he says. "One is that Henri Jayer, who hated stems, is dead. And the other is that with global warming, the stems are more often riper than they used to be." Jayer, a tremendously high-profile grower, influenced many to move away from stems and, until recently, this was the direction

being taken across the region. And the popularity of destemming was linked with a corresponding reduction in greenness and rusticity in many red Burgundies, so there was a good reason for doing it. In a sense, in the past people used stems by default, and the results weren't always good. Now the choice to use stems is an active one, so the people doing it are doing a better job with it.

Jeremy Seysses at Domaine Dujac uses between 65% and 100% whole-cluster fermentations depending on the *cuvée*. "We have the feeling that we get greater complexity and silkier tannins with whole-cluster fermentation," he says. "In high-acid vintages, it helps round things out, and in high-ripeness vintages, it brings a freshness to the wines." For Seysses, the decision about whether or not to destem depends on a number of factors. "Some terroirs don't seem to do so well with whole-cluster. The whole-cluster character rapidly becomes dominant and can appear 'gimmicky;' it doesn't mesh well with the wine, and can give the illusion of complexity, but it feels superficial," he explains. "Of our holdings, I like destemming a little more for the Gevrey vineyards than the others." He also tends to destem more frequently the grapes from younger vineyards with bigger clusters, and in vintages with rapid end-of-season ripening, where the ripening may be a little more uneven.

Mike Symons, winemaker at Stonier in Australia's Mornington Peninsula region, also finds that terroir is the biggest determinant of whether or not he uses stems in his ferments. "We have a couple of vineyards where we like the stems; [these] are north-facing and produce nice ripe stems" he explains. "We pretty much know the vineyards where we like the stems.

One of them is the Windmill vineyard, and another is the vineyard near the winery. There are some vineyards where we don't include the stems, such as the Lyncroft vineyard, which is very cool, or the KBS vineyard Pinot. They would just be awful if we included the stems."

"I normally find a strong correlation between the better sites and the amount of stem/whole bunch I am able to use," says Mark Haisma, an Australian working as a *micro-negociant* (someone who specializes in limited-production wines) in Burgundy and Cornas. "The stems from the best sites are generally cleaner and richer in character."

To decide whether or not to use stems, Mike Symons eats them. "If they taste like broccoli we don't use them." His experience from regular workshops with Victorian Pinot Noir producers is that winemakers are increasingly talking about using stems, but he thinks that you need the right vineyard. "Some people get on bandwagons and they include stems where they shouldn't. It is something you have to be careful of," says Symons. "With a blend where we include stems, we will do it over three days or more to make sure we get it right."

Another well-known Victorian winemaker, Tom Carson of Yabby Lake, admits that he likes to play around a bit with whole bunches in his Pinot Noir ferments. "I am still experimenting, and I'm reluctant to go in too hard. When it's good, whole-bunch fermentation gives fragrance and perfume, and adds a bit of strength and firmness to the tannins. But when it's not good it can dull the fruit, adding mulch and compost character," says Carson. "We want to highlight the fragrance of the Pinot. We don't want complexing elements that are not vineyard-derived." Carson did 8% whole bunch in 2009 and 20% in 2010, but then backed off a lot in 2011 because it was a wet year and the stalks were very green. "We are still learning what is the right amount."

Nick Mills, of Rippon, in New Zealand's Central Otago, uses some whole bunches in the Pinot Noir ferments, but decisions are made based on the fruit. "We do some whole bunches," says Nick, "but this is all done on the sorting table." He adds that, "the sorting table isn't about taking stuff off, but it's for me to taste pips and skins, and figure out what raw material we have.

If we can chew the stems through then we'll put them in. I'd put in 100% whole clusters if we could. It's a better ferment." Overall, Rippon Pinot Noir has 25–40% whole clusters. "The vineyard is incredibly parcelated," says Nick, "with all these small micro-ferments. If we get something really good, then we'll put the whole lot in and do 100% stems, but if grapes come in that I don't like the taste of we'll use no stems."

It seems that lots of growers, like Mills, will use as many stems in the ferment as they can, with the limiting factor being the suitability of the stems. "When you are choosing whether or not to use stems, some people do a positive selection on the sorting table (*tri au positif*) rather than a negative one," says Jasper Morris. "So when they come across bunches with lovely bronze stems, they use them."

Until recently, Eben Sadie of South Africa's Swartland region didn't use any stems in making his celebrated Columella wine, but he decided to change this with 35% of stems included in the 2009 vintage. "For the next 10 years we will work with 20–40% whole bunch," he says. For Sadie, stems are a way to achieve freshness in his wines, but he uses them on a vineyard-by-vineyard basis. Of his eight vineyards, five are destemmed and three are 100% whole-bunch ferment.

Another warm-climate region where some growers are experimenting with stems to help achieve more elegance is France's Roussillon. "I know that with our own wine, Le Soula, in the early years we had problems of over-extraction and rusticity in the red wine despite not practicing any *pigeage* [stomping grapes in open fermentation tanks]," says UK wine merchant Roy Richards, talking of the *domaine* he owns in partnership with Gérard Gauby. "Incorporating the stalks allowed us to restrict that extraction, and added peony to the existing range of perfumes, a quite dramatic transformation!"

Renowned Australian consultant winemaker Tony Jordan has been using whole-bunch fermentations with Shiraz. "What we have been doing with Heathcote and Barossa fruit is trying to get a little more restraint in without moving away from Shiraz structure and character," he says. "There, the use of whole bunches at 10, 15, or maybe as high as 20% can be good. In the wine I made from Heathcote, the way I was doing this was to put 15% whole bunches in the bottom and then crush the rest with wide-open rollers, so you probably get some whole berries in this as well. Then I am obsessed with working the cap very gently to try to let that maceration go through in those bunches. Then you really get impact of it." This is working with warm-climate wines; most experiments with whole-bunch typically take place with Pinot Noir. "The reason I am doing this is because it came out of all my Pinot making," explains Jordan. "It was so successful with Pinot Noir, but why not for other varieties? The *maceration carbonique* character is a fruity element. You can tell it is fruity, but you can't say, 'That's *maceration carbonique*.' It is almost a brightness to the fruit."

CARBONIC MACERATION COMPARED WITH WHOLE-BUNCH

Here we need to take a side journey to Beaujolais and consider the Gamay grape. The winegrowers in Beaujolais practice the most extreme and purest version of whole-bunch fermentation, a technique referred to as *maceration carbonique* or carbonic maceration. Carbonic maceration isn't usually thought of when reference is made to whole-bunch ferment—it produces rather different results—but it is an important element of what takes place during such fermentations, and understanding the technical details of carbonic maceration will help shed light on a part of the flavor impact of whole-cluster fermentation.

Carbonic maceration is the process that occurs when intact bunches of red grapes are fermented in a sealed vessel that has first been filled with carbon dioxide. In the absence of oxygen, these intact berries begin an intracellular fermentation process, during which some alcohol is produced, along with a range of other compounds that can affect wine flavor. Once the level of alcohol reaches 2%, which is after about a week at typical fermentation temperatures, the berries begin to die. They then release their juice, or are pressed before this happens, and a normal fermentation (carried out by yeasts), takes place. The result is typically a relatively pale-colored red wine with low tannin levels and enhanced fruity aromatics.

The basis of carbonic maceration is the biochemical process of anaerobic fermentation,

the breakdown of sugars to release energy in the absence of oxygen. Yeasts use this pathway even when oxygen is present, and the result is that sugar is broken down to alcohol and carbon dioxide. The cells in grapes can carry out a form of anaerobic fermentation, but they are less able to cope with the resulting alcohol than yeasts are. When whole bunches of grapes are placed in an atmosphere of carbon dioxide, they are able to break down sugars, but also malic acid, which is one of the main acids present in grapes. This malate degradation is perhaps the most significant step that takes place during anaerobic fermentation, and it's broken down sequentially to pyruvate, acetaldehyde, and then ethanol. Typically, half of the malic acid is degraded in this way.

In carbonic maceration there is therefore a fall in acidity levels that can be quite significant, with titratable acidity (TA) declining by as much as 3.5 g/liter and pH increasing by up to 0.6 units. However, bear in mind that there would be some loss of acidity during the malolactic fermentation that almost always occurs after alcoholic fermentation in red wines. During carbonic maceration, polyphenols (such as tannins and anthocyanins) migrate from the skin to the pulp inside of the grapes, turning their flesh pink. This process produces various compounds that are important for flavor (or which are flavor precursors). For example, extra amino acids are liberated from grape solids, which increase the nutrient status of the juice and open up the potential for these amino acids to act as flavor precursors. The ethanol produced can esterify some grape components, and one ester produced this way, ethyl cinnamate, gives strawberry and raspberry aromas. Another compound that increases is benzaldehyde, which adds cherry/kirsch aromas. The berries eventually die when alcohol reaches a level of 1.5–2.5%.

The traditional winemaking method in Beaujolais (known as *maceration traditionelle*) is not a strict carbonic maceration. Here, the entire clusters are transported in 13-gallon containers and dumped in wooden *cuvées* (vats) or cement or steel tanks. Some of the berries on the bottom are crushed by the weight of those above them, so they start fermenting and the tank fills up with carbon dioxide. The intact berries begin internal fermentation and then when they die they release their juice, which still has quite a bit of sugar in it, keeping the fermentation process going. The higher pH that results from the intracellular degradation of malic acid means that malolactic fermentation can begin more easily after alcoholic fermentation finishes. In reality, most carbonic macerations are not pure carbonic maceration. "Any break at all in the grape skin and yeast will get in and ferment those berries," explains Tony Jordan, who has been working with whole-bunch ferments in some of his consulting projects. "What we find sometimes is that when we press off we get a kick up in sugar; other times we don't. That is normally because in a ferment over 10 days, you probably have true *maceration carbonique* going for a few days, but by then you start to work it and you are breaking a few berries, and there aren't many left that are truly intact with the stalk."

THE EFFECT OF INCLUDING STEMS IN FERMENTATION

Stems have a number of effects on fermentations, but this is where the story becomes complex and somewhat unclear. There are many different ways of using stems in the fermenter, and the stems themselves can be quite different in terms of how green or lignified they are. "There is an immense difference in flavor profile from all the people who do use stems," says Jasper Morris, referring specifically to Burgundy. "You also have to look at the techniques involved. Here it gets very complicated." Morris adds that, "The stems in the fermenting vat will have perhaps a chemical impact, and certainly a physical impact."

"In small vats, like those used in Burgundy, stems are useful because they drain the juice in a more homogeneous way and keep the temperature of fermentation one or two degrees lower," says French wine commentator Michel Bettane. Jeremy Seysses agrees: "The cap is far more aerated, meaning that it doesn't get quite as crazy hot as it would without any rachis (the main part of the bunch stem) in there, letting some heat escape. It also drains much better when you punch down or pump over as you get no clumps." Nick Mills of Rippon in Central Otago adds that the presence of stems allows the yeasts to move around more easily, and the pressing is better. And Rhône winemaker Eric Texier

claims that in whole-bunch fermentation, the conversion factor of sugar to alcohol is slightly different, resulting in wines with lower alcohol.

In addition to these benefits, Bettane also adds that stems in the fermentation can also help diminish the negative influence of any fungal infection on the grapes. "In 1983, for instance, curiously the whole-bunch-made Burgundies were less flawed by rotten berries than destalked ones." But if large tanks are being used, he points out that it is impossible to keep the stems, because they make the cap too resistant to mechanical pressure. Seysses also says that whole-bunch ferments are harder to punch down. "You have to do it by foot or by piston, you can't do it by arm. All these things change your extraction profile."

Another physical effect of stems in the ferment is a loss of color. "The stems also absorb color, leaching the color of the wine," explains Eben Sadie. "These days everyone wants to make more powerful, impressive wines, so whole-bunch is an unfashionable move because your wine looks weaker. For many people, color is an important property of the wine." But Sadie doesn't see this as a big problem. "I'll lose some color to gain freshness and purity. The wine has more vibrancy and life in it. Where we work in South Africa, the biggest flaw is our wines are often too ripe. It's good to get our wines fresher and more vibrant."

In addition, stems raise the pH of the wine slightly (making it less acidic, usually a bad thing), because of the presence of potassium released by the stems. The potassium then combines with tartaric acid and precipitates the acid from the wine. However, Seysses points out that there is less potassium in the stems these days and so this not so much of a problem. (The potassium got there in the first place because a lot of Burgundy growers used too much fertilizer with high levels of it in the 1950s and 1960s.) Some of this pH rise might also have to do with the *maceration carbonique* element that is a part of whole-bunch fermentations.

There are many variables involved in how the actual stems are added to the vat. "One question is, if you don't use all stems—and there is probably quite a lot to be said for using just some stems— when do you put them in?" asks Jasper Morris. "Do you put them in first, as a sort of base to the *cuvée*, and then you put your destemmed grapes on top?

Or do you put them in last, so the stems slowly float down through the juice? Or do you do some sort of lasagne-like layering between stems and nonstems, which I have heard some people do?"

"With many cuvees and the destemmed fraction being so small, I inevitably co-ferment," explains Jeremy Seysses. "The practicalities of harvest don't always allow it, but I usually like putting the destemmed fruit at the bottom and the whole cluster on top, so that it really stays whole. And as it can take a few days for the ferment to get going, I don't want my healthy whole clusters to be covered with juice as they sit waiting for the yeasts to get going."

Tim Kirk of Clonakilla, from Australia's Canberra District, uses some whole-bunch to make his Shiraz-Viognier, but unlike Seysses' preference, the bunches go in first. Whole bunches are put into two-ton fermenters, only part full. Some Viognier is typically crushed and destemmed and put on this, and then some Shiraz is destemmed and put on top. Kirk reports that some of the grapes in the whole-cluster portion stay attached to the rachis and don't burst: he estimates around 20% of them. Instead, these berries begin fermentation from inside, as in carbonic maceration. If you take these whole berries out partway through fermentation, their pulp is colored red, so they are extracting color from the inside. They are also still a little sweet and, on pressing, these berries release sugar, which acts to prolong fermentation.

Blair Walter, of Felton Road in New Zealand's Central Otago, uses a little whole-bunch to add complexity to his wines, and he notes this delayed sugar release from intact berries. "We typically put in a quarter whole-bunch and destem the rest of the bunches. And then when we punch down we don't go to the bottom of the tank. After 28 days you can still pull out whole bunches. They have fermented inside [the intact berries] and there is still some sweetness that is pulled out." He thinks this remaining sweetness is important because it keeps fermentation ticking along for a while. "Burgundians typically chaptalize in six to eight small additions," claims Walter. "This results in a slightly stressed fermentation producing more glycerol. This changes the texture and adds some fruit sweetness. It surprises me that more people don't use whole bunches."

This partial carbonic maceration character is likely to contribute significantly to the enhanced texture and aromatics often seen in wines made by whole-bunch fermentation. But Michel Bettane thinks that some of this benefit can also be derived from very careful destemming. "Don't forget that new destemmers are so precise and delicate that they allow winemakers to put 'caviar' destemmed berries in the vats with almost the same effect as whole bunch fermentation," says Bettane. "The beginning of the fermentation takes place inside the berry, helping to preserve the best quality of fruit, delicacy of texture, and capacity to age, keeping the youth of fruit and avoiding barnyard undertones."

Mark Haisma is a winemaker with broad experience across different hemispheres. In his previous employment he was at Yarra Yering, in Australia's Yarra Valley, but he's now a micro-négociant in Burgundy, also making a wine in Cornas in the northern Rhône. At Yarra Yering he developed an innovative approach to stem use, which he calls a "macerating basket." "The fruit would be completely destemmed, and I had some stainless steel mesh cylinders made," he explains. "These would be stuffed with the stems. I could take them out when I felt I had what I wanted." I asked him about the results of using stems this way: "I find it adds a great spicy complexity to the wine and also builds your tannin profile. And this way I have absolute control." Haisma is working on this in Burgundy, with some interesting results, but he is unaware of anyone else doing it this way.

"Whole-bunch for me is about controlling the ferment: slowing it down, with a slow release of sugar," says Haisma. "It is a great way to build loads of complexity and savory characters, and still keep a lush, creamy feel to the palate. I think of velvet. This is especially noticeable with my Cornas. As for burgundy, it's all about the complexity and finesse. In the big appellations I feel it adds a structure to the fruit, without adding coarseness or bitterness: characters I hate in Pinot Noir."

The other variable here is the length of time the stems stay in the ferment. Is a cold soak employed, or a post-ferment maceration? This could increase the extraction of flavor compounds from the stems. Dirk Niepoort from Portugal's Douro uses stems to make his

Charme wine, which is known for its finesse and elegance. For him, the length of time the stems are macerating in the *lagar* is critical. In one vintage he says that he misjudged a *lagar* by five hours, and that was enough for the wine to be excluded from the final blend of Charme.

THE NEGATIVE EFFECTS OF WHOLE-BUNCH

What about the negative aspects of whole-bunch fermentation? Blair Walter of Felton Road says that he used to do one fermenter with just whole bunches each year, but has now given up. "For us it is too much," he says. "It is interesting but the wine becomes too herbal—it is like a hessian-sack character." But he still uses smaller proportions of stems in many of his fermentations. "With stems, people expect the wines to become angular. I find the opposite. Destemmed wines taste more angular. A lot of people don't have the courage [to use stems]; they aren't willing to tolerate earthiness and herbal characters in the wine." Tom Carson finds that using too many stems gives his wines a mulchy, herbal character.

"Whole-bunch is a very important part of my ferments," says Mark Haisma. "But the stems need to be clean. Any mold and it really shows in your wine, worse than moldy fruit."

Greenness is the problem that is most often associated with stems. While there has been increased interest in the use of stems in red wines worldwide, one region stands out as an exception, and that is Bordeaux. This is likely because the main Bordeaux varieties of Cabernet Sauvignon, Cabernet Franc, and Merlot all share a degree of greenness in their varietal flavor signature, something that most winemakers will seek to minimize, and won't want to risk exaggerating by including stems. However, Paul Pontallier at Château Margaux has looked at the impact of stems in the course of the extensive in-house research program that this famous estate has established. This stem trial was with 2009 Cabernet Sauvignon from a plot that, in good years, makes it into the first wine. "We wanted to see how important it is to destem," recalls Pontallier. "Our tradition has been to almost totally destem. From the early twentieth century at Margaux destemming was a standard procedure." He points out that some are now suggesting that using some stems could be

a good thing. And on the other side, some estates have become more fastidious about removing even the tiniest traces of stem. The destemming regime in practice at Margaux leaves some tiny pieces of stems in the ferment, such that 0.03–0.05% of the ferment is stems. In this trial, the standard Margaux destemming was compared with 1% stem additions, and 1% stem additions but with the stems cut into tiny pieces. To Pontallier, the results from this trial are obvious. His view is that the current approach produces the best wines, and the 1% stems in pieces the worst. But he is still cautious about generalizing the result. "We shouldn't draw too general conclusions. For this wine I think destemming works, but for other plots, such as a rich wine with soft tannins, it might be different."

In California, Paul Draper at Ridge also avoids using stems with Cabernet Sauvignon. "We have never used stems with the Bordeaux varietals as we have more than enough tannin in any year," says Draper. "In addition in our cool climate—which is as cool as Bordeaux during the growing season but with cooler nights and warmer days—we are sensitive to any green character which, of course, is a risk with stems." Draper also chooses not to use stems for his Zinfandel: "Though this is not as tannic as the Bordeaux varieties, is well-balanced without any additional structure." However, he does include the stems when he ferments the few tons of Petite Sirah that Ridge has at Lytton Springs. Draper thinks there is good reason that stems are most widely used for Pinot Noir. "Given that Pinot Noir has less tannin and fewer different kinds of tannin than virtually any other well-known variety, the use of stems when needed makes more sense."

CONCLUDING REMARKS

There are many different ways of doing whole-bunch fermentations. Combine these different techniques with the variability in the state of ripeness of the stems, and it creates a complex matrix of factors liable to result in different flavors in the final wine. So it is with some trepidation that I'm going to attempt to sum up the way that whole-cluster ferments affect the flavor of red wines.

The state of ripeness of the stems seems to be very important, and this is likely to be determined primarily by the vineyard site, with vintage variation playing a role. In some warmer regions with a shorter ripening time, the stems may still be very green at harvest and thus unsuitable for inclusion at all.

An element of carbonic maceration is an important part of whole-bunch ferments. The intracellular fermentation that occurs in any intact berries will produce interesting aromatic elements, and the slow, gradual release of sugar into the ferment will change its dynamics. Together with this, the reduced temperature of whole-bunch ferments is likely to have some effect on the resulting wine, usually in a positive direction. There may also be some direct flavor input from the stem material to the wine, which can be both good and bad, depending on the state of the stems. And the slight rise in pH that occurs with whole-bunch may increase the susceptibility of the wine to *Brettanomyces*, but at the same time may improve the mouthfeel.

What are the benefits of whole bunch? One is textural—it seems to deliver a textural smoothness or silkiness that is really attractive, especially in Pinot Noir. Along with this, the tannic structure may be increased. I find that young whole-bunch reds often have a grippy, spicy tannic edge that can sometimes be confused with the structural presence of new oak. Frequently cited as a benefit of whole bunch is the enhanced aromatic expression of the wine, and it's common to find an elevated, sappy, green, floral edge to the pronounced fruity aromas, which is really attractive. Freshness is another positive attribute associated with whole-bunch. Done well, whole cluster can help make wines that are more elegant than their totally destemmed counterparts. I would add that whole-bunch wines sometimes start out with distinctive flavors and aromas that can be a little surprising (tasting terms associated with whole-bunch include broccoli, soy sauce, compost, mulch, forest floor, herbal, green, black tea, cedar, menthol, cinnamon), but these often resolve nicely with time in bottle.

"The wines of the 1990s were the 'Parkerized' wines," (monster wines, of the sort that American critic Robert Parker was widely believed to prefer) says Tony Jordan, referring to the move at this time in Australia to make monster wines. "Everyone seemed to think bigger was better

and the wines seemed to be getting bigger in every way. Now there is a big step back from that. And yet if you are in a warm climate, the wines are going to be robust. That's the terroir speaking. But you can still aim for freshness, a bit of brightness of fruit, more elegance on the palate." This is one of the reasons why there is so much interest in whole-bunch fermentation at the moment, because it does represent a tool for making more expressive, elegant red wines, even from sites not known for this attribute.

And even commentators such as UK retailer and Burgundy expert Roy Richards, who used to be opposed to whole-bunch fermentation, are softening their attitudes. "I no longer have an ideological view on this question, and understand that it is rather more complicated than I used to believe," says Richards. "As a disciple of the late Henri Jayer, I followed his mantra that stalks led to green tannins and that new oak to creamy, soft ones. And it is true that in his time his wines stood out for their vibrancy and sensuality whereas those wines from more illustrious *domaines* seemed a little delicate and pasty alongside." Richards adds that, "He is doubtless turning in his grave, seeing his protégés, Jean-Nicolas Meo and Emmanuel Rouget experimenting with whole-bunch fermentation in his beloved Cros Parantoux." Richards thinks that this could in part be attributed to changing weather patterns. "Burgundy is no longer such a marginal climate," he says. "I can understand from the results I have seen that stalks lend finesse and some floral perfume to wines that might otherwise be a little butch—say Corton, Clos Vougeot, Pommard, and certain Morey *premiers crus*."

It seems that the circle has turned. What was once regarded as an outmoded practice—including the stems in red-wine ferments—is now becoming a fashionable winemaking tool for those seeking elegance over power.

12 Barrels and the impact of oak on wine

Barrels are probably the earliest form of wine technology, and their use is still vital in the production of many of today's wine styles. Despite their importance in the winemaking process, the positive effects of barrels were likely discovered providentially. They just happened to be the best way of storing and transporting liquids, and until the advent of epoxy-lined cement and stainless-steel tanks, winemakers lacked alternatives. However, oak's accidental association with wine has been a critical one. The majority of fine red wines are dependent on oak barrels for a vital component of their flavor, as are a good number of whites.

Without oak, wine would be very different. Even where older, larger barrels, which don't have such a direct impact on flavor, are used, their ability to allow exposure of the contents to small amounts of oxygen is important in the development of the wine. In this chapter we focus on the science of barrels and their role in the *élevage* (the French term for "breeding" or "upbringing") of wine.

OAK

Let's begin with a slightly tangential biological perspective. Taking a somewhat simplistic conceptual view, there are four organisms crucial to wine production: two microbes and two woody plants. The microbes are the yeast *Saccharomyces* and the lactic acid bacterium *Oenococcus*. The two plants are the grapevine *Vitis vinifera* and the oak tree, *Quercus*. Of these, two are essential for all wine styles (grapes and yeast) but red wines and some whites need lactic acid bacteria, and many red and white styles wouldn't be possible without oak.

The genus *Quercus* can be split into many hundreds of species. There are four that are principally relevant to wine, three of which are used to make barrels (*Quercus alba*, *Q. sessiflora*, and *Q. robur*) and one of which makes corks (*Q. suber*). *Q. robur* also goes under the name *Q. pedunculata*, while *Q. petraea* is a synonym for *Q. sessiflora*. Taxonomy is a confused branch of science.

Why is oak so good for barrel construction? It is strong yet still quite easy to work. It has the capacity to make containers that are watertight. Its wood is rich in structures known as tyloses. Wood is mostly composed of large, open, water-conducting vessels, the xylem vessels, which travel up the trunk. The tyloses are outgrowths from neighboring cells that block the xylem vessels. American oak is particularly rich in these structures, and thus the wood can be sawn in a number of planes and still be impermeable, while French oak has fewer and has to be split in specific planes for it to make watertight staves. But perhaps most significantly, oak facilitates and is also directly involved in chemical interactions with wine that have positive effects on its flavor and structure. That's why, in this technological age, barrels still haven't been replaced in the cellar.

BARRELS: THEIR FLAVOR IMPACT ON WINE

I remember the first time I sampled through the barrels in a producer's cellar. I was struck by the differences among samples of the same lot of wine that differed only in the barrel they were being aged in. These barrels varied in their toast level, manufacturer, and source of oak. For those who use them, the choice of barrel is an important winemaking decision, and skilled producers will be as fussy about the barrels they use as they are about the condition of their grapes. There are several factors that influence the way that barrels affect wine flavor.

THE TREE

Typically, oaks used in barrel production are classified according to geographic origin. The first, and most important distinction is between French and American oak. American oak is a separate species (*Q. alba*) with different characteristics from the French species (*Q. robur* and *Q. sessiflora*). Within the category of French oak further subdivisions are made according to the forest region, which closely (but not completely) correlates with species used. The situation is further complicated by the fact that each cooper has his or her own style, and the characteristics of the staves will differ according to factors such as age of tree, the part of the trunk they were taken from, the seasoning process and the amount of toast. This interplay between oak species, environment, and human intervention makes the science of barrels almost as complex as that of the viticultural and winemaking processes.

COOPERAGE

Coopers are interested in the central part of the trunk, the dead, tough heartwood that is also known as "stave wood." This is split along medullary rays (horizontal structures that run radially through the wood). Splitting is essential for French oak because if it were sawn, it would be porous. American oak can be sawn because of the presence of tyloses, structures that block the vertical-running fibers at regular intervals. The splitting continues until single staves are produced. The fact that French oak must be split while American oak can be sawn in part explains why French oak barrels are more expensive than their American

oak counterparts. The house style of various coopers is probably the most significant factor influencing the effect of the barrel on the wine.

SEASONING

Before oak is used for barrel construction it must be seasoned. This is to bring its humidity levels into line with where it will be used, and to allow some important chemical modifications to occur. This typically takes two or three years, depending on the thickness of the staves. Seasoning, which normally takes place outdoors, is a balancing act. You want to leave the wood long enough, but not too long. It results in a number of changes to the wood: ellagitannins are reduced, as are levels of bitter-tasting compounds, called coumarins. At the same time, there is an increase in some aromatic components, such as eugenol. It is possible—and much less expensive and quicker—to age staves artificially in ovens, but these important chemical changes don't occur and the oak has fewer aromatic properties and more bitter compounds ready to be leached into the wine.

TOASTING

The barrel-manufacturing process involves heating the staves over a brazier so that they can be bent into shape. Somewhat fortuitously, this slight charring – referred to as "toasting"— coupled with the chemical properties of the

Different species of oak used in barrel production[a]

Name	Origin[b]	General characteristics
Quercus robur, known as pedunculate oaks	French forests, principally Limousin, Burgundy, and South of France	High extractable polyphenol content; makes wines that are more structured, less aromatic
Quercus sessilis, known as sessile oaks	French forests, principally in the Centre and Vosges regions:	Contributes more aroma and less structure
Quercus alba	American	Low phenol content, very high concentration of aromatic substances[c]

[a]Because of the high cost of French oak, central European alternatives are now being considered. Tonnellerie Lafitte is a sister company of Vicard that specializes in Hungarian oak.

[b]The three most commonly encountered French barrel styles are Nevers, Allier, and Tronçais. These encompass the wood from these areas, but are also terms used to 'type' wood from other regions.

[c]American oak has extremely high concentrations of oak lactones. For example, on study by Pascal Chatonnet showed that French sessile oak gave concentrations of methyl octalactone of 77 µg/liter, French pedunculate 16 µg/liter, while American oak delivered a whopping 158 µg/liter.

1 Barrel manufacture: here staves of seasoned oak are being cut to size.

2 Here the staves have been assembled in order, ready to be formed into a barrel.

3 The barrel is toasted over a flame. The degree of toasting is specified by the customer, and affects the ability of the barrel to impart flavor to the wine. This is a stylistic choice.

4 The barrel is being finished off, with metal rings hammered into place to ensure it keeps its shape and is watertight.

5 Oak barrels in the cellar at Kumeu River, West Auckland, New Zealand. This is one of the New World's most celebrated Chardonnay specialists.

6 Stainless-steel barrels are sometimes used where winemakers want to play with lees contact as you would get with a barrel, but without the effect of oxygen or oak.

7 ETS "radar" analysis of oak flavor impact on wine.

2003 Chardonnay Reserve: Barrel Trial

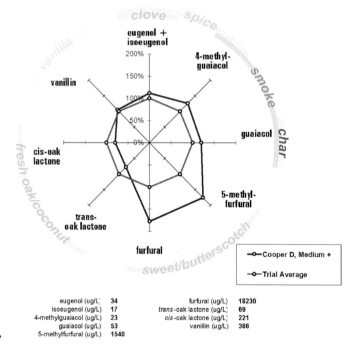

eugenol (ug/L) 34	furfural (ug/L) 18230
isoeugenol (ug/L) 17	trans-oak lactone (ug/L) 69
4-methylguaiacol (ug/L) 23	cis-oak lactone (ug/L) 221
guaiacol (ug/L) 53	vanillin (ug/L) 386
5-methylfurfural (ug/L) 1540	

7

wood, means that the interaction of the wine with the inside of a new barrel imparts pronounced flavor characteristics to the wine. When used appropriately, new barrels can have a significant beneficial impact on the wine that is aged in them. These are summarized in the table.

MICROOXYGENATION
—THE TRADITIONAL WAY

Barrels don't just impart flavor directly. Another equally important, but less talked-about, effect of aging wine in barrels is that this allows a very slight and controlled exposure to oxygen. Normally, winemakers do all they can to avoid exposing their wines to air, but in this case the low-level oxidation that barrels permit is beneficial to the structure and character of many wines.

Wines stored in average-sized barrels (around 225 liters [60 gallons]) typically receive between about 20 and 40 mg of dissolved oxygen per liter per year, but this is difficult to measure precisely because some is consumed by ellagitannins in the oak. Some oxygen passes through the wood itself, the majority passes through gaps between the staves, and the remainder comes through the bunghole. This low-level exposure to oxygen has a number of important effects. Color is intensified because of reactions between tannins and anthocyanins, and tannins are typically softened by polymerization, which eventually causes them to precipitate from the wine. But in the shorter term, barrels can help build a wine's structure in much the same way the micro-oxygenation does.

Some companies, such as California's ETS Laboratories, are now offering analytical techniques that will allow winemakers to test the performance of barrels by analyzing their chemical imprint. To explain the results in ways that winemakers can visualize, they use a graphical representation known as a radar plot. Dr. Eric Hervé of ETS gave me an example to explain how they do this. "Imagine that a winemaker wants a better understanding of how barrels from various origins influence the aroma of one of his top-end wines. The trial is simple: the same lot of wine is aged in groups of barrels from various coopers, with various specifications (e.g. toast level). During the aging, the wine is tasted and analyzed. We report concentrations of oak aroma compounds for each

sample, in micrograms per liter (μg/L). Winemakers can then build a database for future references and comparisons. We also generate a 'radar plot,' for each sample, by comparing concentrations found versus the 'trial average' (mathematical average of concentrations found in all samples from the trial). This allows us to see immediately what aroma compounds are below or above average, and therefore if associated aroma descriptors are likely to be less or more intense, compared to the other wines in this trial." In this way it is possible to compare barrels from different coopers, different species of oak, and even with different toast levels.

A strong relationship of trust has to exist between winemakers and their barrel suppliers, because once wine has gone into a bad or unsuitable barrel, it is too late to reverse the decision. But this sort of analytic technique is likely only to describe a small fraction of oak-imparted wine flavor compounds. Do these sorts of analyses measure all the significant molecular contributions of barrels, or just a few? I asked Hervé about this. "Although the compounds we measure are widely considered the main contributors to the oak flavor, there are many others that may play a role in wine," he responded. "Luckily, most of these compounds belong to a 'family' of related compounds (same origin and similar odors), and measuring one or two molecules only gives a good idea of the 'family' as a whole. This is the concept of 'indicator compounds.' One good example is the volatile phenols family: degradation of oak lignin by heat gives hundreds of these molecules, several of them playing a sensory role in wine, in synergy. Measuring only the two most important ones (guaiacol and 4-methylguaiacol), however, proves to be a very reliable indicator of the 'smoky' character in wines."

OLDER BARRELS

Not everyone likes or wants the flavor imprint that new oak barrels stamp on a wine. While certain styles of wine have the substance to absorb flavor compounds from new oak without being dominated by them, many wines are best aged in second-, third-, or even fourth-use barrels. Older barrels impart progressively fewer flavors to wine, but they still allow the controlled oxygen exposure that is important for resolving structural elements in the wine. While it is possible (and important) to clean barrels between uses, it is impossible to sterilize them. If they harbor spoilage organisms such as *Brettanomyces*, there's virtually no way to remove the potential inoculum because of the porous nature of the wood. There's always somewhere for the bugs to hide. These barrels can still be used, but they need to be monitored very carefully.

ALTERNATIVES TO BARRELS

Barrels are expensive. Oak chips, staves, and even liquid oak extract have been used to give cheaper wines some oak complexity and flavor. The results are mixed, and rarely replicate the characteristic imprint of barrels. However, suspending oak alternatives in tanks during fermentation of red wines offers a means of adding some barrel fermentation character to red wines. Usually, barrel fermentation isn't possible for red wines because of the fact that they are fermented on their skins, and oak interaction is usually restricted to the post-alcoholic fermentation period. This may offer some interesting options for experimentation. It could increase structure development and enhance color stability. Micro-oxygenation, a technology that aims to replicate the slow oxygen exposure wine experiences in barrel, is discussed in chapter 10.

HAS THE WINE WORLD FOCUSED TOO MUCH ON SMALL OAK?

Over recent years there seems to have been a move away from small oak barrels (of the 225- and 228-liter [about 60-gallon] capacity common in Bordeaux and Burgundy) toward larger barrels and even the elimination of barrels altogether in *élevage*. It has been suggested that the increase in demand for oak has led to some quality issues in barrel production. More significantly, many winegrowers have decided that they don't like the flavor of new oak in their wine, and that the oxygen exposure offered by smaller oak may not suit all wine styles. Chris Alheit, a winegrower in South Africa, likens new oak flavors in wine to retsina, where pine resin is added to wine.

Another South African winegrower, Eben Sadie, has been moving away from new small wood to bigger-format containers, such as big old oval casks and concrete.

"Cabernet family grapes do really well in small barrels," Eben maintains. "All varieties with certain tannins need a rapid evolution in order to be bottled at 18 months, but it has become clear that Mediterranean varieties are completely different. They already have advanced tannins and more fragile fruit." This move has proved to be a massive logistics exercise. He is getting big oak from Austria, and buys one or two large barrels a year.

"I have begun to dislike wood. But it is because of where I am," says Sadie. "I was in Burgundy and the northern Rhône last week tasting the 2010s, and the people making wine in continental climates can really use wood and complement their wines. But for us, working with very mature fruit, it is more difficult. The barrel is essentially an incubator for ripening the wine. You ripen the wine in the wood. For us, with a Mediterranean climate, where the grapes are inevitably mature, the last thing we need to do is mature the wine more in the wood. The tannins are mature. What we have to do is move away from oxygen and protect the fruit and freshness of the wine. So I am moving away from wood. Also, on a generic profile across the world, people are moving away from wood. I think it will be an interesting move because we will taste more wine."

"If you look at a vineyard that is 100 years old, why in the world would you put these grapes into 100% new wood, and make this wine taste like a tree that grows in France for at least 10 years of its life?" he asks. "80% of consumers will taste the tree in France and not your terroir, because most people drink the wines before they are 10 years old. So I have moved completely away from oak."

CONCLUDING REMARKS

I have touched on the importance of barrels in winemaking, and some of the significant effects they have on wine flavor and structure. It's clear that this is a complicated, multifaceted subject. The availability of noninvasive analytical techniques for examining the potential of barrels will allow winemakers to make a more informed decision as to which oak sources and *tonnelleries* (barrel-makers) will best suit their wines. Any conscientious winemakers are taking a huge gamble if they aren't meticulous about the provenance of the barrels they are using.

Flavors from oak[a]

Flavor compound	Characteristics	Influence of the manufacturing process
Lactones	The most important oak-derived flavors in wine are *cis* and *trans* isomers of β-methyl-γ-octalactone, known as the oak lactones. On their own, these oak lactones smell like coconut, but in wine they can smell quite oaky, too. The *cis* isomer is described as having an earthy, herbaceous character, as well. The *trans* adds spice to its coconut aromas.	Seasoning barrels affects the ratio of *cis*- to *trans*- forms of oak lactone, and toasting is thought to reduce overall lactone levels. American oak contains much higher lactone concentrations.
Vanillin	The main aroma component of natural vanilla. This is present in significant quantities from oak wood, and contributes significantly to the aroma of oaked wines. If the wine is actually fermented in oak barrels, yeast metabolism reduces the vanillin concentration by reducing it into the odorless vanillic alcohol. Therefore, barrel-fermented wines smell less oaky that the same wines would that were fermented in a tank and then transferred to a barrel, even though they've been in oak for longer.	Levels can be increased by toasting, but decrease at high toast levels.
Guaiacol	Guaiacolol and the related 4-methylguaiacol have a char-like, smoky aroma. 4-methylguiacol is also described as spicy.	Formed by the degradation of the wood component lignin during toasting, and is therefore increased at high toasting levels.
Eugenol	With a clove-like smell, this is the main volatile phenol associated with wood. The related isoeugenol smells similar.	Increased during the seasoning process, and reported to increase during toasting.
Furfural, 5-methylfurfural	Produced by the heat-induced degradation of sugars and carbohydrates. They have sweet butterscotch and caramel aromas, with a hint of almond.	Made when carbohydrates in the wood are degraded by the heat of the toasting process.
Ellagitannins	Tannins absorbed by the wine from the wood are known as ellagitannins. They are capable of modifying the structure of the wine, as well as combining with anthocyanins and increasing color. They have an astringent taste. Ellagitannins belong to a class known as hydrolysable tannins, and the two main isomers are vescalagin and castalagin.	Their concentration decreases at heavy toasting levels.
Coumarins	Derivatives of cinnamic acids that are present in oaked wine at low concentrations but which still affect flavor: their glycosides are bitter and their aglycones acidic.	

[a] Many of these compounds occur at levels below their individual detection thresholds in wine. Yet they can still have an effect on wine flavor and aroma through important synergistic effects. For example, the perception threshold for oak lactone is reported to be 50 times lower in the presence of vanillin. In addition, the combination of more than one of these can produce complex flavor/aroma sensations. Whether or not any of these flavors will be positive depends on the context of the wine. It's a complicated business. Recent research also points toward a volatile sulfur compound of the thiol (mercaptan) class that can contribute to oak flavor. This is also why white wines are more "oaky" when they are fermented in steel and transferred to a barrel.

13 Reverse osmosis, spinning cones, and evaporators: alcohol reduction and must concentration

Californian wine consultant and industry commentator Clark Smith is a controversial figure. He comes across as articulate, shrewd, and, at times, very funny. To some he's an inspired innovator, to others he's the devil incarnate. Why? It's because he obtained the first patent for alcohol reduction by a technique known as reverse osmosis, which is increasingly used worldwide as one of the latest high-tech tools in the winemaker's arsenal. To traditionalists, these machines, and other newer ways of reducing alcohol in wine, represent a very real threat to the "soul" of wine itself.

A key concept underlying Smith's message is that grape phenolic maturity is independent from sugar levels. That is, while grape sugar accumulation and acid respiration are highly dependent on the climate, the color, aroma synthesis, and tannin evolution occur at the more or less the same rate wherever grapes are grown. This means that in warm regions the big problem is that the rapid accumulation of sugars can impose a premature harvest, even though the grapes have not reached phenolic maturity. The traditional compromise has been that if you want to pick grapes at optimum flavor maturity then you have to tolerate alcohol levels that are often excessively high, and in many cases are to the detriment of wine quality. In cooler climates the problem is quite different. For producers in the classic European wine region achieving phenolic ripeness often means leaving grapes to hang well into the onset of the fall rains, risking dilution. Vinovation's reverse-osmosis technology offers solutions to both problems. If you have too much alcohol you can simply remove it from the finished wine. Grapes got rained on? You can concentrate the must before fermentation.

This chapter takes a close look at reverse osmosis, and the related technologies of spinning cones, Memstar, and vacuum concentration. I'll look at how widespread the use of these techniques is, and try to assess whether they are compatible with fine-wine production.

HOW IT WORKS

So how does reverse osmosis work? You might remember the principle of osmosis from school science lessons. If a semipermeable membrane separates two liquids, water will flow across the membrane from the less concentrated to the more concentrated solution. This is not what you want to do with wine, of course, because the must or finished wine would get more dilute. The answer, or at least part of it, is to increase the pressure of the more concentrated solution, which then reverses the flow. It's not quite as simple as this, though. As the pore size of the filter membrane decreases to the levels needed for wine applications, dissolved molecules begin to foul it, making it unusable. The key to the success of reverse osmosis is the use of what is known as tangential or crossflow filtration. So, instead of the flow being pushed through the filter, the process mimics the blood capillaries in the kidney by directing the flow across the filter. As a consequence, the turbulent fluid flow keeps the filter clean. Yes, a lot of energy is wasted just keeping this membrane from getting clogged up, but it works. There are currently three key uses for this technique: removing water from grape must and removing alcohol or volatile acidity from a finished wine. A more recent development is the possibility of removing 4-ethylphenol (a major contributor to the negative flavor impact of *Brettanomyces*, see Chapter 17) from finished wine. This has led some researchers to joke that various consulting companies will first give you a "brett" problem (a shorthand name for *Brettanomyces*) in your wine by encouraging the use of micro-oxygenation, and then offer to clean the wine up for you with reverse osmosis.

ALCOHOL REDUCTION

If you grow grapes in hot climates—as in most New World regions—you usually have to make a compromise with the time of harvest. Pick too soon and you'll have ideal sugar levels and poor phenolic ripeness. Wait for optimum phenolic ripeness (flavor maturity) and the danger is that the sugar levels will be so high that you'll end up with overly alcoholic wines. This is a problem that is on the increase. The choice to harvest later to make riper, richer styles of wines may shoulder some of the blame, but so does global warming. In many regions it is getting tough to make wines with alcohol levels that don't impair the quality of the wine.

For this reason, high alcohol is a hot topic. For example, in Australia wines are getting increasingly alcoholic. A survey by the Australian Wine Research Institute (AWRI) showed a steady increase in red-wine alcohol levels over a 20-year period from 1984–2004, with a rise from 12.3% to 13.9%. This is just the average, and many ambitious red wines are nudging 15% alcohol. Is it a problem? Some winemakers argue that to pick any earlier would risk green, unripe flavors in their reds, and that high alcohol levels are a price willingly paid for sweet, rich, fruit flavors. Others would counter that many grapes are now simply picked too late, in the chase for super-ripe flavors that have been popular with critics. What no one can deny is that high alcohol has become a talking point.

Is high alcohol such a problem? Evidence suggests that it is, principally because high levels mask aroma. Ethanol has been shown to modify the solubility of many of the aroma compounds found in wine, making them reluctant to leave solution. This makes the wine less aromatic. It can also make some wines taste "hot." An analytical chemistry study in 2000 showed that as alcohol rose from 11% to 14%, there was reduced recovery of typical wine volatile compounds.

In 2007, Vicente Ferreira's group at the University of Zaragoza (Spain) identified a range of esters responsible for the fruity berry flavors in a series of red wines. But when they added more of these to the wine, it didn't increase the fruity impact, because of the suppressing effect of other wine components, including alcohol. They showed this in another experiment in which they added increasing levels of ethanol to a solution of nine esters at the same concentration they are found in wine, and discovered that the fruity scent quickly falls as alcohol rises, to the point that when alcohol reached 14.5%, the fruity aroma had been totally masked by the alcohol.

In part, these rising alcohol levels could be reflecting a choice by winegrowers to pick grapes later. This may be driven by perceived demand on the part of consumers for bigger, riper wines, with sweet fruit flavors. But there's convincing evidence that in some regions at least, high alcohol may also be a consequence of increased average temperatures caused by climate change. In this case even winegrowers who don't want to make riper, more alcoholic wines are forced to wait until grapes have very high sugar levels, because the consequence would be picking grapes that haven't reached appropriate flavor development.

In a 2008 study, Petrie and Sadras looked at the advancement of grapevine maturity in Australia over a 13-year period from 1993. They studied three different varieties from a range of regions across the country. The results were striking. Even over this short timescale, the various regions showed advancement in date of designated maturity of between half and three days a year. So it does seem that climate change is having some measurable effect

Left Partly raisined berries that have been left on the vine late in the season. The result is very ripe, slightly jammy fruit flavors, and high sugar levels that will translate to high potential alcohol levels.

on grapevine phenology across Australia's vineyards. But why is it that this is resulting in more alcoholic wine? Wouldn't it be possible just to pick a little earlier, advancing harvest in line with the advances in phenology?

Not so, it seems. Sadras and Moran published a very interesting paper looking at how warming trends could result in the decoupling of flavor and sugar ripeness. This is a very interesting concept, and needs some explaining.

Grape ripening is split into two processes, which are known commonly as sugar ripeness and flavor ripeness (this is also referred to as phenolic or physiological ripeness; in the Sadras and Moran study, anthocyanins are used as an indicator of its progress). In an ideal world, the process of sugar accumulation (which happens in tandem with lowering of acid levels) goes hand in hand with the chemical transitions that result in changes in levels of flavor compounds and their precursors in the grapes. Simply put, sugar accumulation occurs through photosynthesis, which takes place during hours of daylight. The development of flavor ripeness occurs via the process of respiration, which is independent of light, but is dependent on temperature.

The ideal scenario sees grapes reaching flavor ripeness at a point where they still have adequate acidity and the right amount of sugar to give moderate alcohol levels. Should sugar ripeness proceed too fast, then by the time flavor ripeness is achieved potential alcohol is very high and acidity low enough to make supplementation in the winery necessary.

In this new study, Sadras and Moran looked at the effect of elevated temperature on the coupling between sugars and anthocyanins. The study took place in experimental vineyards in the Barossa Valley, and involved three different experiments on neighboring rows of vines in both the 2010 and 2011 growing seasons. In the first, Shiraz and Cabernet Franc vines were exposed to elevated daytime temperatures using a passive open-top chamber device that didn't change the night temperatures or the humidity. These were compared with controls. In the second, the elevated-temperature experiment was repeated with reduced crop load. In the third, the elevated-temperature experiment was repeated with deficit irrigation (as compared to the normal irrigation level).

The data suggested an interesting relationship between sugar and anthocyanin accumulation. This has two phases: in the first, sugars increase with no change in anthocyanin, followed by a linear phase where both increase in parallel. Elevated temperatures delay the onset of this linear phase. The result is that with warmer temperatures, flavor ripeness lags behind sugar ripeness. At these higher temperatures the onset of flavor ripeness (the linear phase) is being delayed, but the actual rates of accumulation of both sugar and anthocyanins in the linear phase is unchanged.

Interestingly, experiment 3 showed that water deficit increases the ratio of anthocyanins to sugar in elevated temperature, which, along with similar results in other studies that have addressed this, suggests that regulated deficit irrigation is one way of restoring the flavor: sugar ripeness balance disrupted by higher temperatures. In contrast, lower crop levels didn't seem to help restore this balance in experiment 2.

This is an important paper, and it explains why the recent rises in alcohol levels might not just be a result of fashions in preferred red-wine styles. The warming trends experienced in Australian vineyards are making it harder to get proper flavor development at sensible potential alcohol levels. Deficit irrigation appears to be one way to counter these changes, but there may well be other viticultural interventions that could help shift the balance back.

TECHNOLOGICAL SOLUTIONS

So, if your alcohol levels are too high, don't worry. Companies such as Conetech and Wine Secrets in California can reduce the alcohol levels in your wine. Or, there is the old trick of adding water to must. It's really common in warm climate regions, but it is illegal (although the regulations in California do allow for a reasonably generous water addition leeway when winemaking processing additives are added). This is problematic. While there's little risk of being caught, there are techniques, such as isotope analysis, that can show if this has occurred.

Legally, though, two technologies are used to reduce alcohol. The first is reverse osmosis, which relies on a technology called crossflow filtration. A portion of the blend is put through a crossflow filter, which acts something like the kidney tubules in that the liquid being filtered isn't forced through a membrane, but instead runs through a tube under pressure, the walls of which are made of the filtration membrane. The advantage of this technique is that the flow of liquid keeps the membrane pores from clogging, but it does mean that the surface area of membrane required is enormous. This is achieved by having a column consisting of numerous small tubes. As the wine passes through, a mixture chiefly composed of water, acetic acid, and alcohol is removed from it. This is taken away, and the alcohol is then removed either by distillation (this is the reverse osmosis process that Clark Smith gained a patent for), or another membrane process (this is called Memstar). This means that the water can then be recombined with the wine to produce a lower-alcohol wine that can be blended back to the original larger batch to produce wine with the desired alcoholic strength.

The second technique, the spinning cone, achieves the same ends by rather different means. Commercialized by the U.S. company Conetech, the spinning cone column contains around 40 upside-down cones, half of which are fixed while half spin. In a vacuum environment, the cones spin the wine into thin liquid films, and a cool vapor rises off the wine, carrying the volatiles from the liquid. In the first pass, the ultra-light component consisting of the delicate flavors and aromas is carried off and condensed. This is known as the "essence," and it is saved for later in order to be recombined with the wine. The second pass takes off as much

Above A spinning cone column.

alcohol as you want to remove. Theoretically, you could then recombine the remaining low-alcohol wine with the essence and the alcohol and end up with the same wine you started with.

Currently, Conetech treats wines from around 600 clients worldwide, and last year treated six million gallons. However, because it only treats a small proportion of each wine, around 10%, which is then back-blended, around 50 million gallons of wine will have been alcohol-reduced in this way. As well as a plant in California, Conetech also has plants in Chile, South Africa, and Spain. The cost of the spinning cone machine is around $1 million, whereas a reverse osmosis machine costs around $30,000 and is small enough to be moved around, making it possible to do alcohol reduction in the winery.

The rules are potentially a problem for this sort of technology. In the USA spinning cones have been authorized for alcohol reduction, but in Europe this technique used to be allowed only on an experimental basis. This meant that you were allowed to treat 50,000 hectoliters (about 1.3 million gallons), but the wine couldn't leave the country of origin. But in November 2008 the European Union regulations changed, making it legal to remove up to two degrees of alcohol, where specific local appellation laws permitted this (in many AOPs in France, for example, this is still not permitted).

Tasting through a series of the same wine with the alcohol reduced, both by the spinning cone and reverse osmosis, reveals how much of an impact the presence of alcohol has on the perception of other wine components. "Alcohol is

Above A reverse osmosis machine.
Photo courtesy of Clark Smith

a masking agent," says Conetech's previous head winemaker Scott Burr, "so taking it away reveals what's there. It also adds sweetness to the palate."

How does a winemaker decide how much alcohol should be reduced? "A big client who we've been working with a long time might specify it wants its wine at 13.85%," says Burr. "A smaller client might say it has a Zinfandel at 16.8% alcohol, so we do a run, take the 4% component and blend wines at a bunch of different alcohol levels," he says. "I don't tell the client what the alcohol level should be." In terms of which wines are more successful, Burr says that it is not a curve, but instead there are sweet spots. "There are all kinds of different ones. The more oak, the bigger the variance."

Clark Smith reckons that 45% of premium Californian wines are alcohol-adjusted, either by reverse osmosis or the spinning cone. Alcohol reduction is becoming a widely adopted tool throughout the winegrowing world, but it is not without controversy, simply because it's seen as a rather artificial technique that subjects a portion of the wine to rather dramatic physical forces.

Clearly, such interventionist strategies don't sit well with a concept of natural wine. But there is an interesting, almost philosophical question here. Could it be that if a grower is stuck in the position where he or she are forced to choose between either developing the flavor and achieving sensible alcohol levels, whatever is done in the vineyard, might this justify the use of alcohol-reduction technologies as a last resort? They do sound highly manipulative, and most fine-wine producers would prefer to do as little as possible in terms of winemaking interventions. But the tantalizing possibility remains that a wine that speaks more eloquently of its origins—the vineyard site or terroir it came from—could be realized with the use of these tools. I know it sounds like heresy, but could alcohol-reduction techniques in the wine cellar assist in the terroir expression that traditionally minded winegrowers seek? Could reverse osmosis or the spinning cone have an important role to play in fine-wine production, compensating for the effects of global warming?

The argument would be that technology like this is simply a tool, and tools themselves are morally neutral—it is how they are used that matters. A winegrower who genuinely wants to make a wine that expresses the vineyard site optimally, but finds the wine's ability to do this hindered by the masking, sweetening effect of high alcohol, could use such a tool to produce a better wine that has more of a sense of place to it, (in theory, at least).

"In the past, I confess to having used some technological methods (spinning cone and reverse osmosis) to remove alcohol from our wines," admits Californian winenmaker Randall Grahm, "but I am now quite opposed to the practice (and of adding water as well-known as 'Jesus Units' in the trade). If you are doing the right work in the vineyard, you should not be compelled to resort to these extreme solutions."

He continues: "In the winery, the most practical solution that we have found to keeping alcoholic degree in check has been the use of open-top fermenters, warm fermentation temperatures, and especially the use of indigenous yeasts. The 'wild' yeasts have at least for us been absolutely brilliant. They tend to give us a much longer, more even fermentation (and also don't stick)."

But how do wines treated in this way compare with control wines in terms of wine-flavor chemistry? Clearly more independent work is called for here, a sentiment shared by Roger Boulton, Professor of Enology and Chemical Engineering at the University of California-Davis: "There are no independent published reports of the sensory effects of this treatment compared to a control, only proprietary claims and selective testimonials. There is no published example of a side-by-side comparison across several wines." It should be added that the use of crossflow filtration for reducing alcohol levels in wine is not new; it has been trialed since the mid-1980s, but it is only in recent years that it has been the focus of much attention.

Reverse osmosis is a subtractive rather than an additive process, taking something out but not adding anything that wasn't there in the first place. The loop is intact. In Australia, the rules seem to be a little more flexible. They allow winemakers to replace the permeate removed from the wine with water evaporated from juice—this water is the "low-baume juice." The exception that stops this from being an outright misdemeanor in the eyes of purists is that the water involved would once have been inside a grape in the appropriate appellation, and thus carries its appellation with it. It saves having to bother with stills and all the rules and restrictions that distilling alcohol carries with it. However, some Australian commentators suggested to me that this is not the case, and that there is a strong recommendation by the official committee dealing with label integrity that a wine component must retain at least 8% alcohol or sugar equivalent to 8% potential alcohol for it to be considered wine, unless the reverse osmosis loop is kept intact and water derived from permeate distillation is added back to the wine. This illustrates the complexity of adapting rules to new technologies.

The time-honored method of reducing alcohol levels in wines is the hose method—adding water to high-sugar musts. This is of course illegal, but it is widely practiced. It has the added drawback that it dilutes all the components of the wine. Some winemakers use reverse osmosis to remove the water and alcohol from wine, and then add water back in that has no connection with the grapes. While this is a cheaper method than distilling the permeate or using low baume juice, it is still illegal.

MUST CONCENTRATION

Reverse osmosis (RO) has also found a home in the classic wine regions of Europe. Here the problem is that harvest often coincides with the onset of rainy autumnal weather. If it rains during harvest, you can end up with a dilute wine, and many potentially good vintages are ruined this way. Smith explains, "In Bordeaux, true ripeness often means hanging into the rain, and a reverse osmosis can squeeze this rain back out so we can obtain flavor concentration and alcohol balance." He points out that the widely used "traditional" technique of chaptalization ignores this dilution and corrects just the alcohol imbalance with added sugar. One problem with using reverse osmosis for must concentration is that it requires clarified juice to avoid fouling the filtration device, so the smaller the amount to be concentrated, the better. This is a factor that limits the uptake of the technology for red wines because not everyone wants to have to clarify their grape must prior to fermentation. A modern reverse osmosis machine operating at 1500 psi can concentrate just a portion of the must to 42 Brix, which can then be used for blending.

An alternative method of removing water from the must is by means of vacuum distillation. These concentrators heat the must to temperatures of around 77–86°F under vacuum, and can treat from 265–2,100 gallons must per hour with an evaporation capacity of 40–320 gallons per hour. These machines have been popular in the past, and their use preceded that of reverse osmosis. However, they have drawbacks. The heating of the must can introduce aroma losses and apparently a butterscotch-caramel character to the fruit, and they are much more expensive than today's reverse-osmosis machines. One point worth mentioning is that the must usually has to be heated to temperatures above those claimed by machine manufacturers. To work one of these devices at 77°F requires a vacuum of around 85%, which is hard to achieve. Some industry people have told me that many of the machines need must at 115°F to work effectively, which could substantially change its character.

So far so good. In principle the use of reverse osmosis or vacuum distillation to remove excess water from the must is simply a correction of a vintage anomaly, with the winemaker removing just the rainwater that otherwise would have caused a diluted wine. But the problem is that in a market dominated by critics who award high scores to super-concentrated wines, the temptation to do a little more than just remove the rainwater is strong, even if there are drawbacks to using both. In effect, they aren't a million miles away from the technique of *saignée*, which aims to increase the ratio of juice to skins in red-wine fermentation by bleeding off some juice after a brief maceration.

Must concentration by reverse osmosis or vacuum distillation is allowed in the European Union, but the regulations limit its use to a 20% maximum volume decrease and a 2% volume maximum alcohol potential increase. It's illegal to chaptalize and concentrate the same batch of wine. Still, even within these limits, it is possible to make a super-concentrated wine that some critics may have trouble differentiating from one produced by the more traditional route of low yields, good vineyard sites, and careful fruit selection.

How widespread is must concentration? It's hard to get an accurate picture, because companies that supply these services are unwilling to name names, and most properties owning machines don't tend to boast about it. I've spoken to quite a few people, and the only clear answer I can get is "more than you might think." James Lawther, an expert commentator on the Bordeaux scene reckons there are more than 60 reverse-osmosis machines in operation in the region, and about as many vacuum concentrators again. Then there are a number of contract companies that offer RO, all of which are pretty busy at harvest time. Must concentration is also being practiced in Burgundy, but on a much smaller scale.

Elsewhere in Europe, the picture is one of a technical revolution in the offing. In Germany, more than 100 wineries have experimented with must concentration, mainly in Baden and the southern areas. It is easier to achieve with white wines, because juice clarification prior to fermentation isn't problematic the way it is in reds. In Italy, there are reverse-osmosis machines in Piedmont, Tuscany, and Alto-Adige. It seems that things are just beginning in Spain, too.

So far not mentioned is concentration of finished wines by reverse osmosis. This has been tried, but is not widely legal. Richard Gibson, previously with Southcorp but now running a consultancy called Scorpex, recalls trials with dilute Riverland fruit. At 40% reduction the wine started to become really interesting, with various aromatic compounds reaching threshold level. At 60% they had what he mischievously referred to as a "Grange blender." There are two potential drawbacks with this technique. First, the wine must be sound to start with: any unripe or vegetal characters will be concentrated along with positive flavor components. Second, it is subtractive and there is a loss of volume, which may not be economic in certain circumstances.

IS IT A SIN?

A crucial question regarding these winemaking interventions (alcohol reduction, must concentration, and micro-oxygenation [covered in Chapter 10])—is whether they are appropriate or honest manipulations for making high-quality wines. In other words, is it cheating?

As Clark Smith puts it, "The central debate about reverse osmosis and other high-tech wine-production innovations is not about whether they work anymore. It is about whether we will go to hell if we use them." Smith adds, "That they work is taken more or less for granted, in the same way that procreative ability is a side issue in assessing one's daughter's suitors."

Randall Grahm makes an important point here. He thinks it's a question of context: "If a producer produces a *vin d'appellation*, there is an implicit contract that he enters into, whereby he effectively promises a wine of some degree of 'typicity', which I suppose would include the characteristics of the vintage. If he utilizes certain techniques in the winemaking process to wipe out vintage characteristics, even though he is perhaps producing a wine that most people would prefer, I believe that he is acting in bad faith." Grahm adds, "If he is producing a *vin de table*, *vino da tavola*, or producing wine in the New World, I think that a different set of criteria apply. His contract is simply with the consumer to make the best wine that he can and that will offer the consumer vinous *jouissance*."

Perhaps the ethos and motivation of the winemakers themselves are more important than the issue of technological innovation. If they are passionate people whose goal is to produce the very best wine their terroir permits, then there's little danger that, should they choose to employ them, they will use techniques such as these irresponsibly.

In the meantime, open discussion of these technical issues can only be a good thing. Consumers need to be kept informed so they can participate in the debate about how much "intervention" they would like in the wines they are purchasing. Finally, Smith is quite critical of wine writers, who he thinks have failed to educate consumers properly about the role technology has to play in winemaking. "There's a major disconnect between what's being done to improve wine quality and what wine writers choose to let consumers look at. The idea is that they are protecting the notion of wine. They feel it is so fragile that if we tell people what is really going on then the mystery will go away."

14 Sulfur dioxide

Sulfur dioxide in wine is one of the most frequently discussed and yet simultaneously one of the most frequently misunderstood issues in winemaking. Winemakers, retailers, writers, and even consumers talk about it, but with the exception of the first group listed here, I suspect that most don't have a clear understanding of the issues involved. It's undoubtedly a technical sort of subject that fits firmly into the category of the chemistry of winemaking, but I'm going to try to keep this chapter readable and interesting without sacrificing depth of content. This is because I'd like people actually to read this chapter. Sulfur dioxide is an important subject, so it's a good idea to have a decent grasp of the issues relating to its use.

First I'll look at why sulfur dioxide is such an important component of winemaking, and how it acts as a "chemical custodian" of wine quality. Then I'll turn to the important subject of how sulfur dioxide can best be used, and why it is generally not a good idea to use too much or too little of it. Finally, and perhaps most interestingly, I'll report on some brave souls who, opposed to any sort of winemaking additives at all, have tried to make wines with very little or even no added sulfur dioxide.

Above Peter Godden of the Australian Wine Research Institute.

WHY SULFUR DIOXIDE IS JUST ABOUT INDISPENSABLE FOR WINEMAKING

Sulfur dioxide (SO_2) acts as a guardian of wine quality in two ways. First, and most importantly, it protects the wine from the ill effects of oxidation. Secondly, it acts as an antimicrobial agent, preventing the growth of unwanted spoilage bugs in the wine. Peter Godden of the Australian Wine Research Institute (AWRI) describes SO_2 as a "magical substance" because it has these effects at very low concentrations. "We're mostly talking about a maximum of 150 parts per million (150 mg/liter) and much less in most Australian wines," he explains (Godden showed data indicating that of a representative sample of Australian wines,

at least 70% have levels below 100 mg/l, and close to 60% have 80 mg/l or less). Correct use of SO_2 is a subject that he and his colleagues of the AWRI industry services team have been spending a lot of time advising Australian winemakers on in recent years. "A wide range of the problems that we see have relatively few root causes, and SO_2 use, or misuse, has been the most common," although he adds that since they began this work, data from the Institute's commercial Analytical Service Laboratory, which analyses many thousands of Australian wines each year, indicates that things have improved noticeably.

FREE AND BOUND

Key to understanding the effects of SO_2 is the ratio between the free and bound forms. When SO_2 is added to a wine, it dissolves and some reacts with other chemical components in the wine to become "bound." This bound fraction is effectively lost to the winemaker, at least temporarily, because it has insignificant antioxidant and antimicrobial properties. Various compounds present in the wine, such as ethanal, ketonic acids, sugars, and dicarbonyl group molecules, are responsible for this. Winemakers routinely measure total SO_2 and free SO_2, with the difference between the two being the amount in the bound form. Importantly, equilibrium exists between the free and bound forms such that as free SO_2 is used up, some more may be released from the bound fraction. It's slightly more complicated than this, though, because some of the bound SO_2 is locked in irreversibly and the remainder is releasable, and of the free portion most exists as the relatively inactive bisulfite anion (HSO_3) with just a small amount left as active molecular SO_2. From the winemaker's perspective, molecular SO_2 is the interesting part. Typically, levels of 0.8 mg/liter molecular SO_2 will be aimed for in white wines, and this will probably require there to be 15–40 mg/liter of free SO_2 present. Reds can get by with a little less.

THE IMPORTANCE OF PH

One of the key factors affecting the function of SO_2 is pH. For the benefit of those who have long forgotten their school chemistry lessons, pH is a measure of how acidic or alkaline a solution is (technically it relates to the concentration of hydrogen ions in solution), where a pH of 7 is neutral and below and above this the solution is progressively more acidic or alkaline, respectively. Thus a wine with a lower pH is more acidic. All wines are acidic (with a pH of less than 7), but some are more acidic than others. In two respects, pH is important here. First, at higher pH levels more total SO_2 is needed to get the same level of free SO_2. Second, SO_2 is more effective, that is, actually works better, at lower pH. So, as well as having more of the useful free form for the same addition, what you have works better as well. It's a double benefit. This is shown in the table below.

The most useful attribute of this wonder molecule is that it protects wine against the effects of oxidation. Oxidation is discussed in depth in Chapter 10, so I won't go into too much detail here. There are two sorts of oxidation processes in wine. The first, which only really happens in grape must, is enzymatic, and is caused by oxidases. The levels of these enzymes are much increased in damaged or rotten grapes, so where these are likely to be present it is especially important to use sufficient SO_2. The message here for winemakers is that it is important to use grapes that are as

clean as possible, with the absolute minimum of fungal damage. It follows that sweet wines made from botrytized grapes need substantially higher levels of SO_2 to protect them against oxidation. Significantly, botrytized wines are also very high in compounds that bind free SO_2, with the result that winemakers can end up adding enormous levels and still not have significant free SO_2.

The other types of oxidation reactions that occur in wine are chemical. Oxygen itself isn't terribly reactive, but it is made reactive by the presence of reduced transition metal ions, principally iron and copper. (If this is all sounding a little too technical and complicated, that's because it is.) The next stage is the production of quinones from phenolic groups, and also the production of hydrogen peroxide, a powerful oxidizing agent. While SO_2 doesn't react with oxygen itself, it is present in the free form that can bind with the quinones and peroxide, and take them out of the equation. Otherwise they would go on to react with other wine components. SO_2 can also bind with the products of oxidation such as ethanal (also known as acetaldehyde), which otherwise would make the wine taste and smell oxidized. So it isn't preventing oxygen from having an affect on the wine, but it is limiting

Percentage of free SO_2 in the molecular form at different pH levels

pH	% molecular SO_2
2.9	7.5
3.0	6.1
3.1	4.9
3.2	3.9
3.3	3.1
3.4	2.5
3.5	2.0
3.6	1.6
3.7	1.3
3.8	1.0
3.9	0.8

the damage and clearing up some of the mess. White wines generally need higher levels of SO_2 than reds to protect them. This is because red wines are richer in polyphenolic compounds, which give the wine a natural level of defense against oxidation. White wines that have been handled reductively (that is, protected against oxygen exposure through the use of stainless steel and inert gases in the winemaking process) are especially vulnerable to oxidation and need careful protecting.

Aging of wine is what is known as a "reductive" process. It works properly in the absence of oxygen, which is why a good, tight seal by the closure, whether a cork or a screw cap, is important. While there's some debate about whether tiny traces of oxygen might be needed to ensure optimum evolution of wine in the bottle, it is universally recognized that any significant influx of oxygen will rapidly oxidize the wine, that is, the oxygen will combine chemically with compounds present, negatively affecting the flavor.

But SO_2 is also microbicidal. It prevents the growth—and at high enough concentrations kills—fungi (yeasts) and bacteria. Usefully, SO_2 is more active against bacteria than yeasts, and so by getting the concentration right winemakers can inhibit growth of bad bugs while allowing good yeasts to do their work. SO_2 is usually still added to the crushed grapes in wild-yeast fermentations. While it kills some of the natural yeasts present on grape skins, the stronger strains survive and thus are selected for preferentially. Sweet wines and unfiltered red wines are at higher risk of rogue microbial growth, so with these it is especially important that correct SO_2 addition is practiced.

It follows from all this that if you don't use enough SO_2 in your winemaking you run the dual risks of oxidized wine and off-flavors and aromas from unwanted microbial growth, together with potentially considerable bottle variation.

BEST PRACTICE IN SULFUR DIOXIDE USAGE: GETTING THE RATIO RIGHT

But while this might encourage some winemakers to add in more SO_2 just to be on the safe side, Godden suggests that the best way to ensure wine quality is not by using more SO_2, but by using it smarter. His idea is that the key measurement for

winemakers is not the free SO_2 level, but the ratio of free to bound SO_2. That is, the key to effective SO_2 usage is getting the ratio of free to bound SO_2 as high as possible, to maximize the benefits of the amount added. He sent me data gathered by the AWRI Analytical Service on a typical cross-section of Australian wines which shows that, in a range of reds, free SO_2 has been steadily increasing in recent years while the total SO_2 has actually been decreasing. Thus the ratio of free to total SO_2 has improved. "I consider the use of the ratio of free to total SO_2 as one of the most useful quality-control measures during winemaking," says Godden. He has similar data on Australian white wines, although he feels that there is probably some more room for improvement with these.

How is a good ratio achieved? Starting with healthy grapes is important. Grapes suffering from rot have significantly higher levels of compounds that will bind SO_2 and also enzymes that encourage oxidation. Judicious filtration, where necessary, will also help make SO_2 additions more effective by reducing microbial populations to a level where the SO_2 is more effective against them. General cleanliness in the winery is also helpful.

Perhaps most important, though, are two critical winemaking interventions. First, controlling turbidity by careful racking, fining, and filtration (if necessary), and second, the timing and size of additions. There are three points where wine is likely to be subject to considerable oxygen stress or risk of bug growth: at crushing; at the end of malolactic fermentation (or alcoholic fermentation where malolactic is discouraged); and at bottling. At each of these points a healthy dose of SO_2 is highly recommended. Crucially, for the same total addition, it is much more effective to add your SO_2 in fewer relatively large amounts rather than many small additions. In the latter case you run the risk of never getting your free SO_2 levels high enough for it to do its job properly.

HEALTH EFFECTS

Now we turn to the consumer's perspective. Is SO_2 healthy? Not completely, is the simple answer. SO_2 can cause adverse reactions in some asthmatics, which can be quite dangerous at ingestion levels as low as 1 mg. For this reason some doctors have even gone as far as suggesting that asthmatics avoid wine altogether. For most people, it is probably fairly harmless at the levels used in winemaking, but anyone drinking wine on a regular basis is probably taking in more than medical experts recommend (although it could be debated that these levels are set a little low in the name of caution). SO_2 levels in wine are subject to regulation by various authorities. The EU has set a maximum permitted level that varies with wine type from 160 mg/liter (or parts per million) for dry red wines to 300 mg/l for sweet whites and 400 mg/l for botrytized wines. In Australia, the regulations permit 250 mg/l for dry wines and 350 mg/l for those with more than 35 g/liter of residual sugar. In the USA the maximum level allowed is similar, and any wine with more than 10 mg/l SO_2 (this level can be reached naturally even if no SO_2 is added) has to be labeled "contains sulfites." On the basis of animal experiments, the World Health Organization has set the recommended daily allowance (RDA) of SO_2 at 0.7 mg per kg bodyweight. Doing some simple sums, this would permit a 70-kg (154-pound) human to take in 49 mg SO_2 a day. Half a bottle of wine with an SO_2 level of 150 mg/l would provide 56 mg of SO_2, thus exceeding the recommended daily allowance.

Many people who suffer adverse reactions to wine, such as headaches and flushing, blame SO_2, partly, one suspects, because it seems an obvious candidate as an added chemical substance. After all, the bottle will have "Contains Sulfites" written on it, usually with no other ingredient labeling. The issue of adverse wine reactions is a complex one, and the scientific literature offers few clear indications of the culprit compounds. However, many foodstuffs contain higher levels of sulfites than wine—the worst offenders being dried fruit, which typically contains about 10 times the level in wine.

VINS SANS SOUFRE: THE QUEST FOR NATURAL WINE

So if the presence of SO_2 is just about essential for winemaking, why would anyone want to do without it? There are two reasons. First, because of the increasing concern people have about what they put in their bodies, they are anxious not to consume anything that has been chemically manipulated. To many people unaware of the issues, SO_2 use sounds like a gratuitous addition of unnecessary chemicals. This creates a potential market for "additive-free" wines. Second, there exists a band of passionate winemakers who see wine as a natural product. Eliminating SO_2 usage is seen as the final hurdle in the quest for fully "natural" wines. The desire for naturalness runs strong, and it's not just fringe winemakers who pursue this goal.

When I wrote the first edition of this book in 2005, the natural-wine movement was a tiny group, with just a few way-out winegrowers working without added SO_2. Now, eight years later, it has grown into a sizeable, lively coalition of hundreds of winegrowers. And while many do add a little SO_2 at bottling, there are many who avoid this additive at all.

In France there is now a definite market for these wines, with dozens of natural wine bars and restaurants. (The more conservative London wine scene is even blessed with a handful of wine bars focusing on natural wines.) This market really began in the 1980s, with a circuit of wine bars in Paris, all of which wanted to serve fresh wines with a purity of fruit to them. An instrumental figure in this trend has been Jacques Néauport, whose inspiration was the late Jules Chauvet. Chauvet was a small *négociant* in the Beaujolais region with an enquiring mind who tried out a number of novel ideas, one of which was making wine without added sulfites. Since Chauvet's

Above Frank Cornelissen, winegrower on Mount Etna, Sicily. Frank works without any added sulfur dioxide.

death, Néauport has consulted for a number of growers, the first of whom was Overnoy in the Jura, and whom included the late Marcel Lapierre in Morgon. Néauport developed a vinification method specially adapted for working without SO_2, involving carbonic maceration under very cold conditions. Catherine and Pierre Breton, Thierry Allemand, Jean Foillard, and Pierre Frick are other well-known pioneers of working without SO_2 for at least some of their wines. A common claim is that wines made in this way have a greater purity of fruit and are aromatically more interesting. In addition, the claim is made, if you overindulge, you are less likely to suffer a headache later.

Indeed, storage is the key problem with totally SO_2-free wines: they need to be kept below about 57°F at all points in the supply chain. It's easy to see how consumers buying their wines direct from a grower and putting them straight into a temperature-controlled cellar might achieve this, but this degree of protection is hard to envisage in the modern retail environment. For this reason alone, even if SO_2-free wines were to prove sensorially superior (and many dispute this claim), it's unlikely they will ever become widespread. Other keys to successful no- or very low-SO_2 wines include using good-quality fruit and keeping a spotless winery.

It must be pointed out, though, that even if no SO_2 is added during the winemaking process, there will still be some present in the wine because it's a by-product of fermentation. Yeasts produce small quantities or around 5–15 mg/liter quite naturally, so the notion of a totally sulfite-free wine is illusory.

Why focus on what is, in many respects, a winemaking movement on the periphery, of little commercial significance? The late Joe Dressner, who imported these wines into New York, made a good point. He thought that the *sans soufre* producers have been more generally useful to the wine world in that pushing the extreme has moved everyone along in the direction of using less SO_2 overall, and certainly less than some of the heroic levels that used to be applied in many parts of Europe. He likened these pioneers to those who have advocated unfiltered wines in the past, in that, while not everyone bottles unfiltered, these days, quality-minded winemakers will generally look not to filter unless it's really necessary. Now if the drive to use lower levels of added SO_2 can be coupled with a better understanding of how this wonder molecule works, so that it can be used smarter, as Peter Godden advocates, then everyone is likely to benefit.

15 Reduction: volatile sulfur compounds in wine

It's difficult to know how to pitch a chapter like this. While every winemaking expert I've consulted agrees that reduction is an important topic to cover, there's no doubting that the subject is a horrendously technical one. My goal here will be to make this rather technical subject understandable to nonexperts, while at the same time not being scared of the hardcore wine chemistry where it's necessary for the story.

"I feel reduction is a key topic for the trade," says Sam Harrop, a consultant winemaker. Why? "Because there is a lot of confusion over just what it is. The term 'reduction' is used extensively in tastings and I feel that in a lot of circumstances people don't really understand what it means."

What is a "reduced" wine, and what do people mean by "reduction character"? First, what does the term "reduction" mean? It's a term that is used to describe wine's sensory characteristics caused by the presence of a suite of volatile sulfur compounds (VSCs), whose presence is commonly referred to as SLOs for "sulfurlike odors". "Reduction" is actually something of a misnomer: the term was coined because these characteristics frequently develop in wine that has a very low redox potential (hence, is "reduced"), such as occurs when barrels are left for a long time without racking, but these sulfur compounds can theoretically also be present in wine that is oxidized. "'Reduction' is a simplification, a language abuse," says Dominique Delteil, consultant winemaker. "As often occurs in wine vocabulary, tasters have been willing to link sensory sensations to chemical or physical states, without being sure they are real or not. Reduction is typical of this." Delteil continues, "I prefer to call this concept 'sulfur flavors' rather than 'reduction.' 'Sulfur flavors' are commonly associated with descriptors such as 'burned match,' 'garlic,' 'onion,' 'leek,' 'rotten egg' according to

Left Consultant winemaker Sam Harrop MW, who has developed a reputation as a specialist on wine faults, and runs the fault clinic at London's International Wine Challenge each year.

the intensity and the culture of the tasters (how often they have met those aromas, how intense were they when they built their references)." These sulfur compounds still exist in oxidative conditions, and this leads to misunderstandings when they are then also described as "reduction." "As a winemaking consultant," says Delteil, "I have seen winemakers making technical mistakes because of this false concept. For example, some have continued to aerate an already oxidized wine because it had 'reduction' characters, so it was thought to need oxygen."

Despite this nomenclature issue, reduction is such a well-recognized, convenient, and widely used term, so I'll stick with it. Sometimes the term "sulfides" is used as a catch-all to describe reduction issues, but this is technically incorrect and carries with it a serious risk of confusion, because as well as sulfides and disulfides, mercaptans (thiols) and thioesters are also involved in reduction.

There's also some confusion about the difference between sulfides and sulfites. "I get this all the time," says leading New Zealand winemaker Matt Thomson. "Someone will say. 'I think this wine has got a bit of sulfur.' And I say to them 'Sulfite or sulfide?' They say, 'Sulfur.' I say, 'No, there is a big difference.' This is something that is prevalent throughout the entire industry. People don't know how to define the difference between sulfite and sulfide." Sulfite is the presence of excess sulfur dioxide, whereas sulfide refers to a subset of volatile sulfur compounds. They are completely different.

These sulfur-containing compounds are the responsibility of the yeasts. The chemistry involved is rather complex and still not all that well-understood, but here's a simplified summary of what is currently thought to be the situation. Grape musts are generally deficient in organic sulfur compounds, principally the sulfur-containing amino acids methionine and cysteine. Their absence triggers the yeast cells to begin a process called the sulfate reduction sequence (SRS) pathway, which uses inorganic sulfur compounds to generate these amino acids. The SRS pathway forms the HS^- ion from sulfate, which is transported into the yeast cell by sulfate permease and then reduced to sulfide by two enzymes, ATP-sulfurylase and sulfite reductase. The HS^- ion is then sequestered by either O-acetylserine or O-acetylhomoserine to form methionine or cysteine. However, if there is insufficient nitrogen present in the must in the right form, this doesn't occur, and hydrogen sulfide (H_2S) is released from the yeast cell. H_2S is reactive and can then combine with other wine components to form the volatile sulfur compounds that we are discussing. In addition, high levels of sulfite (SO_2) can diffuse into the yeast cell and bypass regulatory mechanisms, causing H_2S production. Elemental sulfur in the must can also cause the same problem. Theoretically, when nitrogen is in short supply, yeasts can also break down the sulfur-containing amino acids and the result would be the liberation of volatile sulfur compounds in the must. However, the concentrations of these amino acids available to the yeast make this mechanism unlikely.

It seems that stressing yeasts in various different ways—for example, by temperature changes—can also cause reductive problems, although the mechanisms involved don't seem entirely understood. The ability of yeasts to produce H_2S is highly strain-dependent, and is therefore at least partly genetic. Some yeast cells are naturally much lower H_2S producers than others.

There's another important story taking place in yeasts involving sulfur compounds. In a separate pathway, yeasts produce a range of volatile thiols, known as polyfunctional thiols, from must precursors that add a fruity quality to the wine. These are particularly important in Sauvignon Blanc aroma, where they contribute passionfruit, boxwood, and grapefruit aromas (at higher levels they can make the wine taste a little "sweaty"). Again, some yeast strains are much better at liberating thiols from precursors than others, so there is a strongly genetic component to this.

One reason that reduction has become such a hot topic of late is the growing use of screw caps. Corks allow a little oxygen transmission, which means that the redox potential of the wine will be higher than in bottles sealed with screw caps, which allow much less oxygen transmission. The low redox potential in wines sealed with screw caps can trip winemakers up: they can bottle a wine that is free from noticeable SLOs, only for the SLOs present to be modified by the low-redox environment into a smellier form, with the result being "reduction" noticeable to tasters. One such reaction would be the reduction of disulfides to mercaptans (thiols). As a result, winemakers need to change the way they prepare their wines for bottling if they are intending to use screw caps. The wine needs to be free of any reductive tendencies at all. Frequently, copper-fining trials will be done before bottling. Copper removes mercaptans, but not disulfides, and winemakers will look at using as little as necessary to clean up the wine.

New Zealand winemaker and chemist Alan Limmer PhD has written widely on the subject, bringing his knowledge of wine chemistry to bear. Limmer has pointed out that screw-cap reduction is not a problem that can be completely eliminated simply by more careful winemaking. "In essence we are talking about thiol accumulation, post-bottling, from complex sulfides which do not respond to prebottling copper treatment," claims Limmer, in response to the assertion that fining with copper removes reduction defects. "This reaction occurs to all wines containing the appropriate precursors, irrespective of closure type. But the varying levels of oxygen ingress between closures leads to significantly different outcomes from a sensory point of view."

Volatile sulfur compound production by yeasts

The main way in which yeasts produce volatile sulfur compounds is through the sulfate assimilatory reduction pathway. Sulfate is taken up and used in the biosynthesis of organic sulfur-containing compounds such as the amino acids methionine and cysteine. Low concentrations of these two amino acids in grape must cause the induction of the sulfate assimilation pathway, in which it is taken up, reduced to sulfite and then reduced to sulfide by sulfate reductase. In addition, extracellular sulfite can also be reduced to sulfide. The sulfide is used to form methionine or cysteine, in conjunction with *O*-acetylhomoserine or *O*-acetylserine, respectively. If these two are in short supply, then sulfide is released and combines with hydrogen to form hydrogen sulfide. Methionine can be de-animated and reductively decarboxylated to form methional. Methanethiol can be formed from methionine by demethiolase. And sulfide can react with acetaldehyde (aka ethanal) to form ethanethiol. All these volatile

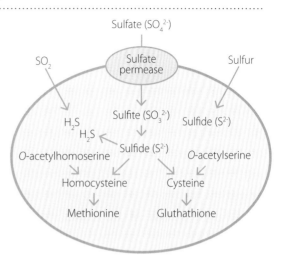

sulfur compounds are quite stinky. But there is still a lot we don't know about how volatile sulfur compounds end up in wine and cause the problems that they do.

Limmer's explanation for screw-cap reduction is that sulfides present in the wine at bottling need a small level of oxygen ingress through the closure, otherwise they can become reduced to thiols. So the use of a closure, such as cork, which allows a little oxygen ingress (but not too much), is a necessary concession to the vagaries of sulfur chemistry. Limmer points out: "Controlling ferments not to produce the complex sulfides is beyond our means currently. This sulfide behavior of the ferment is more controlled by the yeast genetics than the winemaker."

How is reduction recognized? At worst reduction is easy to spot: hydrogen sulfide gives eggish, drain-like aromas that spoil the wine. Any such wine is clearly faulty. But the most common forms of reduction are subtler. The form seen most often is "struck match/flint", presumably due to mercaptans. This form of reduction can be attractive in the right sort of context, but many people find it objectionable. In its most subtle form, this enhances the minerality of white wines, and is often encountered in white Burgundy, where it can add to complexity. Increasingly, New World Chardonnays are being made with a deliberate hint of this complexing matchstick reduction.

Related to this is the cabbage/cooked vegetable character that comes from either mercaptans or disulfides. Again, this is sometimes found in white Burgundies, and although it's initially repellent, it can be complexing. Reduction also manifests itself by means of a slight rubbery quality, which is most often detected in red wines. There are also strong, smoky, roast-coffee aromas in some reds that suggest high-toast oak, but which are actually due to volatile sulfur compounds. It should be added that low-level reduction may have only a small sensory impact on the aroma of wine, but can affect the palate, obscuring

Left A bag of nitrogen supplement diammonium phosphate (DAP) at a winery. This is routinely used to correct low yeast available nitrogen (YAN) levels, to avoid problems with volatile sulfur compound correction, but in some cases can be counterproductive.

the fruit expression and adding hardness to it. Thus it is also an issue of mouthfeel.

One of the difficulties with this discussion is that the perception of SLOs seems to depend quite highly on the context of the wine. Also, at different concentrations the various SLOs will have a different sensory impact, which makes it hard to be sure what the signature impact of a specific sulfur compound is. In turn, this makes specific diagnosis of "reduction" faults difficult for tasters.

While the development of SLOs in wine is normally best avoided, there are some circumstances where they can contribute something positive. New Zealand winemaker James Healey (winemaker at New Zealand winery Dog Point, previously with Cloudy Bay) points some of these out. "In Champagne the bready/brioche character from aging for a period on lees is a result of a certain type of reduction in association with autolysis and liberation of the contents of yeast cells into the wine. The reductive characters from fermentation of Chardonnay juice containing highish solid concentrations result in accentuated nuttiness and improved texture after aging on lees for some time." Dominique Delteil also has examples of the different manifestations of reduction. "First, a very ripe Languedoc Syrah macerated to reach licorice aromas, and then aged in oak. In that wine, hints of 'burned match' could be very interesting from a sensory point of view. They will match the ripe fruit/vanilla style. Most wine-drinkers will appreciate that because those aromas are in a very sweet, aromatic environment, so they won't express as dominant. Second, a cool-climate, unripe Cabernet Sauvignon. Let's suppose that this wine has exactly (chemically speaking) the same amount of the sulfur compounds that gave the interesting light 'burned match' in the above Syrah. In that wine those chemicals will give a different sensation that the same taster will translate as 'leeks,' 'green bean,' and eventually 'garlic.'"

It is because of this that reduction is one of the most interesting and intriguing topics in the wine trade. A fascinating idea is that what is often thought of as minerality in wine could actually (in some, if not in all cases) be due to reductive characters. Portuguese winemaker Dirk Niepoort thinks that what often ended up as minerality in

some famous wines was actually due to certain imperfections in the must due to the terroir. "For example, because of a lack of a certain nutrient in a must from a certain terroir-fermentation would create off-flavors during the fermentation, which in the end would give the so-typical character of that area," he hypothesizes. "Certain treatment habits in the vineyards (for example, sulfur treatments done late) can create some residues in the must that influence fermentation, producing some reduction, which are cleaned up by the time the wines are bottled but give a lot of character."

THE DIFFERENT SULFUR COMPOUNDS IN WINE

Time to talk to an expert on wine flavor chemistry. Dr. Leigh Francis is a senior research chemist with the Australian Wine Research Institute (AWRI), and he has a particular interest in sensory analysis. I quizzed him about how he would define "reduced" in the context of wine-tasting. "The word is usually reserved for aromas arising from what are presumed to be sulfur compounds," he explained. "Often tasters at the AWRI would have sufficient technical knowledge that they are aware of how particular sulfur compounds smell, and often use the chemical name rather than an evocative descriptor." Dr. Francis then described the descriptions of some of these compounds in sensory analysis work. "Reduced" would encompass aromas such as:

- hydrogen sulfide (and only rarely would a taster go the extra step of delineating the aroma as "rotten-egg gas" or similar);
- mercaptans or thiols: and for these aromas tasters would define with words such as cabbagey, rubbery, burned rubber;
- sometimes a more general term would be used such as "sulfidic," which would encompass both of the above;
- struck flint.

There is at least one other specific sulfur compound that tasters are aware of, and this may sometimes be included in the term "reduced" DMS (dimethyl sulfide) which has its own normally distinctive aroma of cooked vegetables, at high levels reminiscent of cooked corn or canned tomato, at lower levels similar to blackcurrant drink concentrate. Sometimes wines showing an apparent DMS aroma will have a "reduced" stinky note as well.

Left Famous French wine scientist Denis Dubourdieu who has done pioneering work on the polyfunctional thiols that are important in wine aroma.

In addition, Francis points to recent research carried out by Denis Dubourdieu and colleagues at the University of Bordeaux. "Relatively newly identified compounds 4-mercapto-4-methylpentan-2-one, 3-mercaptohexan-1-ol and 3-mercaptohexyl acetate are responsible for the tropical fruit/passionfruit aroma at particular concentrations, and cat's urine aroma at higher levels. These compounds have been found in Sauvignon Blanc, but also in other white varieties and even in red wines, where they probably don't provide a tropical-fruit aroma but may contribute to blackcurrant character. Work has been done to indicate that yeast strains will strongly influence the levels of these compounds during fermentation." And there's more: "Benzenemethanethiol has recently been implicated in smoky/gunflint aromas, and oak-derived thiol compounds have been implicated in coffee-oak character," he adds.

But it's not a simple case of avoiding these reduced characters altogether. "Most commonly reduced aromas would be considered a negative by our panelists, but not always," he explained, "and often our panelists are simply instructed to rate the intensity of particular attributes rather than give a quality judgment. Usually the term in a tasting note would indicate that a wine had a 'stinky' cabbagey/rubbery/sewage note that, depending on the intensity, could render it faulty and unpleasant, but it may be only a slightly noticeable character. If the compound responsible is hydrogen sulfide, it can disappear with swirling the glass and make a final decision more difficult for the taster."

Researcher and consultant Pascal Chatonnet also sees reduction as a double-edged sword. "An excess of reduction can produce off-flavors coming from an excess of volatile sulfur compounds. But a good equilibrium permits the maintenance of some very sensitive aromatic products involved in the 'flowery' and 'fruity' character of young wines."

TOYING WITH REDUCTION

So it seems that while the development of reduced sulfur compounds in wine is normally best avoided, there are some circumstances where they can contribute something positive. Indeed, some winemakers have deliberately set out to use reductive-winemaking techniques to encourage the development of these compounds as complexing agents in their wines. This is currently quite a hot topic in winemaking, it seems, although few research studies have addressed it directly.

James Healey comments that, "Toying with reduction is a risky business that requires a bit each of courage, experience, and knowledge mixed in appropriate portions." He continues, "Reduction can be a very positive thing in certain wine types and not in others. The reductive characters from fermentation of Chardonnay juice containing highish solid concentrations result in accentuated nuttiness and improved texture after aging on lees for some time. And the 'cat's-pee' or sweaty character that develops during fermentation of Sauvignon Blanc from cooler climates is the result of a certain reduction-related compound. In fact, the sweaty character is so close to hydrogen sulfide that one must be careful that the sweat doesn't mask the sulfide in a young wine about to go into bottle. I have seen wines spoiled as a result of the sweat diminishing with time and sulfides becoming more evident. Obviously the sweaty story could be seen as good reduction or bad reduction, depending upon your liking for Sauvignon."

Leigh Francis agrees that reduced characters aren't always negative. "Overall, as always with flavor chemistry and sensory properties, it is hard to be too definitive about negative/positive. A small amount of a particular character in a wine that has other fairly strong aroma attributes will likely be accepted and liked. It's when the aroma dominates that the [reduced sulfur] compounds could be considered negative, but this is probably the case with any aroma you can think of: too much of a good thing can be too overwhelming and make the wine too simple and not attractive to drink much of. For sulfur compounds in general it does seem, however, that, when they're at very high concentrations, an unpleasant note becomes evident, no matter how pleasant they seem at lower levels, which is not the case for many other flavor compounds. It is likely that hydrogen sulfide will be negative, no matter what level."

Delteil gives two explanations, one from a sensory perspective and one from a chemical one. "Sensory answer: the same sensation (the molecules reaching the sensory bulb in the palate cavity) will produce different perception (neurosensorial unconscious feeling) and finally a different translation (the conscious expression, such as the pronounced word 'garlic') according to the other compounds reaching the sensory bulb at the same time. This phenomenon is very well-known in cooking. So some vegetal character in the unripe Cabernet will emphasize the sensation and translation. On the contrary, the ripe aromas of the Syrah make more acceptable and even interesting the same amount of chemical compounds.

Chemical answer: recently there has been significant progress in understanding the chemistry of macromolecules in wine, particularly with the work carried out in the INRA in Montpellier by Professor Moutounnet and his team. An interaction has been identified (nonclassical chemical links between molecules) between volatile compounds and some polysaccharides from the grape, the yeast, or the lactic acid bacteria. It is also known now that ripe grapes release more macromolecules into the wine. So, in our examples, maybe there was the same amount of sulfur compounds (from a classical analytical chemistry point of view) but the volatilities were different because of a higher macromolecule concentration in our ripe Syrah."

Delteil thinks that with the kinds of wines he is making in the Mediterranean regions, from very ripe grapes, "It is interesting to manage sulfur characters closer to the razors edge, once we are sure that we are managing the basic sulfur off-flavors risks." Oxygen is a useful tool in this sort of manipulation. "Once one gets the first level of security with yeast, nutrients, and so on, with a lower amount of oxygen one can play around with the 'burned match" character. Working this way one also helps preserve native fruit characters. So,

there is a double effect: hints of burned match with rich, fresher fruit. If the wine is backed with very rich, ripe fruit, it is a style change that can be attractive. In hot-climate areas, it recalls the good characters of cooler areas (when they reach ripening). It is trendy today not to push too far the natural ripe Mediterranean style. Oak management is also a tool to complement ripeness with mineral/smoky/fresh-fruit characters."

Winemaker Matt Thomson points out how many of the best *domaines* in Burgundy are masters at getting just the right level of struck-match mercaptan characters in their Chardonnays. He thinks that Michael Brajkovich of Kumeu River in New Zealand is also very good at this. I asked him if yeasts produce it. "I am not sure about the biochemistry of it, but we have found it comes early, so I don't think it's produced after fermentation," says Thomson. "Once you get it, it seems to remain there. It seems stable with copper, too. I have also found it stable to ascorbic acid and copper. I have had it before in my wines and I think it's great." Thomson adds, "Mike has told me how he gets it, through solids, and we have used solids. But what he told me is, that instead of filling your barrels out of tank, when you get all the lees in the last couple of barrels and they stink like hell, he keeps the must homogeneous so there is a gentle mixing process going on while he fills the barrels. But there is more than this: I think the yeasts he has resident in the winery are very good at producing this mercaptan. This is the other part of the story. This is why wild ferments work for some and not for others."

I asked Thomson whether reduction could also work in reds as a complexing factor. "Sometimes, but at a low level. Often, sulfides shut down the fruit, and this can be the problem. Often the wine can appear just slightly oxidized, with something shutting it down, and then you'll see some reduction. I have seen some struck-match notes in Pinot that I don't object to, and occasionally in Barolo, and also in cool-climate Syrah. That bacon/juniper thing can work nicely with the struck-match note. I don't really like egg!"

Managing sulfur compound flavors in wine

How can the negative effects of sulfur compounds be managed? "By prevention," says Dominique Delteil. "For prevention, a minimum knowledge of the phenomenon is necessary." He explained the good practice he tries to develop with the wineries he works with. "Sulfur compounds are produced by the yeast, either during its life or after its death: this is 99% of sulfur flavors. Then, those compounds are chemically extremely reactive. So, what we identify with a word such as 'garlic' comes from yeast native molecules (molecules liberated by the yeast in the chemical form they have when they reach our sensory bulb) and from sulfur compounds that are produced through different chemical reactions during fermentation and aging. But almost everything comes from the yeast, so prevention is based on the yeast." Delteil explains that avoiding the production of excessive sulfur compounds is a complex matter. "The key point is the yeast strain. There are huge differences between the different yeasts. That's true among the hundreds of selected oenological natural yeasts available today, and also among the indigenous yeasts. This is then amplified by the yeast nutrition, the second key point. The higher the sugar in the juice, the lower the natural content in complex nitrogen compounds (particularly amino acids), the higher the yeast stress and higher the risk of sulfur compounds excess. Those conditions are typically found in the Mediterranean areas. Once the yeast strain and the nutrition are OK, oxygen will be an extra tool for the winemaker. If the first two key points are not managed it is hazardous to manage the problem just with oxygen. Why? Stressed yeast cells produce sulfur compounds continually, and 'high-oxygen' winemaking processes add oxygen once a day in reds and once or twice during the whole fermentation in white or rosé. So the oxygen added arrives to find some already-stabilized sulfur compounds. The third key point is yeast management during aging. Once a winemaker has made a fermentation without a sulfur problem, he must manage the risk due to dead yeast. Many practices are consistent. If one keeps a certain level of dead yeast (we call them light lees), it is recommended that they should be stirred regularly into the wine to avoid reductive zones in the tank or the barrel. Here reductive is used in its proper physicochemical sense. This movement is the reason for the traditional stirring (known as *batonnage*) during aging with lees. The right amount of oxygen is also important, either through the stirring or through direct managed injections.

So, how should winemakers enhance the positive effects of sulfur compounds? "They should amplify the other positive characters in the wine," says Delteil, "that is, ripe aromas (fruity, spicy, vanilla, etc., according to the grape, the place, the style goals, the market goals) and round mouthfeel sensations. Among ripe aromas, sulfur flavors will express as mineral characters and in mouth they won't provoke too much dryness and bitterness in the after taste." Why talk about mouthfeel? Some common sense: volatile compounds that are aggressive to the olfactive bulb through the called ortho-olfaction (direct smell versus retro-olfaction) are not 100% volatile. Most of the compounds (in quantity) stay in solution in the wine (at tasting temperature) and so will come into contact with the mouth, and will come across as aggressive there just as they are in the olfactive bulb. That impact of sulfur compounds on mouthfeel is not often known and integrated in winemaking although it is very important for consumers."

DECONSTRUCTING TERROIR?

Sam Harrop has a fascinating idea, albeit one that may prove difficult to test. In chemical terms, minerality is an ill-defined quality in wine, but when it does occur it is commonly explained as being a terroir character. But what if it is actually a consequence of reduction, caused by a combination of volatile sulfur compounds at low levels? "Wines from many of the best wine regions in France show mineral/reductive qualities," says Harrop. "Perhaps these qualities are derived in the winery and not the vineyard."

"I do believe that minerality and reduction are related," says James Healey. "It could be a result of struggling ferments coupled with nutrient deficiency/vine stress, but I don't think that this is why great white Burgundy or Riesling achieve this character. I think these wines derive this character because the producers understand how to get it from their vines and vinifications.

Volatile thiols identified in wine

Name[1]	Sensory descriptor	Perception threshold	Comments
3-mercaptohexyl acetate (3MHA) [3-sulfanylhexyl acetate]	Grapefruit zest, passion fruit (*R* enantiomer); more herbaceous, boxwood (*S* enantiomer)	9 ng/liter (*R*), 2.5 ng/liter (*S*)	Very important in the fruity aroma of Marlborough Sauvignon Blanc. During aging of wine, this hydrolyses to 3MH.
3-mercaptohexan-1-ol (3MH) [3-sulfanylhexan-1-ol]	Passion fruit (S enantiomer), grapefruit (R enantiomer)	60 ng/liter (S); 50 ng/liter (R)	Always present in Sauvignon Blanc, at concentrations ranging from several hundred ng/liter, and even s high as mg/liter. Wines with highest levels of 3MH tend to have the highest levels of 3MHA.
4-mercapto-4-methylpentan-2-one (4MMP) [4-methyl-4-sulfanylpentan-2-one]	Boxtree, broom	0.8 ng/lite	Present at up to 100 ng/liter in fresh box leaves; can be found at up to 40 ng/liter in some Sauvignon Blancs.
3-mercapto-3-methylbutan-1-ol (3MMB)	Cooked leeks	1500 ng/liter	Rarely found in wine above perception threshold
4-methyl-4-mercaptopentan-2-ol (4MMPOH)	Citrus zest	55 ng/liter (20 ng/liter in water)	
2-furanmethanethiol (also known as 2-furfurylthiol and 2-furfurylmethyl mercaptan)	Roast coffee	0.4 ng/liter	Identified in sweet whites from Petit Manseng variety, red Bordeaux wines, and has been found in toasted oak. Also found in Champagne aroma.
Ethyl-3-mercaptoproprionate	Meaty		Found in the aroma of Champagne.
3-mercaptobutan-1-ol	Onion, leek		
Mercaptopropyl acetate	Meaty		
2-methyl-3-mercaptopropan-1-ol	Broth, sweet		
2-methyl-3-mercaptobutan-1-ol	Raw onion		
3-mercaptopentan-1-ol	Grapefruit		
2-methyl-3-mercaptopentan-1-ol	Raw onion		
3-mercaptoheptan-1-ol	Grapefruit		
4-methyl-4-mercaptopentan-2-ol	Citrus zest	55 ng/liter	Rarely found in wine above perception threshold.

Volatile thiols identified in wine (continued)

Name[1]	Sensory descriptor threshold	Perception	Comments
Methanethiol (methyl mercaptan)	Rotten cabbage, stagnant water	1.5 ng/liter	
Ethanethiol (=ethyl mercaptan)	Rotten onion, burned rubber at threshold levels; at higher levels this is skunky or fecal	1.1 ng/liter	
2-mercaptoethanol	Barnyard		
Ethanedithiol	Rubber, rotten cabbage	0.3 ng/liter	Potentially responsible for reductive aromas in some wines. Found as a component of Champagne aroma.

[1]Terminology can differ. For example, "mercapto" is sometimes interchanged with "sulfanyl" in these names. Here I have chosen to use the most widely adopted name for each of the thiols, which is currently the "mercapto" version; technically, the correct IUPAC name is the sulfanyl one, so I have given this in brackets as well in a couple of incidences, although this terminology is not yet widely used in the literature. However, this could change over the lifespan of this book.

I suppose that someone could fluke it from time to time, but there are many great producers that consistently hit the nail year after year."

Delteil agrees that flavors from sulfur compounds are often misidentified as terroir characters. He recounts his experience with a client in Friuli. "According to the commercial manager, the flinty character was too high in the wine, even though they were applying classical prevention practices." Delteil looked to see how the flintiness could be reduced, and found the best method was through amplifying the ripe-grape character. "It was a classical pendulum effect between ripe fruit and sulfur, and we now have an accepted level of flint hints for the market goal. I personally think that excessive sulfur characters are too often presented as a terroir expression (so nobody can say a word about it because it's terroir!) although they are a nonunderstanding of the risks created by a situation, and the winemaker. And when one applies a better-adapted process, that famous terroir appears to be a luscious fruit source."

CONCLUDING REMARKS

So, questions, few solid answers, and lots of room for further research. The chemistry of reduced sulfur compounds is complicated, but an increased understanding of their evolution in wine would permit winemakers to manipulate their levels with a view to increasing complexity. There's even the tantalizing prospect that some of the "terroir"-like characters in Old World styles could be understood in terms of sulfur chemistry, and perhaps allow the production of more "terroir"-like New World styles.

This leads to a final—I think, crucial—question. Would a technically perfect wine be boring? Wine is bewilderingly chemically complex. Do there exist components of wine that, at low levels, impart complexity, at higher levels are considered as faults, and vice versa? And how can winemakers walk the tightrope of encouraging the development of these complexing factors while avoiding faulty wines?

16 Microbes and wine: yeasts and lactic acid bacteria

Yeasts don't get enough credit. When it comes to wine, grapes get all the glory. But without yeasts, all we'd have is grape juice. The choice of yeast, or indeed the decision of whether to use indigenous yeasts or cultured strains to carry out fermentation, is an important part of winemaking. Not only do yeasts convert sugar to alcohol, they are also able to influence the flavor and aroma of the final wine. Of the estimated 1,000 or so volatile flavor compounds in wine, at least 400 are produced by yeast. Take some freshly crushed grape must. It doesn't smell like much of anything. In contrast, the fermented wine is often rich in aroma and flavor, thanks to the action of yeasts.

We've agreed that yeasts are undervalued, so this chapter seeks to redress the balance a little, looking at their role in winemaking and addressing the science of fermentation. I'll also cover the controversy that surrounds wild-yeast ferments and attempts to engineer beneficial traits into yeast strains by genetic modification.

And if yeasts are the underdogs, bacteria are almost completely ignored. But they, too, are important in modifying the flavor of wine in those situations where malolactic fermentation takes place—and this includes practically all red wines and a good subset of whites. So we'll be taking a look at bacteria, too.

MICROBES AND WINE: SOME CONCEPTS

Until the nineteenth century, fermentation must have seemed a mysterious process to winemakers. It wasn't until the now-famous studies of Pasteur in the latter half of that century that yeasts were shown to be directly responsible for converting the sugar in grape must into alcohol. He correctly surmised that the particular yeasts carrying out the fermentation could influence the flavor of the wine. As Émile Peynaud, the famed wine

scientist from Bordeaux, author of influential works on winemaking and wine-tasting pointed out, "Before his time, good wine was merely the result of a succession of lucky accidents."

You can't see yeasts. Like bacteria, these unicellular fungi are far too small to be seen with the naked eye. Humans are tremendously visually centered, and it is perhaps because of this that we've found it hard to understand and be comfortable with the microbial world that surrounds us. In medicine, this has no doubt contributed to the current antibiotic crisis. Because we are unable to appreciate that microbes are everywhere, the message that there are good bacteria as well as pathogenic ones has been hard to handle. We're much more comfortable with the idea that bugs are bad and we should zap them all, and that the only good bacterium is a dead bacterium. This sort of attitude has encouraged the crazily irresponsible overuse of antibiotics, and led to the development of widespread antibiotic resistance, a major threat to human health.

So before we get to the gritty details, it's probably worth taking a conceptual look at microbes in winemaking. Yeasts and bacteria are ever-present in the winery. Even in a spotlessly clean environment there will always be some receptive surface, such as an uneven soldered joint in metal pipework, where microbes can hide. Barrels are particularly receptive to microbes because the structure of wood means that it is almost impossible to sterilize it. Because a potential source of inoculation is just about ubiquitous, all yeasts and bacteria need is the right sort of environment, and they will begin to grow. The grape must represents a sugar- and nutrient-rich medium that's ideal for the growth of certain microbes, although with its strong osmotic potential, it also presents challenges for these

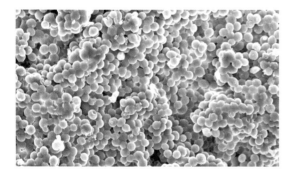

Above Scanning electron micrograph of *Saccharomyces cerevisiae*, the wine yeast.
Picture courtesy of Ann Dumont, Lallemand

microbes to overcome. As the must ferments, it changes, and its suitability for one species or strain wanes as its suitability for another develops.

Let's illustrate this in picture terms. Take the side of a mountain. At the base, vegetation is lush and plentiful. The environment here suits a wide variety of organisms. Move a short way up and the change in climate affected by the altitude difference will mean that a different population of plants will prevail. This will continue up the mountainside, until conditions are such that toward the summit plants can no longer establish themselves. It's something like this in fermenting wine. Create the right conditions and you can select for the population of organisms that you want to be growing at that particular time. Winemakers tend to concentrate on eradicating rogue organisms from the winery. But they might be better off concentrating on ensuring that their musts and developing wines that represent ideal habitats for the sorts of microbes they want to encourage, while at the same time not neglecting winery hygiene. To quote Peynaud "The winemaker should imagine the whole surface of the winery and equipment as being lined with yeasts."

Microbes have short generation times, so they can be fiercely competitive. If conditions suit one yeast or bacterium a little more any others, it will rapidly outpace the competition and establish itself as the primary fermenting organism. Winemakers have to make sure that the musts they are working with give a competitive advantage to the sorts of bugs they'd like to see growing.

"NATIVE" YEASTS

Yeasts are widespread, not only in the winery, but also in the vineyard. They spend winter in the upper layers of the soil, spreading to the vines during the growing season via aerial transmission and insect transfer. They colonize grape skins during the maturation phase, although they never reach very high levels on intact grapes. Contrary to popular opinion, the bloom on the surface of grape skins isn't made up of yeast populations, but rather a wax-like scaly material that doesn't harbor many fungi.

Only a limited number of yeast species are present on grapes, the so-called "native" yeast populations. These include: *Rhodotorula*, the apiculate yeasts *Kloeckera apiculata,* and its soporiferous form *Hanseniaspora uvarum* (the most common by far), and lesser amounts of *Metschnikowia pulcherrima*, *Candida famata*, *Candida stellata*, *Pichia membranefaciens*, *Pichia fermentans*, and *Hansenula anomola*. Also present may be potential spoilage organisms, such as *Brettanomyces*. It needs to be added that yeast nomenclature can be a confusing business, with various synonyms in common use for the same bug. This isn't surprising, because until the recent development of molecular methods for typing yeast strains and species, it was very hard to tell all of them apart.

The main wine yeast, the alcohol-tolerant *Saccharomyces cervisiae*, is relatively rare in nature. Attempts to culture *S. cerevisiae* from the skins of grapes have largely proved unsuccessful. The only way its presence can be demonstrated is by taking grape samples and placing them in sterile bags, crushing them under aseptic conditions and seeing what happens, an experiment that has been done in Bordeaux. At mid-fermentation, *S. cerevisiae*, which is undetectable on grape skins, represents almost all the yeasts isolated. In a few cases no *S. cerevisiae* is present and apiculate yeasts do the fermentation.

It has been suggested that rather than *S. cerevisiae* originating in the vineyard,it is actually a human introduction, crossing over to winemaking from its ancient and widespread use in baking and brewing. But this has been shown not to be the case.

CULTURED OR SPONTANEOUS FERMENTATIONS?

So you have your grapes and want to make some wine. You want to start a fermentation. There are two approaches open to you. Traditionally, the only option would have been to crush the grapes and leave the yeasts already present to get on with it. This is known alternatively as a "spontaneous," "wild-yeast," "indigenous- yeast," or "native-yeast" fermentation. Since the 1960s, with the ready availability of cultured strains of *S. cerevisiae*, winemakers now have the choice of inoculating the must with a starter culture of their preferred yeast. Estimates are that worldwide, some 80% of wine is produced by inoculated fermentations, but this is a tricky figure to verify.

The choice between native and cultured yeasts has opened up a philosophical divide between winemakers who want to bring fermentation under control and those who prefer to leave things to nature. These days there's a loose kind of Old World–New World divide, with the former largely preferring to use indigenous yeasts (at the high end, at least) and the latter relying on cultured strains, although this divide is far from an absolute.

What happens during a spontaneous fermentation? Because the initial inoculum of yeasts from the winery environment and grape skins is quite low, things can take a while to get going. This introduces an element of risk: if bugs such as *Acetobacter* (the acetic acid bacterium that turns wine to vinegar) establish themselves before the fermentative-yeast species, then the wine will be at risk of spoilage. Also, there's no guarantee that the native yeasts that become established will do a good job. Like *S. cerevisiae*, all the various native yeasts exist in many different strains, some desirable, others not. With a spontaneous fermentation you take what you are given.

Apiculate ethanol-sensitive yeasts, such as various species of *Hanseniasporia* (and its anamorph [asexually reproductive] form *Kloeckera*), which is the dominant yeast found on grape berries, usually dominate the initial stages of a wild ferment. As these decline, yeasts—various species of *Candida*—take over (*Candida stellata*, *C. pulcherrima* [and its teleomorph, the sexually reproductive form *Metschniokowia pulcherrima*], and *C. colliculosa* [and

its teleomorph *Torulaspora delbrueckii*]). It has been estimated that in an uninoculated ferment, as many as 20 to 30 strains participate. But as alcohol levels reach 4–6%, most native species can't take it, and the alcohol-tolerant *S. cerevisiae* will take things onward from here. The only two non-*Saccharomyces* yeasts that can tolerate higher alcohol levels are *C. stellata* and *C. colliculosa*. Other species of yeasts may be present in wild ferments. These include *Cryptococcus*, *Debaromyces*, *Issatchenkia*, *Kluyveromyces*, *Pichia*, *Rhodotorula,* and *Zygosaccharomyces*.

There are twists to this story, however. Most winemakers will add some sulfur dioxide (SO_2) on crushing. This will slant things in favor of *S. cerevisiae* and the more robust of the native species, eliminating some of the more problematic wild yeasts and spoilage bacteria that tend to be more sensitive to the microbicidal actions of SO_2. Temperature also affects the balance of yeast species in the fermentation. Cooler temperatures (below 57°F) favor wild yeasts, such as *Kloeckera*, whereas higher temperatures shift things in favor of *S. cerevisiae*. Added to this, as harvest gets underway, the winery equipment will be a ready source of inoculum, and fermentations will get going a lot faster, with *S. cerevisiae* establishing itself sooner. Studies have shown that after a few days of harvest operations, half of yeasts isolated from the first pumping over of a spontaneously fermented red-grape tank are *S. cerevisiae*. Aside from the actual properties of the wild yeasts themselves, spontaneous ferments cause a delay the onset of vigorous fermentation. This will allow oxygen to react with anthocyanins and other phenolics present in the must, enhancing color stability and accelerating phenolic polymerization.

Why take the risk of a spontaneous wine fermentation? In many cases the motivation will be ideological—this is the traditional approach in certain areas and there is a reluctance to adopt alternative methods, or a disbelief in the integrity or efficacy of these alternative methods. Others do it for quality reasons because native yeasts are thought to produce wines with a fuller, rounder palate structure, and the ferments tend to be slower and cooler, burning off fewer aromatics. There is also a cost saving because cultured yeast has to be paid for. It should be pointed out that even where cultured strains of yeast are used, wine must is not a sterile medium, and even though a good dose of sulfur dioxide (SO_2) will kill off the more susceptible bacteria and yeasts already present, some indigenous strains will likely play a small role in the fermentation.

AN IDEOLOGICAL DIVIDE: WILD YEASTS VERSUS CULTURED YEASTS

Winemaker Nicolas Joly sums up well the objection that some winemakers have to using cultured yeasts to initiate fermentation. "Re-yeasting is absurd. Natural yeast is marked by all the subtleties of the year. If you have been dumb enough to kill your yeast, you have lost something from that year." He's not alone. Many winemakers, particularly those in the classic Old World European wine regions, see the use of cultured yeasts as unnecessary and even plain wrong. They argue that the native yeasts present in the vineyard are part of the terroir.

"We are very big fans of wild-yeast ferments," says Pierre Perrin, of Château de Beaucastel in France's Châteauneuf-du-Pape region. "It can be risky if the fermentation doesn't begin quickly. But if you do a *pied de cuve* two or three days before your first harvest day, you can mix this with your crop after a few days of maceration and then the fermentation will go. The key is to have a fermentation departure that isn't too fast (which eliminates the maceration process) or too slow (which risks acetic problems)." However, most New World winemakers, and a growing band of Old World producers, now initiate winemaking with cultured yeasts, seeing the control of

fermentation parameters this affords as being key to quality. The choice of specific yeast strain is also seen as an important winemaking decision, because the various properties of the yeast can be chosen to complement or add to the wine style being made. Yeast expert Sakkie Pretorious points out that the outcome of spontaneous fermentation is highly unpredictable and describes the risk involved as "potentially staggering."

I asked Rui Reguinga, a Portuguese consultant winemaker, about whether he preferred cultured or indigenous yeasts. "I always use cultured yeasts," he explained. "While we might lose a little typicality, with the cultured yeasts we see more of the potential of the variety. We have a correct fermentation that starts at the right time and finishes without problems. If things go well with wild yeasts you have better typicality, but if you have bad yeasts you have a problem." However, Reguinga doesn't see the choice of yeast strain as a way to alter the intrinsic characteristics of the wines. "We use neutral yeasts in order to achieve a successful fermentation without influencing the aroma." There are specific regional considerations that alter his decision. With his work for Quinta dos Roques in the Dão region, he is dealing with a harvesttime where there can be sudden changes of temperature. This can make it problematic to get fermentation going. Additionally, there's an element of risk management, especially when you are dealing with large quantities of someone else's wine. "It is too much to play with 200,000 liters [52,835 gallons]," says Reguinga.

"Success with wild yeasts depends on the right combination of yeasts and cellar. On occasion the wine is more complex," says University of California Fresno Professor Ken Fugelsang, "but in the majority of years the dangers outweigh the benefits." It is now even possible to buy commercial cocktails of wild yeasts plus cultured *S. cerevisiae*. "You do see the succession," says Ken Fugelsang, who has used these experimentally. "*S. cerevisiae* doesn't take off immediately."

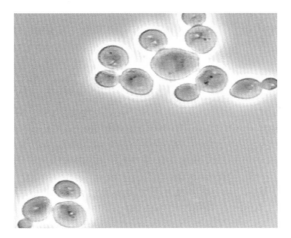

Left *Torulospora delbruckii*, a wild yeast species that is now available as a cultured, commercial strain, viewed under a light microscope.
Picture courtesy of Ann Dumont, Lallemand

How does he feel native yeast ferments affect the wine? "They add mouthfeel and structure, with an interesting bouquet." Fugelsang points out that it is hard to separate the wild yeast action from the fact that the native yeasts work slower, allowing oxygen to have more effect on the wine.

MATT GODDARD'S WORK ON WILD YEASTS

New Zealand-based yeast researcher Matthew Goddard has shown that in a natural ferment, *S. cerevisiae* engineers its environment to give itself a competitive advantage through the production of alcohol and heat. Fermentation of sugars is actually less energetically favorable than respiration of these sugars (which requires oxygen), but even in the presence of oxygen *S. cerevisiae* will still choose to ferment these sugars, producing alcohol in the process. Goddard speculates that because *S. cerevisiae* is a specialist at consuming ripe fruits, it chooses to make ethanol to protect this valuable food resource. The ethanol acts as an antimicrobial, reducing competition, and also deters most vertebrates. Thus *S. cerevisiae* is choosing a less efficient pathway, which you'd have expected evolution to eliminate unless there was some collateral benefit. This production of ethanol is therefore an example of niche construction.

Goddard's work makes use of new genetic tools that are able to differentiate between different strains of yeasts. "It is not an easy thing to do. We use exactly the same technique that is used in forensic criminology situations—genetic fingerprinting—to distinguish different strains of *S. cerevisiae*. This affords us a powerful insight into the variation and relatedness of different strains that we have isolated from different areas."

His most important findings relate to a series of studies that took place in West Auckland, New Zealand. New Zealand is geographically remote, and was only inhabited by humans as recently as 700 years ago. The human introduction of wine occurred just 100 years ago. In the first study, Goddard looked at a spontaneous Chardonnay fermentation at boutique winery Kumeu River, when 800 random isolates were taken from the ferment. Initially, *S. cerevisiae* was present at very low frequency (1/1,500 of the population), but by day 11 it dominated the fermentation. Using genetic fingerprinting, Goddard found

88 different genotypes from 380 isolates of *S. cerevisiae*. Using some statistical techniques to analyze the genetic data, he concluded that there were around 150 different genotypes present in the fermentation, and these derived from six very distinct subpopulations.

But were these "wild" populations of *S. cerevisiae* actually escaped commercial strains? Goddard and his team compared them with a large database of commercial yeast strains and found that they were quite different. They then combed the winery, sampling winery equipment and walls before harvest, and failed to find any *S. cerevisiae*. The conclusion was that the strains in the wild ferments they analyzed were brought into the winery with the grapes. "For a long time the impression in the industry has been that *S. cerevisiae* doesn't live in the natural environment; it only inhabits wineries, and that these are escaped domestic/commercial strains," says Goddard. "When I was at the Oregon Wine Board in February to talk about some of this, many of these people had been schooled through the UC Davis system, and this is what they had been taught. They were astounded by our results."

In the next phase of the experiment, Goddard and his co-workers sampled soil, bark, and flowers from Matua Valley vineyard, just 3.75 miles from Kumeu River and which is surrounded by bush. They found 122 different colonies of *S. cerevisiae*, with 22 different genotypes. Two were found in vine bark, two in buttercup flowers, and the remainder in the soil. None of the Matua Valley genotypes matched any commercial isolates, nor did they match the isolates in the Kumeu River ferment. The inference is that it is these local strains, from soil bark and flowers, which end up carrying out the indigenous fermentations.

It makes sense that *S. cerevisiae* must be living somewhere in the environment, because ripe fruit is only available as a habitat for a couple of months a year at most. But how are these strains spreading among the soil, bark, and flower niches, and then getting onto the fruit? The obvious explanation seems to be by insect transfer, so Goddard took 19 samples, over a five-month period, from an apiary with hives near both West Auckland vineyards that he has studied. From the 67 colonies of bee-borne *S. cerevisiae* analyzed, two were almost identical to isolates from the Matua Valley vineyard.

But one question remained. How similar were these New Zealand isolates to yeasts found in other countries? The answer was they were unique. On average, the New Zealand isolates were found to share less that 0.4% of their ancestry with international strains of *S. cerevisiae*, and therefore were unique populations. "This was the first demonstration that strains in spontaneous ferments derived from the local environment," says Goddard, "and these strains appear to be unique (or at least distinctive) when compared with strains from elsewhere on the planet."

There was a further twist to the story to come, however. Kumeu River winemaker Michael Brajkovich suggested that Goddard should examine a new French oak barrel to see whether there was any possibility that yeasts could be transferred internationally in this way. So they looked at a new barrel imported from Chagny in Burgundy, and found 50 different isolates of *S. cerevisiae* with 40 different genotypes. This is the first time that *S. cerevisiae* has been shown to be present in new oak barrels. Significantly, one of these isolates was shown to be the same as one of the genotypes present in the Kumeu River wild ferment. "This was a bit of a shock," says Goddard. "We were pretty stunned to see this. Again, it is applying ecological population biology techniques to this system in a rigorous way, and bringing the right tools to the question."

Goddard has extended his study to other wine regions. One of these is Central Otago, where he has looked at yeast populations at Felton Road. Nigel Greening of Felton Road reports that the results show exactly the same thing as found at Kumeu River: the yeasts carrying out fermentations are local. "The biggest single cohort is unique to place, and this is 30–35% of the yeasts," says Greening. "The second-largest cohort tracks back through the barrels to the forests the oak was grown in. This is really interesting for Chardonnay if you are barrel-fermenting. Then for the cohorts after this it gets harder to define where they came from. Just as with Kumeu River, Matt Goddard couldn't find a single strain that tracks back to a laboratory yeast, so we were clean."

One interesting question is whether systemic fungicides have any effect on the microbial populations on grapes, and this is a topic that Goddard is currently looking at with a master's student. "We know a bit about what Mancozeb and copper oxychloride do to 'nasty' fungi, but what do they do to good fungi? There's a huge amount we could investigate. The anecdotal evidence from winemakers is that if you spray, it is harder to conduct a spontaneous ferment. Killing off the microbes in the vineyard, be they yeast or otherwise, has a knock-on effect."

Goddard's studies are hugely interesting and have important implications for how we see microbial populations as part of terroir. They also answer some long-disputed questions in wine microbiology. "If different regions harbor different populations of wine yeasts, then the use of these region-specific yeasts in winemaking means that the resulting wine more faithfully reflects the sense of place—or terroir—of the wine," says Goddard. His studies should really be extended to other vineyard regions (for example, famous areas such as Bordeaux, Burgundy and Champagne) to see whether they hold true there also. It would also be interesting to see whether regions that are wall-to-wall vineyard, with extensive herbicide use to create effectively a monoculture, are also harboring unique populations of yeast. And it would be good to see whether there's a difference between warm-climate vineyards and cooler-climate vineyards.

CULTURED WILD YEASTS

What about winemakers who want some control over their fermentations, but like the qualities provided by cultured yeasts? In recent years, a number of cultured non-*Saccharomyces* yeasts have become available. The yeast company Lallemand has a strong research program on the aromatic potential of non-*Saccharomyces* yeasts, combined with the optimization of the production of these yeasts in dry form. This allows winemakers to take advantage of fermenting with non-*Saccharomyces* wine yeasts while maintaining control over the fermentation process. The yeast *Torulaspora delbrueckii* strain 291 is now available to the winemaker in combination with specific *Saccharomyces* yeast in sequential inoculation, which results in wines with a distinct sensory profile and mouthfeel. Other yeasts are also being investigated, such as *Hanseniospora orientalis*,

Schizosaccharomyces pombe, Metschnikowia pulcherrima, Hansenula anomala, Candida stellata, and *Pichia anomala,* and also *Kluyveromyces wickerhamii* to control the development of *Brettanomyces* yeast through their production of specific compounds. All those non-*Saccharomyces* yeast are found in the natural flora of vineyards and wineries, similarly to cultured *S. cerevisiae* used for alcoholic fermentation. CHR Hansen, another yeast company, has a reasonably broad portfolio of non-*Saccharomyces* products. "Prelude" is a single species culture of *Torulaspora delbrueckii,* and they also offer blends. "Melody" and "Harmony" are both blends of *Kluyveromyces thermotolerans, Torulaspora delbrueckii,* and a strain of *S. cerevisiae.* "Rhythm" and "Symphony" are both blends of *Kluyveromyces thermotolerans* and *S. cerevisiae.*

Some important flavor compounds produced by yeasts during fermentation

Class of compound	Examples	Notes
Acetate esters	Ethyl acetate	Esters are largely responsible for fruity characteristics of wines. They hydrolyze during aging, but often remain at concentrations above threshold. Synergy between the esters present in wine determines how they are perceived
	2-methylpropylacetate	(also known as isobutyl acetate)
	2-methylbutylacetate	(also known as amyl acetate)
	3-methylbutylacetate	(also known as isoamyl acetate)
	Hexylacetate	
	2-phenylethylacetate	
Branched chain esters	Ethyl-2-methylpropanate	These are among the most powerful esters
	Ethyl-2-methylbutanoate	
	Ethyl-3-methylbutanoate	This has an aroma like strawberries, which contributes to the fruity character of some red wines
Higher alcohols (also known as fusel alcohols)	2-methylpropanol	(Also known as isobutanol)
	2-methylbutanol	(amyl alcohol)
	3-methylbutanol	(isoamyl alcohol)
	2-phenylethanol	This has a pleasant rose-like aroma.
Volatile sulfur compounds	Hydrogen sulfide (H2S)	The most common volatile sulfur compound in wines—smells like rotten eggs
	Ethanethiol (ethyl mercaptan)	Aroma of rotten onions or burned rubber at threshold levels; at higher levels it is unpleasantly fecal
	2-mercaptoethanol	Aroma like a barnyard.
	Methanethiol (methyl mercaptan)	Rotten cabbage

SELECTING FOR YEAST STRAINS: DESIRABLE PROPERTIES

Yeasts do a whole lot more than just convert sugar into alcohol. They are responsible for the metabolic generation of many wine flavor compounds from precursors in the grape must. Because of this, the use of strains of cultured yeast with specific properties has become an important winemaking tool, although one that is not universally welcomed—some traditional producers see this as a way of cheating.

Wine microbiologists see the development of yeast strains with enhanced abilities as an important goal in furthering wine quality. There are two ways of doing this. The first is by more "traditional" genetic techniques that don't involve the direct introduction of new genes. These include selection of variants (choosing the best of a range of natural genetic variants); mutation and selection (using a mutagen to increase the frequency of genetic variation and then selecting for those mutants with enhanced properties); hybridization (mixing together different species); and spheroplast fusion (a special way of joining together yeast cells to produce progeny with enhanced properties). The second is by transformation: the precise introduction of new, specified genes into the genome of the yeast strain of interest. This is also known as genetic modification (GM).

Both have their benefits and drawbacks. If the trait of interest is polygenic (multiple genes are involved), then non-GM methods are the best way to select for this. However, they are less precise, which means you run the risk of losing the beneficial traits of the starting yeast strains. GM methods are much more precise and elegant, but things get complicated if more than one gene is involved, and there is still the huge hurdle of public antipathy. It is likely that both types of strategies will prove important.

What are some of the desirable properties that microbiologists would like to engineer into yeasts? There's actually a fairly long list, summarized below.

Goals for genetic improvement of wine yeast strains

Fermentation performance	Fermentations are often faster than optimum, so lower temperatures often control them. Sometimes, though, they are sluggish and can even become "stuck," with disastrous consequences for wine quality, and improved yeast strains could help here. Targets include increased stress resistance, improved grape sugar and nitrogen uptake, resistance to high alcohol levels, and reduced foam formation. It would also be beneficial to have yeasts that can utilize the nitrogen sources in wine that they currently don't (this would help avoid stuck fermentations), and for them to be resistant to toxins produced by wild yeast strains.
Biological control of spoilage bugs	Spoilage microbes are a constant threat, and are countered by the addition of sulfur dioxide. A new development has been the use of antimicrobial peptides and enzymes as an alternative to chemical preservation. It would be ideal if these could be synthesized directly by yeasts.
Processing efficiency	The fining and clarification of wine consumes time and resources, and risks the removal of flavor components. Wouldn't it be great if yeasts could do this job? They could be engineered to secrete proteolytic and polysaccharolytic enzymes that would remove proteins and polysaccharides, which can form haze and clog filters. Another avenue of research is the regulated expression of flocculation genes, which would enable winemakers to encourage the yeast to enter into suspension for fermentation, and then settle quickly as a residue on completion.

Goals for genetic improvement of wine yeast strains (continued)

Flavor and sensory qualities	Wine is a complex mixture of hundreds of different flavor compounds, many of which are synthesized by yeasts. Yeast-derived compounds can be both positive and negative in terms of flavor impact. Therefore, yeasts should be selected for which have a positive impact. Yeasts that possess enzymes that liberate color and aroma have been selected for, as have those producing ester-modifying enzymes. Elevated alcohol levels are becoming an increasing problem in many regions and attempts have been made to produce yeasts that ferment to dryness at lower alcohol levels by diverting more sugar to glycerol production. Yeast strains are also being developed to adjust acid levels biologically.
Healthy properties	Sometimes wine contains elevated levels of undesirable compounds such as ethyl carbamate and biogenic amines. It is ideal to minimize the levels of these. Yeasts could also be developed with enhanced production of supposedly health-enhancing compounds such as resveratrol.

Modified yeasts: a tool in the battle against rising alcohol levels?

One of the major problems facing winemakers in warm-climate regions is the high level of alcohol. This has become more acute now that phenolic ripeness rather than sugar levels commonly dictates harvests. Is it possible to select for or engineer yeasts that are less efficient at converting sugar to alcohol, possibly by making different end products from some of the sugar? Typically 16.83 g sugar becomes 1% of alcohol. In a simulated white wine, one study showed that 56 commercial yeast strains produced alcohol levels ranging from 11.75% to 12.09%, a difference of just 0.34%. One proposal has been to make yeasts that still ferment to dryness, but which divert more of the sugar they use to glycerol, rather than alcohol. It sounds like a promising strategy, and it is one that has succeeded to an extent. Yeasts with elevated levels of the glycerol-3-phosphate dehydrogenase (GPD) enzymes do make more glycerol and thus will result in wines with lower alcohol levels. There are two main problems with this strategy as a way of countering high-alcohol wines. First, there is a lack of consumer acceptance of wines made with genetically modified yeasts. Second, glycerol isn't a neutral flavor component in wine. Ramp up your glycerol levels and your wine will taste different. "Wines with elevated glycerol levels are peculiar in flavor presentation," says Ken Fugelsang. "To produce enough glycerol to reduce alcohol levels will make a red wine with a decidedly sweet finish." So, it would seem a different solution is needed.

It is clear that there is a lot of scope for the manipulation of yeast strains to enhance wine quality. However, such developments are unlikely to appeal to the traditional wine producers who see the wild yeasts that dominate the early phases of spontaneous fermentations as part of their terroir, and an important factor in fine-wine production.

MALOLACTIC FERMENTATION

The term "malolactic fermentation" is actually a bit of a misnomer, because what we're referring to is the transformation of one acid, malic acid, into another, lactic acid, with the release of carbon dioxide. This is just one of the effects that the malolactic bacteria have on wine, but it's the most important one.

THE BACTERIA INVOLVED

The bacteria we are concerned with, known as lactic acid bacteria (often abbreviated to LAB), are present in the vineyard on the grapes, much like the yeast species that carry out spontaneous fermentations. The microbial population on the grapes gets slimmed down a bit at crushing, because of the hostile conditions of grape juice, with high acidity and high sugar content. Yeasts are better at dealing with these conditions than bacteria, and the bacterial populations tend to tumble a little, especially when fermentation gets going.

It's quite common for winemakers to add some sulfur dioxide at crushing. This is toxic to

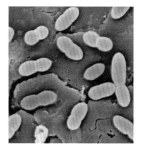

Left *Oenococcus oeni,* the main lactic acid bacterium. Picture courtesy of Ann Dumont, Lallemand

both yeast and bacteria, but bacteria are more susceptible, and so this addition tends to clear the field for the yeasts to do their thing. Typically, the population of lactic acid bacteria present in must is 100–10,000 viable cells per milliliter, expressed as CFU/ml (where CFU stands for colony-forming units, referring to cells that can form a colony when plated onto a culture medium). This initial population dips as fermentation progresses because of the increasingly hostile conditions encountered—alcohol levels rise and nutrients are depleted from the must. Also, some yeast strains are believed to release antimicrobial substances to eliminate the microbial competition.

By the end of alcoholic fermentation, there will usually be only 10–100 CFU/ml of these lactic acid bacteria remaining, but they are ready and waiting to start a rearguard action. By this stage other things will have shifted in favor of bacterial growth: the nutrients, depleted during the active phase of fermentation, will have restored their levels a degree, and free sulfur dioxide levels will have dipped to around zero. The other factor important for bacterial growth is temperature, and if this increases a little, for example, through the cellar warming up in spring, then bacterial populations can build up again.

There are four genera of lactic acid bacteria important in wine. The main one is *Oenococcus* (a single species, *O. oeni*). There are also 12 species of *Lactobacillus*, three species of *Pediococcus*, and a single species of *Leuconostoc*. As with yeasts, it is not just the species that matters, but also the strain: *O. oeni*, for example, shows important strain differences. Many malolactic fermentations are spontaneous, but since the 1980s it has been possible to buy cultured strains of *O. oeni* to inoculate with and start malolactic fermentation to order. I will explain more about this later.

WHEN IT OCCURS

There's a joke that I've heard told by winemakers. It goes: "How do you initiate malolactic fermentation?" The answer? "Bottle your wine." Traditionally, malolactic fermentation has been rather unpredictable. It happens when it happens, and sometimes it happens after bottling, with disastrous consequences.

To initiate the process, the population of lactic acid bacteria has to reach a threshold level of 1 million viable cells per milliliter (10^6 CFU/ml). Factors influencing this population growth include pH (it happens more easily at higher pH, but then the species and strains taking part will likely be a little different), the nutrient status of the wine, the temperature, and the strains of bacteria present. In many classic European regions, harvesting takes place in fall and cellars become increasingly cold as fermentation finishes, such that there is a long delay before malolactic fermentation, which takes place the following spring as the cellars warm up. This means a delay of several months between the completion of alcoholic fermentation and the onset of the malolactic conversion.

This delay between the end of alcoholic fermentation and the onset of malolactic fermentation can be a risky time for red wines. This is because the wine is relatively unprotected from microbial growth because free sulfur dioxide levels are low or zero, creating ideal conditions for the growth of the spoilage yeast *Brettanomyces*, and also risk the growth of acetic acid bacteria. Winemakers are not able to use sulfur dioxide at this stage because they risk inhibiting malolactic fermentation altogether.

This is why some choose to inoculate with bacterial cultures and reduce this risk period. Once malolactic fermentation is finished then it is possible to add sulfur dioxide to protect the wine. But it's a complicated business, because wines respond differently to the ingress of oxygen that barrel maturation permits, depending on whether or not free sulfur dioxide is present at appreciable levels. *Élevage* is a complicated business, and perhaps as much an art as a science.

Another factor is whether or not to complete malolactic fermentation in tank or barrel. A recent trend for high-end red wines has been to move them to barrel before alcoholic fermentation is fully complete, and then do malolactic fermentation in barrel. Some winemakers, however, believe they get better results and more authentic wines by carrying out malolactic in tank before going to barrel. The existence of frozen or freeze-dried cultures of *Oenococcus oeni* for starting malolactic fermentation makes this choice a matter of winemaking style. Such frozen or freeze-dried cultures are added directly to the wine—in the past some cultures required a reconstitution stage—but they are faced with quite a task, because they are being added to a fairly hostile environment. These cultures of *O. oeni* are selected for the ability to tolerate low pH and high ethanol, but they aren't selected for their resistance to sulfur dioxide because this would then make it tricky to stabilize the wine at a later stage. Single strains are best because combinations of strains may end up competing with one another.

A recent shift has been experimentation with co-inoculation of cultured yeast and lactic acid bacteria. There are clear advantages of completing alcoholic and malolactic fermentation simultaneously for commercial wines, including reduced time in tank or barrel and reduced risk of unwanted microbial activity. But the strains of yeast and bacteria need to be carefully matched for this to work.

Recently, a genetically modified yeast strain, ML01, has been available to winemakers where the use of genetically modified yeasts is permitted (currently this is only true in the USA). This strain has a gene from lactic acid bacteria engineered into it, so it is capable of carrying out the malolactic transformation at the same time as alcoholic fermentation. This is achieved by integrating the malate transport gene (*mae1*) from the yeast *Schizosaccharomyces pombe* and the malolactic gene (*mleA*) from *O. oeni* into the yeast genome at the *URA3* locus. No winemakers have gone on record admitting using this yeast because of the negative publicity that this could create among opponents of genetically modified organisms in food and drink production. It's a tricky issue. From the scientific viewpoint, this is an elegant solution to avoiding the problems of malolactic fermentation, but from the societal viewpoint, while there are no safety issues, consumers in many countries are simply not comfortable with genetic modification.

THE TRANSFORMATION: MALIC TO LACTIC ACID

The transformation at the heart of malolactic fermentation is that of malic acid (*malum* is Latin for apple), with its sharp, green, appley taste to lactic acid (*lac* is Latin for milk), which tastes softer and more appealing. The reaction involved is a decarboxylation, and it releases carbon dioxide. For this reason, wines that are undergoing malolactic fermentation often have a bit of a prickle to them when you taste them, and if a wine is unstable and malolactic fermentation starts in the bottle, then the wine will end up being slightly fizzy. This transformation results in a pH shift of 0.1–0.3, which is significant. But it is not just the pH change that makes a sensory difference because the difference in taste between the two acids is also important in changing the perception of the wine.

THE FLAVOR EFFECTS OF MALOLACTIC FERMENTATION

The sensory effects of the work of lactic acid bacteria are quite complex, and still far from being properly understood. Some of the effects are positive, while others are negative.

First, the negatives. One of the concerns about the work of lactic acid bacteria is that they are able to produce compounds that are negative to health. In fact, this was one of the motivations for developing the genetically engineered yeast ML01 mentioned above. Lactic acid bacteria sometimes make the biogenic amines histamine, ethylamine, isoamylamine, cadaverine (phenylethylamine), putrescine (diaminobutane), and diaminopentane. Their production depends on the strain, and they are a health concern to many consumers, even though it is not clear whether the levels typically found in wines are responsible for some of the symptoms claimed, which range from nausea and hot flashes to headaches. The production of biogenic amines seems to be worse in wines with higher pH, although selected lactic acid bacteria have been chosen for low biogenic amine production in all cases.

Lactic acid bacteria can metabolize the amino acid arginine, but incomplete metabolism can result in the production of ethyly carbamate, a (mild) carcinogen, via the production of citruline as an intermediate. Yeasts can also produce ethyl carbamate. Another potential carcinogen, acrolein, can be produced from glycerol by lactic acid bacteria: acrolein combines with phenolics in the wine to give a strong bitter taste, and is undesirable in wine. It is produced mainly by *Lactobacillus*, which tends to predominate malolactic fermentations at higher pHs.

Citric acid present in wine can be degraded by lactic acid bacteria, resulting in elevated levels of acetic acid (volatile acidity). It is degraded to pyruvic acid and then acetic acid, and this process also results in the production of diacetyl, butanediol, and acetoin. Diacetyl production is a major issue for malolactic fermentation because of its pronounced sensory effects. It yields smells of butter, popcorn, and yogurt. At low levels (below 4 mg/liter) it can be quite nice in some wines, but at higher levels (typically 5 mg/liter and above) it isn't.

The positives? Lactic acid bacteria have β-glucosidase enzymes that can help to release positive flavor compounds from their precursors. An example would be the floral-smelling monoterpenes. They can also release esters, which smell fruity, although some reports suggest that lactic acid bacteria have esterase enzymes that can remove these compounds from wine. It depends on the strain and the wine. And, of course, there is the beneficial sensory impact of the malolactic conversion to consider.

The pH level of the wine makes quite a difference to the species/strain of lactic acid bacteria that get involved. Above pH 3.5 malolactic fermentation occurs faster, and *Pediococcus* and *Lactobacillus* are more likely to be involved. This can be a bad thing, because it increases the risk of cheesy, buttery or milky off-flavors developing. High pH is also a huge risk factor for *Brettanomyces* in red wines, in part because sulfur dioxide is much less effective at higher pH, and because malolactic raises pH this is quite an issue. Lower pH (below 3.5) is desirable for malolactic fermentation with lower risk of off-flavors developing. However, if the pH is really low, then malolactic fermentation may struggle to get going.

WHICH WINES NEED MALOLACTIC?

Almost all red wines need to complete malolactic fermentation. The only exception I am aware of is red Vinho Verde, from the north of Portugal. These red wines are made predominantly from the Vinhão variety, which is a *teinturier* (red-fleshed grape). They are inky, dark, and vivid, and taste like cask samples, made to be drunk young and often with a bit of spritz. Many of them never make it to bottle, but are served from carafes, and the fact that they don't undergo malolactic fermentation gives them a nice green, appley edge that kind of works. But the vast majority of reds do undergo malolactic fermentation. Wines made in warm climates may be made from grapes that have very low malic acid levels to start with, so malolactic fermentation makes relatively little difference. Many whites undergo malolactic fermentation, but here this is more of a style issue. Those who want to preserve freshness and acidity in their wines may choose to carry out partial malolactic fermentation.

So, for white wines, malolactic fermentation is in part a stylistic choice. If winemakers want to prevent it from taking place, the usual method is to discourage the growth of bacteria by a number of interventions, including chilling, adding sulfur dioxide, racking the wine, and taking clarification measures. In addition, in many countries the use of the enzyme lysozyme, which specifically attacks bacteria, has been approved.

From this brief overview we see that malolactic fermentation is one of the most unpredictable, and least understood of all the stages of winemaking. Not only is the conversion of malic to lactic acid important in the flavor development of many wines, but also there are other flavor compounds produced by lactic acid bacteria that may be significant, in both positive and negative senses. It is only relatively recently that reliable cultured strains of *Oenococcus oeni* have been available to winemakers, and their use represents a valuable control step for those winemakers who choose to inoculate for malolactic. The choice of whether to permit malolactic fermentation or not is an important decision for white-winemakers. Where spontaneous malolactic fermentation is chosen, winemaking options in terms of temperature, nutrient availability, pH, and concentration of sulfur dioxide will affect the timing of this stage, as well as the risks involved.

17 *Brettanomyces*

Of all wine faults, *Brettanomyces* is one of the most multifaceted—and also one of the most fascinating, partly because it is one of those "faults" that in some contexts can be regarded as a positive. Indeed, there are many sought-after, expensive wines that owe some of their character to *Brettanomyces* (or, "brett"). It's also a controversial topic, with some arguing that brett is to be avoided at all costs, while others think a less dogmatic approach is more in order.

"*Brettanomyces* is very complex," says New Zealand winemaker Matt Thomson, an expert on the subject. "It is not a simple story at all. People want to make it a simple story, and you just can't. To make it a simple story is to make it inaccurate."

First, here are some basics. *Brettanomyces* is a genus of yeast (that is, a unicellular type of fungus, not a bacterium) also known as *Dekkera* (a technical point, this—the yeast can exist in two states, and this latter name is used for the sexual, spore-producing form). While several species names are commonly used, the current classification has the wine-relevant brett as just two species, *B. bruxellensis* and *B. anomala*, with the former by far the most important. Brett was first discovered by the brewing industry as an important component in British and Belgian beer styles in the early years of the twentieth century. Indeed, when the first single-culture *Saccharomyces cerevisiae* yeasts (this is the species of yeast used for making wine) were used to make British beers, people noticed that something was missing. This was the imprint of brett, which in the context of a good bitter can add real interest. Interestingly, brewers commonly refer to brett character in beer as being "vinous."

The reason Brett is a problem in winemaking is that it is annoyingly resilient, sitting around, biding its time, and then growing in conditions where virtually nothing else can. In practical terms, this means that it does its real damage after the regular alcoholic and malolactic fermentations are complete. Brett grows slowly, is tough, and doesn't need much to feed on. While it is seen in white wines (albeit very rarely), it's predominantly a red-wine problem. And the reason it is such a problem is that it produces some distinctive flavors that, at higher levels, can ruin wines.

THE MICROBIAL ECOLOGY OF WINE

The microbiology of wine production is a complex business, covered in more depth elsewhere in this book. Let me revisit an analogy from a previous chapter, where we compared the microbial ecology of wine to plants growing on the slopes up the side of a mountain. At the bottom there are hundreds of different types, with the pattern of vegetation changing and progressively decreasing in diversity with altitude (and a corresponding drop in temperature). It's something that happens with fermenting wine, except that here the variation is temporal and not spatial—it is a gradually changing environment for yeasts. In freshly crushed grape must there are many different yeast species present, and these rapidly disappear as fermentation starts and alcohol rises. The environment becomes more and more inhospitable, and after a while the only significant yeast species present is *Saccharomyces cerevisiae*. As alcoholic fermentation finishes, the *S. cerevisiae* population decreases significantly. If by this stage the sugar and nutrient supplies are exhausted, that's the end of things and the wine is stable. But if they aren't, this leaves the way open for spoilage bugs to develop. Brett is one of the worst culprits here. It is a resourceful bug that can make use of a range of substrates, making it hard to control.

WHAT DOES BRETTY WINE TASTE AND SMELL LIKE?
RECOGNIZING BRETT

Before we look in any detail at the factors that contribute to brett problems, let's first focus on perhaps the most important part: what are the key characteristics of this wine fault as it might be encountered in wine? The sensory effects of brett are many. The first sign is reduced varietal character, followed by the degradation of certain fruity aromas by esterases present in this yeast. Esterases are enzymes that cause the breakdown of esters, a chemical group important in conferring fruitiness. Thus grape varieties such as Pinot Noir are particularly badly hit by brett, because it loses its bright cherry and violet characters, and this loss of fruit can be an early cue for the presence of brett while the wine is in barrel. So this is the first action of brett: one of subtraction.

"Brett seems to take away the nonfermentable sugars that would otherwise add a bit of rounded character," says Matt Thomson. "Brett can metabolize the pentoses, and with these low levels of residual sugar (2–4 g/liter) you don't pick sweetness, but you pick body. Brett can take it from 1.5 g/l, which is bone-dry as far as we can see it, and it can end up at 0.5. That 1 g will make a difference," claims Thomson. "The tannins look a bit more rustic. For me, with brett in Pinot and Nebbiolo, what disturbs me most is the loss. It is not what it adds, but what it takes away from those aromatic reds that really upsets me.

"It upsets me when I hear people say that a little bit of brett is OK in Pinot Noir," continues Thomson. "I think, 'Obviously you don't know what it is like without it, or you wouldn't say that.' It takes the magic away. Tasting barrels, you see the different stage of the same wine, and this is crucial in understanding brett."

What does brett add? Volatile phenols and fatty acids are the key molecules responsible for the olfactory defects in wines affected by *Brettanomyces*. According to Peter Godden of the Australian Wine Research Institute (AWRI), "The anecdotal dogma in this area is that 4-ethylphenol (4EP), isovaleric acid (IVA; also known as 3-methylbutyric acid) and 4-ethylguiacol (4EG) are the key molecules, in order of sensory importance." But he adds that he has seen variations in brett character in different bottles of the same wine. 4EP is the most prominent molecule in bretty wines, giving aromas of stables, barnyards, and sweaty saddles (apparently, but I must admit never to having smelled one). Its presence in wine is an almost certain indicator of a brett infection, and this is what most diagnostic labs test for to indicate the presence of brett. 4EG is a little more appealing, known for its smoky, spicy aromas. IVA, a volatile fatty acid, is known for its rancid, horsey aroma. Godden emphasizes that this is a complex area of study: "There is not much of a relationship between overall brett character and 4EP levels, and there are synergistic effects between the three most important sensory compounds."

"The different components [Brett] can have, and even the ratios of 4EG, 4EP, and isovaleric acid change the perception," says Thomson. "It does show differently in different wines. There was some research some time ago that showed that 4EG is produced at much higher levels in Pinot Noir than 4EG is. This is what I find: I tend to get the medicinal, smoky 4EG notes more than I do 4EP. If you have 4EP in a Pinot it is well gone. It is hugely bretty. I find 4EG quite disturbing, to be honest, because the other thing that has gone on is that you have lost your pentose and your body, and more importantly, you have lost the fruit."

As with other volatile odorants, people differ widely in their sensitivity to these molecules, and each individual shows a range of different thresholds. For example, the threshold for detecting an odorant differs from the threshold for recognition of the same odorant. Godden suggests that a useful sensory threshold to use for 4-ethyl-phenol is 425 µg/liter (a microgram, µg, is one millionth of a gram). At this concentration and beyond, a wine will typically be noticeably bretty. Below this concentration, the character of the wine may be changed but people won't, on average, recognize that this is due to 4EP. Pascal Chatonnet, who did pioneering work on the problem of brett in Bordeaux wines, found that 425 µg/liter of 4EP negatively affected the sensory property of the wines from this region that he tested. For 4EG, thresholds are in the range of 100 µg/liter. But specifying thresholds is an inexact science. These may be altered by the style of the wine and the presence of other volatile compounds. Others have suggested levels as low as 300 µg/liter and even 150 µg/liter for 4EP.

Because the threshold for 4EP drops when 4EG is also present—and in brett-infected wine they typically occur together in a ratio of about 10:1—this threshold is best calculated for a 10:1 mixture of 4EP and 4EG. In fact, the ratio of 4EP to 4EG varies among wines, although there is quite a good correlation overall. This ratio is fairly consistent for wines made from a particular varieties or from particular regions, varying from 8:1 for wines from Bordeaux, to 3:1 for Pinot Noir, and 24:1 for Shiraz where this has been tested. The speculation is that regional and varietal differences alter the composition of precursor compounds in the wines.

It's actually quite difficult to teach people how to spot brett, because the characteristics of bretty wines will vary depending on the substrates that were initially available to the brett cells as they multiplied, the precise strain of brett involved, and also the context of the other flavors present in the wine. Depending on the combination of spoilage compounds produced and their relative concentrations, the overall effects of brett will differ. I've experienced some bretty wines that were more earthy and spicy while others have been more at the fecal/animal-sheds end of the spectrum.

HOW COMMON IS BRETT?

The short answer is that brett is highly prevalent, and represents an increasing problem, even in New World countries such as Australia. "We first started raising this as an issue five years ago," says AWRI's Peter Godden. "Since then, AWRI has started a major project looking at *Brettanomyces*." As a scientist, he feels that for such an important issue, this is a relatively under-researched area. "There is a lot of conjecture: anecdotal observations are very important but we have to be careful with them because they can skew people's opinions."

Although brett can and does occur with whites, it is predominantly a red-wine problem. This is because red wines are far higher in polyphenol content, and generally have a higher pH, both factors which encourage brett development for reasons outlined below.

With rising standards of winemaking worldwide, I was a little surprised to hear that brett is on the increase. There seem to be two contributing factors to this rise. First, there is the current trend for "natural" wines. "Minimalist winemaking is a perfect recipe for bretty wine," says Godden. "It's probable that the increase in brett in the 1990s can be traced back to the winemaking fad to stop adding sulfur at crushing." Indeed, the most effective way of preventing brett

is to maintain an adequate concentration of free sulfur dioxide (SO_2). Randall Grahm of California's Bonny Doon comments, "If one is ideologically committed to no sulfitage at the crusher, this increases one's chances of brett dramatically. Likewise, if one uses low or no SO_2 in the *élevage* of the wines, this greatly increases the risk of brett." Preliminary studies by the AWRI show that there is a lot of genetic variability among *Brettanomyces* strains. This makes the correct use of sulfur even more important. If it is added in small, regular doses, winemakers might unintentionally be selecting for SO_2-resistant strains of *Brettanomyces* or, to put it another way, superbrett strains that are then even harder to eliminate. So timing and magnitude of SO_2 additions are important as well as the actual concentrations. The best way to get rid of brett seems to be large SO_2 additions at strategic intervals.

Second, there is the move towards "international" styles of red wine, made in an extracted style from superripe grapes. "These are higher in pH and are richer in polyphenols," explains Grahm. Thanks to its role in modulating the effectiveness of SO_2 additions, it is likely that pH is important. The higher the pH, the less effective SO_2 is, and the more likely that *Brettanomyces* will grow. Polyphenol content is important because these compounds are the precursors for the volatile phenols largely responsible for bretty odors.

A vital risk factor is the presence of residual sugars and nitrogen sources left over at the end of fermentation. With the gradual rise in alcohol levels over the last 20 years, the yeast commonly isn't metabolizing the last bit of sugar. Chatonnet reports that as little as 0.5 g/liter residual sugar is enough for brett to grow to levels where it has a significant impact on flavor profile. Godden suggests that one solution is to try to keep the wines warm while they are being pressed. As well as sugar, a nitrogen source is needed for brett to grow. In fermenting wine, *Saccharomyces cerevisiae* uses amino acids as a nitrogen source. A recent winemaking trend has been to add diammonium phosphate (DAP) as a supplementary nitrogen source for yeasts, to reduce the risk of stuck fermentations. However, fewer than half of musts need actually use this additive, and DAP has been described as "junk food" for yeasts—they'll use this in preference to amino acids, leaving them in the wine as a nitrogen source that encourages the growth of brett. Phenolic compounds are also a substrate for brett growth.

Old barrels are frequently touted as the main culprits of brett, but Randall Grahm adds, "The received wisdom about old barrels, old *foudres* being the great repository of brett, I think, is somewhat mythical and simplistic: dirty barrels, dirty wines, QED." Grahm adds that, "Since brett is largely ubiquitous, a rampant brett infection is often more of a function of a large inoculum coming in on the grapes."

To gauge the extent of the brett problem, Godden and his colleagues undertook a survey of Cabernet Sauvignon wines in five major regions of Australia. "If a consumer were to go out and buy a mixed dozen," he told me, "several bottles would have more than 425 µg/liter 4-ethyl-phenol: if you drink wine regularly, you'll have come across a lot of brett." By the end of July 2003, 228 wines from the vintages 1996–2001 had been analyzed for 4EP and 4EG, but because of difficulties in establishing the analytical method for IVA, just 25 wines had been investigated for this latter compound. The results showed that the 4EP and 4EG levels were highly correlated, although neither level correlated with IVA concentrations.

Godden and his team at the AWRI have also studied the concentrations of brett-flavor components in a range of varietal red wines used in their Advanced Wine Assessment Course in July 2003. Altogether 192 bottles were analyzed for 4EP, 4EG, and IVA—the data are presented in Table 1. In particular, the Cabernet Sauvignon wines seemed particularly affected by brett. The average levels of the Cabernet Sauvignon and Shiraz wines were both above threshold, a startling finding.

Concentrations and ratios of 4-ethylphenol (4EP) and 4-ethylguaiacol (4EG), and concentrations of 4-isovaleric acid, in varietal red wines used for the July 2003 AWRI Advanced Wine Assessment Course

Variety	Number of wines	Mean 4EP (µg/liter)	Mean 4EG (µg/liter)	Mean ratio of 4EP:4EG	Number of wines	Mean IVA (µg/l)
Cabernet Sauvignon	33	771	76	14	30	1264
Nebbiolo	14	368	49	9	13	1155
Pinot Noir	13	120	50	3	11	718
Shiraz	19	495	37	24	16	929
Total	79			70		

Reproduced with permission of the AWRI.

There's a widespread misconception that brett contamination is a hallmark of wineries with poor hygiene. "Brett can occur in the cleanest cellars," says Matt Thomson, who is an expert on the subject. Brett has been identified in every wine region where people have looked for it. Thomson thinks that oak is largely to blame for many infections, because brett can live in the oak and it is almost impossible to get out by cleaning. "If you use new oak, you will get brett: it is not something you can associate just with a dirty cellar."

"If you have more new oak in your cellar, you will get more brett. If you have high pH wines, and harvest riper, you will get more brett. If you harvest riper and end up with high brix, the ferments will struggle toward the end and you'll get more brett. If you go to barrel with some residual sugar, you will get more brett. If you have low SO_2 or leave the wine with no SO_2 after malo [malolactic fermentation] for an extended time, you will get more brett. If you adopt practices that allow cross-contamination of barrels, you will get more brett. If you have a warm cellar you will have more brett. If you do these things that promote brett, brett will become a bigger problem.

"As soon as you have brett in your cellar you have to treat every barrel as a unit that can cross contaminate," adds Thomson. "I'd love some research to be done on coopers and brett. Coopers understandably run a mile if you mention it. Think about the very specific metabolism that *Brettanomyces* has. It can use oak lactones. The chances of *Brettanomyces* evolving in the absence of those substrates and having that ability to metabolize those specific substrates would be pretty low. It would be a huge coincidence. The chances are that they have grown up in the presence of oak in forests."

He continues: "If someone went and had a decent look for *Brettanomyces* in the forests in France, they might find that it is there. If you swab the inside surface of a toasted barrel, of course you won't find it. But if you split the barrel apart and look between the staves where wine leaches in and then leaches out, you'll probably find it. I think there's a decent chance that brett is resident in perhaps some forests. Brett didn't appear to be there in the 1980s. I think the boom in oak usage that occurred because Robert Parker said oak is good and let's have more, meant that the coopers expanded. They probably started recruiting forests that may have had high levels of brett. That is my theory, and I'd love someone to do the research to find out whether this is the case.

"Because we stopped the cross-contamination—and we are pretty rigorous on that—we have found that some coopers tend to have more brett, some where every barrel has brett, and some that don't seem to have brett. We have concluded that the bretty barrels either come with brett, or the oak promotes brett to a greater extent by having more substrates. We can't rule either out. It warrants research."

Brett likes oak. It particularly likes toasted new barrels, and has been found as much as ¼ inch deep in staves. This makes it very hard to remove by steam or ozone cleaning. It can feed off a compound, cellobiose, which is formed when barrels are toasted. Because it is so resilient, it will survive most cleaning attempts, and once it is there it is difficult to eradicate.

Before the 1990s, brett was common in Bordeaux. The wines of several well-known classed growths were renowned for their distinctive "stink." This was almost certainly because of brett infections, but without the data—and most properties would understandably be reluctant to own up to this—I can't name any names. Since the early 1990s, however, brett has become much rarer, and this is mainly due to the groundbreaking work of Dr. Pascal Chatonnet. In 1993, Chatonnet carried out a survey of 100 French red wines, and showed that a staggering one-third of those tested had levels of volatile phenols above the perception threshold.

The conclusion seems to be that *Brettanomyces* is widespread, and virtually every barrel of red wine has the potential to go bretty. Create the right environment for it, and you'll have a brett infection. Thus the key objective for winemakers isn't to create a sterile winery, which will never happen, but to make sure that their barrels aren't a receptive environment for brett to grow in. However, although unclean wineries may not be the main factor in brett prevalence, cleanliness in the winery is likely to reduce the level of inoculum, so is something that should be aimed for.

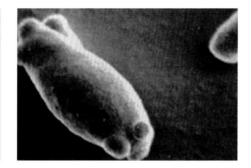

Far left The Sniff' Brett testing kit for detecting *Brettanomyces* in wine.

Left A scanning electron micrograph of *Brettanomyces*. Picture courtesy of Ann Dumont, Lallemand

BRETT, MOURVÈDRE, OR TERROIR? A CASE STUDY

Brettanomyces is a favored topic among wine geeks, who'll often enter into lengthy discussions about whether a certain wine is bretty or not. One wine that keeps appearing in this context is Château de Beaucastel, the highly regarded Châteauneuf-du-Pape estate. To some, the distinctive earthy, slightly animal-like characteristics of many past vintages of Beaucastel have reflected an expression of terroir, or even the higher-than-average Mourvèdre content of this wine. Others think it's because of brett infection. Who is right?

Back in early 1998, Charles Collins, an American wine collector, became so frustrated with the endless wine geek discussions about Beaucastel and brett that he decided to find out for himself. He got hold of some scientific papers on the subject and did some studying. "I realized that the presence of the compound 4EP is a virtually certain indicator of the presence of a brett infection," recalls Collins. He contacted a lab that does testing for 4EP and sent them some Beaucastel from his cellar. "I opted to test two of the most famous vintages, the 1989 and 1990," Collins told me. "These wines are supposed to represent what great Beaucastel is all about." He prepared the samples for shipment in sterilized glass 375 ml (about 12.5 fl oz) bottles and used fresh corks to seal them. The wines were labeled such that the lab had no clue as to their identity.

The results? According to Collins, "They showed indisputable evidence that significant brett infections occurred in both the 1989 and 1990 vintages of Beaucastel." Microscan and plating tests showed only small amounts of mostly dead brett cells, but the 4EP levels were 897 µg/liter for the 1989 and a whopping 3,330 µg/liter for the 1990. Collins concludes, "If you personally like the smell of brett, then none of this should you dissuade you from buying and cellaring Beaucastel. You should, however, give up the myth that the odd flavors are due to terroir—they aren't."

"We believe in natural winegrowing and winemaking, and I must admit that this has led us to have serious debates with scientists spanning three generations," responds Beaucastel's Marc Perrin. "In the mid-1950s, for instance, our grandfather, Jacques Perrin, decided to stop using chemical pesticides or herbicides on the vineyard. At that time, when scientists were recommending the use of such chemicals for productivity or lobby reasons, that seemed crazy and impossible. Now, it seems that people have changed their mind and more and more vineyards are turning organic. I could quote many more examples of opposition between a scientific vision of wine and our traditional/terroir oriented philosophy of wine, and the subject of *Brettanomyces* is just one more," he explained. "There are certainly some *Brettanomyces* in every natural wine, because *Brettanomyces* is not a spoilage yeast (as many people think) but one of the yeasts that exist in winemaking. Some grapes, like Mourvèdre, are richer in 4EP 'precursors' than others and we have a high percentage of these grapes in our vineyard. Of course, you can kill all natural yeasts, then use industrial yeast to start the fermentation, saturate the wine with SO_2, and then strongly filtrate your wine. There will then be no remaining yeasts, but also no taste and no *typicité*. That is the difference between natural wine and industrial wine, between craftsmanship and mass-market product."

ADDING COMPLEXITY?

Beaucastel has been widely acknowledged as one of the world's great wines over recent decades. Yet from Collins' limited sampling coupled with individual tasters' experiences, it seems likely that some of the most successful past vintages of this wines have been marked by high levels of brett. This leads us to a critical, and fascinating, question. Is brett ever a good thing? In small quantities, can it have a positive influence on certain styles of red wines?

If surveys such as those of Chatonnet and Godden are to be extrapolated across all wines, it is likely that many wines with above-threshold levels of brett have received critical acclaim and have been enjoyed by countless consumers. This leads to the conclusion that while most people won't enjoy a really stinky wine, low levels of brett might not be a problem. In fact, a little brett might even add complexity to certain robust styles of wines.

Bob Cartwright, previously senior winemaker of Leeuwin Estate in Western Australia's Margaret River region, acknowledges that, "A lot of winemakers like to have some as a complexing character. The question is, how much is too much?" Randall Grahm is undecided. "I suppose this could theoretically add some complexity to a wine. The problem is that, for now, this is not easily controllable." Pascal Chatonnet is opposed. He sees the problem of brett as a lack of fruit and loss of typicality. "If brett is able to grow in all the red wines of the planet—and this is the case—then all the wines will have the same odor, which is a pity."

Peter Godden is another who isn't keen on the idea. "My view is that if we could eliminate it altogether, we would," but he stressed that he wouldn't go so far as to say it is always negative. "In tests where brett character has been added, it has a severe adverse effect on the palate. 4EG can be interesting and complexing and doesn't have the negative palate effect of 4EP, but with brett infection you get 10 times as much 4EP as 4EG."

Richard Gibson takes up the theme. "There will be endless debate about whether a small amount of brett character enhances the complexity of a wine. Certainly, a little of the spicy character of 4EG may be preferable to cheesy IVA, so it is a case of what you get from brett, not only how much you get. But the key point for me is management and control. If you want a little, or if you want heaps, the key question remains the same. How can I manage my wines and winery environment to deliver the outcomes I want for my style, without threatening the quality of other wines in the cellar?"

Consultant winemaker Sam Harrop conducted a fascinating tasting as part of his Master of Wine (MW) dissertation. He was interested in the contribution of brett to wines made from Syrah. To this end he convened a tasting of 25 leading Syrah-based wines, including notables such as Penfolds Grange 1990, Henschke's Hill of Grace 1996, Jaboulet's La Chapelle 1996, and Chave Hermitage 1997. A dozen or so participants from the wine trade (this author included) were asked to comment on the wines, and specify whether they detected any brett. After the tasting, samples of each wine were sent off for chemical analysis for 4EP and 4EG. The results were telling. Although I don't have permission to specify here the levels in individual wines, 11 of the 25 had above-threshold levels of 4EP (in excess of 425 µg/liter). Eighteen out of 25 had 4EP levels higher than 100 µg/liter and three had levels in excess of

2000 µg/liter. Interestingly, the performance of tasters in detecting brett didn't correlate terribly well with actual levels. These striking results illustrate that brett seems to be an element of the character of critically acclaimed Syrah-based red wines, and it would seem premature to dismiss brett altogether as a wine fault in all circumstances. The context seems to be important.

Randall Grahm has an novel suggestion, though: "It would be very interesting if we could isolate a strain of brett that worked in wine, depleting nutrients but producing very low levels of 4EP. In this way, one could inoculate one's wine with brett, much the same way as one inoculates one's wine with malolactic bacteria, thus depleting nutrients and rendering the wine safe from further microbial degradation." Richard Gibson echoes more revolutionary sentiments: "Currently in Australia the move is to eradicate brett, but with more understanding, management of brett may move to developing the ability to produce blenders with high proportions of the right brett compounds for judicious back-blending, managed in a way that does not threaten the total winery environment with infection."

This brings us around to the topic of control of brett. As we've already seen, maintaining high free SO_2 levels is probably the most effective means of limiting brett growth. For two reasons, pH is an important factor here. First, lower pH increases the effectiveness of free SO_2, such that less is needed, and second, more SO_2 is likely to be in the active molecular form at lower pH. Other preventive steps include completing fermentation so there is no residual sugar left in the wine to act as a substrate, and maintaining lower temperatures in the barrel, vat, or tank. Low cellar temperatures seem to make quite a difference.

High grape-sugar levels are a risk factor for brett, because *Saccharomyces* yeasts are made to struggle to complete primary fermentations. Sam Harrop suggests that part of the problem may lie in the vineyard. "A winemaker who is serious about managing *Brettanomyces* should work their vineyards to get fruit maturity at lower levels of alcohol. This will support a healthy fermentation, reducing residual sugar content, and limiting the possibility of *Brettanomyces* growth. In addition, fruit harvested at lower sugar levels will have lower natural pH levels, which aids greatly in managing *Brettanomyces* throughout the winemaking process by way of its relationship with SO_2."

There is a chemical way of eradicating brett. Dimethyl dicarbonate (DMDC, known commercially as Velcorin) is an effective sterilant that works by deactivating enzymes in microbes, and while it is hazardous to apply, it is completely safe once in wine. A winemaker with access to DMDC can kill off any active *Brettanomyces* present, effectively stabilizing the wine before blending or bottling. In the absence of DMDC use, sterile filtration would seem to be a wise precaution for stabilizing wines at risk from brett before bottling. The downside is the possibility that filtration will remove flavor compounds from the wine. It is clear, though, that unless you are a winemaker who feels very lucky, that playing with low levels of brett necessitates accurate tools for its measurement and management.

A new product has recently become available for the control of *Brettanomyces* in wine. It's called No Brett Inside, and it is a preparation of chitosan, isolated from fungus *Aspergillus nigrans*. The mechanism of action isn't totally clear, but it is thought to bind the outer cell wall of brett cells and disrupt metabolism in some way, as well as causing the brett cells to stick together and fall to the bottom of the tank or barrel. While this product will not remove any 4EG or 4EP already produced by brett, it will prevent any further problems. Chitosan is a linear polymer of D-glucosamine and N-acetyl-D-glucosamine, and is made by a process called chitin deacetylation.

"Sniff Brett" is another tool for winemakers hoping to control brett. It comes in the form of small vials, which contain a liquid culture medium containing coumaric acid. Wine is added, and after two or three days' incubation you can use your sense of smell to pick up the presence of brett, whose aromatic impact is vastly enhanced by the culturing process. Sniff Brett is useful for smaller wineries without proper lab facilities.

Descriptors used to describe *Brettanomyces* characters

Band-aid	Barnyard	Horsey	Wet dog
Ammonia	Stable	Medicinal	Rancid
Mouse droppings	Smoky	Pharmaceutical	Sweaty
Manure	Spicy	Leathery	Cheesy
Burned beans			

Compounds identified as contributing to *Brettanomyces* odors

Volatile phenols	*Medium-chain fatty acids*	Others
4-ethylphenol	octanoic acid	2-phenylathenol
4-ethylguaiacol	dodecanoic acid	isoamyl alcohol
	isobutyric acid	cis-2-nonenal
	isovaleric acid	trans-2-nonenal
	(3-methylbutyric acid)	β-damascenone
	2-methylbutyric acid	ethyl decanoate

18 Corks, screw caps, and closures

Forty years ago, all wine bottles were sealed with natural cork. There was no debate about closures simply because there was no other practical way to seal wine bottles. The wine trade certainly knew about the issue of musty taint (referred to as "corked" wine), but seemed to tolerate it. The cork industry was under no pressure to do a better job because it had no competitors.

From this near-monopoly position, cork has seen its market share shrink in the face of alternative closures. Estimates are that of a global closures market of some 20 billion units (sealing 75 cl or 37.5 cl [25 or 12.5 fl oz] wine bottles, but excluding smaller single-pour formats), screw caps are now around 2.5 billion, alongside synthetic corks, which are around 2 billion. This still leaves natural cork (including technological cork-based closures) with a healthy share of the market. However, the march of alternative closures continues. As yet, fine wines have remained largely sealed by corks but the notable exception here is Australia and New Zealand, where screw caps are by far the dominant closure type for all levels of wine.

CORK: A REMARKABLE NATURAL SUBSTANCE

In the debate on closures, cork has frequently been cast as the bad guy. What's often forgotten is that it has remarkable natural properties. It may be highly unfashionable to say this, but cork is a gift from nature ideally suited for sealing bottles. It comes from the bark of the cork oak, *Quercus suber*, which is an unusual and useful tree. If you stripped the bark of most trees, they would die, because you'd be removing the cambium, the cylinder of dividing cells just inside the bark that is responsible for new growth in the stems or trunks of woody perennials. The cork tree has such a thick bark that it can be stripped from mature trees without harming them. Unusually, cork trees have two cork cambium layers. The first, which has its origin in the epidermis, is removed when the tree is about 20 years old, and a new cork cambium then forms a short distance below the site of the first. From then on, new cork tissue accumulates rapidly and can be harvested every nine to ten years, until the tree reaches a venerable old age of 150 or so.

The key to cork's mechanical properties is that it is formed of a honeycomb network of densely packed cells, whose walls have been "suberized." The molecular composition of suberin is still a bit of a mystery. The latest view is that it is formed by a hydroxycinnamic acid–monolignon (poly) phenolic domain embedded in the cell wall, which is linked to a glycerol-based (poly)aliphatic domain located between the plasma membrane and the inner face of the cell wall. "Suberin" is therefore a term that should be used carefully, because it doesn't refer to a single molecular entity.

Suberinization also involves the deposition of a number of waxes in this inner wall region. A wine cork consists of hundreds of millions of these suberized cells, rendered inert and impermeable. Because these cells are filled with gas, the whole cork structure is compressible and elastic. Cork can be compressed to about half its width without losing flexibility and has the remarkable property of being able to be compressed in one dimension without increasing in another. It can resist moisture for decades, and will stay compressed, thus maintaining a seal, for equally long periods.

Because of this composition and structure, cork is good at sealing wine bottles. A decent cork will provide a good seal for 30 years, possibly longer, allowing the wine to develop and mature. And, despite the tightness of the seal that corks provide, it is relatively easy to extract them using one of a wide array of different designs of corkscrew. Added to this, taking the cork out has become a valued part of the tradition of wine. It may sound silly, but there's something special about uncorking a bottle.

CORK'S ACHILLES' HEEL

But before you begin to wonder whether this chapter is actually an "advertorial" paid for by the cork industry, let me try to put things into perspective. Cork has an Achilles' heel. As a natural substance it is variable, and is prone to failure. Most significantly, it harbors a contaminant that is able to spoil wine at fantastically low doses. Meet TCA, curse of the wine industry. TCA is the commonly used abbreviation for a chemical called 2,4,6-trichloroanisole. The dirty secret of the wine trade is that around one in 20 bottles of wine is ruined as soon as it is bottled by this cork taint, which is the name for when a wine takes on a musty odor. The main culprit is TCA present in some corks, although recently other related anisoles have also been implicated in cases of musty taints. In extreme cases it's hard to miss a corked wine because the mustiness can sometimes be overpowering. In other situations the taint is subtler, reducing the fruitiness of the wine, giving it a subdued aroma, usually with a faint whiff of damp cardboard or old cellars in the background.

The problem with TCA is that it is incredibly potent, and so most people can detect it at concentrations as low as 5 parts per trillion (ppt, the same unit as nanograms per liter), and some are even more sensitive. This makes it hard to eradicate. To give you a better idea of this figure, it's equivalent to one second in 64 centuries. Where good data has been collected, the frequency of cork taint hovers around 2–5% of bottles sealed this way, a rather contested figure that we will address later.

TCA: ITS ORIGINS

Where does TCA come from? It's a compound produced primarily by interaction between microbial metabolites and chlorine in the environment. The use of chlorine in washing steps in cork production was thought to contribute to this, but now that chlorine-based products have been replaced by alternatives, such as hydrogen peroxide, cork taint is still with us, suggesting that an exogenous chlorine source may not be needed. In fact, in a study published in the *Wine Industry Journal* in 1987, some researchers from Australian wine company Southcorp analyzed cork trees *in situ* from four regions of Portugal. They detected TCA in 58 out of 120 trees analyzed. Microbes such as mold-forming fungi live in the small pores (called lenticels), which run throughout cork bark. The lenticels are areas of the cork where cells have divided faster than elsewhere, forming a looser structure that allows air through this otherwise impermeable barrier. They can be seen in corks as darker, colored lines or imperfections. In addition, the processing steps used in making corks from sheets of barks may encourage fungal growth and thus TCA production. Chloroanisoles can also be produced in the absence of microbes. All that is needed are the phenolic precursors and a chlorine source. It is a complicated subject, but the message here is that TCA is endemic to cork.

It needs to be emphasized that TCA isn't confined to corks, nor is it the only compound responsible for musty aromas. Musty off-odors are a major problem in the food industry. Chloroanisoles other than TCA are also potential contaminants, especially TeCA (2,3,4,6-tetrachoroanisole), which is detectable in wine at concentrations of 10 ng/liter. A 2004 study by Pascal Chatonnet and colleagues identified a further potential musty contaminant, 2,4,6-tribromoanisole (TBA). This causes musty off-odors in wine at concentrations of 4 ng/liter, and is formed from its precursor TBP (tribromophenol). TBP is used as a pesticide inside buildings, and barrels, corks, and plastics are all susceptible to TBA contamination from the environment in situations where TBP has been used. Old wooden structures are especially prone. In support of this hypothesis, Chatonnet cites the results from the analysis of wines carried out in Canada by the Liquor Control Board of Ontario (a monopoly supplier owned by the provincial government) on wines it intends to list. In 2002, 2,400 wines were tested. Of those that were considered to have musty taint, only 49% had significant levels of TCA (>2 ng/liter). Other chloroanisoles or TBA may have affected the other 51% of tainted bottles. Because these weren't tested for, we don't know how much influence they had. Barrels can also be a source of musty taint, although there is some heated debate about just how much contamination of barrels occurs. Cooperages claim it is a much lower rate than those offering barrel-screening services.

Several wineries have had problems with TCA or TBA contamination of their premises, resulting

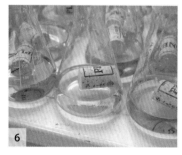

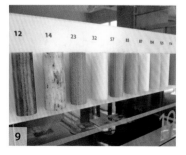

1 A freshly harvested cork oak. This is one of the few trees you can strip the bark from without killing it.

2 Planks of harvested cork bark are bundled together, ready for processing.

3 The cork planks are then boiled to soften them and wash away impurities.

4 A strip of bark with corks punched out of it.

5 High-grade corks, hand-selected. These will be very expensive, in excess of €1 a piece.

6 A soak test of corks being conducted by a winery, in a bid to identify any tainted batches before they are used.

7 Corks from some high-end Bordeaux wines. While Australia and New Zealand have almost entirely shifted to screw cap for their wines, leading European bottles are still cork-sealed, with very few exceptions.

8 Diam closures, made from cork particles, which are now being used even on high-end wines.

9 At Nomacorc's head office in Zebulon, North Carolina, a range of prototypes are on show. The current Select Series Nomacorcs are quite different to the early incarnations of this successful wine closure.

10 This screw-capped wine is actually from 1977, and it's a Clare Valley Riesling made by Riverina College students. The wine had survived beautifully.

in large volumes of wines suffering from low-level musty taint. Musty taint has also been identified in wine with barrels as the source. But all the indicators are that the vast majority of TCA taint is down to the cork, because in large competitions, such as the International Wine Challenge where in excess of 20,000 bottles are opened and tasted, with very few exceptions all the bottles displaying musty taint are sealed with natural cork.

THE PREVALENCE OF CORK TAINT

How common is cork taint? Each year, the International Wine Challenge is held in London. An enormous number of bottles are opened and tasted systematically by panels made up largely of experienced tasters. Although a tally of cork-tainted bottles has been kept in previous years—in 2001 it was 6% and in 2002 it was 4.6%—in 2003 a "super juror" verified all cases of suspected mustiness as cork taint. The results were that of 11,033 bottles sealed with natural corks, 4.9% were considered to be corked. A further 2.79% were faulty for other reasons. This figure tallies well with results from other surveys. Starting from 2005 onward there was a dedicated faults clinic overseen by Sam Harrop, an expert on wine faults. Over the five years from 2006–10 the average rate of cork taint was 2.8% (individual years: 2.8%, 3.3%, 3%, 3.2%, 1.9%). The weakness of this sort of sampling is that it is likely either to produce false positives, or that low-level TCA contamination will be missed in some cases. Although there is no chemical analysis of bottles judged to be affected, the scale of the sampling is impressive enough to mean that these are useful results, and most wine-trade tasters are pretty good at spotting cork taint. Influential U.S. publication *The Wine Spectator* keeps a tally of corked wines. *Spectator* taster James Laube reported that of the 3,269 Californian wines tasted in *The Wine Spectator* office in 2012, 3.7% were corked. He adds that this is the lowest percentage seen yet, down from 3.8% in 2011 and a high of 9.5% in 2007. According to judge Andrew Jefford, in the 2012 *Decanter* World Wine Awards, some 3.3% of the 14,120 entries were dismissed as being spoiled by TCA. However, 10% of those wines turned out to be either screw-capped or not stoppered with cork, so there may have been some false positives. Judges can sometimes detect cork taint where there is none, or confuse other faults with cork taint.

OTHER ISSUES RELATING TO CORK PERFORMANCE
IS CORK A NEUTRAL CLOSURE?

Are corks neutral in contact with wine? During the cork-manufacture process there are steps such as seasoning, boiling, and stabilizing the cork planks, which are designed to remove various tannic and phenolic compounds, rendering the cork as neutral as possible. The cork cells are relatively inert, but it is likely that the cork is not completely neutral and will interact with the wine chemically, albeit to a rather limited extent, for example, by releasing phenolics into the wine. The process of "flavor scalping," where packaging absorbs aroma components, is a problem in the food industry. Aware from studies that TCA could be absorbed from the wine by the cork, scientists at AWRI investigated potential flavor scalping of aromatic wine components by a range of closure types. The wine used was the same as in the AWRI closure trial, a 1999 Semillon, spiked with the flavor compounds they wanted to study. Closure performance was compared with that of the same wine in a sealed glass ampoule. After two years, the concentrations of many of the flavor compounds had changed significantly, partly through absorption by the closures but also through chemical modification independent of closure. Screw caps didn't absorb any flavor compounds, performing similarly to the sealed glass ampoules. The corks, technical corks, and synthetic closures absorbed relatively non-polar volatile compounds, while the more polar compounds weren't absorbed by any closures. Of these closures, technical corks absorbed a little more than the natural corks, and synthetics absorbed a lot more than either. The conclusion was that synthetic corks are responsible for considerable flavor scalping, and even natural corks are capable of absorbing certain wine aroma components in limited amounts.

SOLVING THE PROBLEM OF CORK TAINT

There are three strategies for combating cork taint. First, eradicate TCA from natural corks, curing the cause of the taint, and rescuing the cork from an otherwise gloomy future. The second is to manufacture synthetic, taint-free corks from alternative materials such as plastic, thus allowing wineries to keep their current bottling lines and consumers their corkscrews. The third, which is a little more radical, is to ditch the concept of corks altogether and turn to different forms of closure, such as screw caps.

THE KEY PROPERTY OF CLOSURES: OXYGEN TRANSMISSION

It was only when wine producers began exploring alternatives to cork that discussion began about the technical requirements of wine-bottle closures. Beginning with the crucial question, one for which we don't yet have a clear answer: what would an ideal closure look like in terms of performance? Results from comparative closure studies show that perhaps the most important property of closures is their level of oxygen transmission. Do we want our closure to give a completely hermetic (airtight) seal that doesn't allow any oxygen transmission at all? That is, would a sealed glass ampoule be the perfect closure? Or do we want some oxygen transmission? If the answer is yes, then how much?

Over the last few years there has been a change in thinking in the closures field. It concerns the old notion of the closure "sealing" the bottle. It used to be thought that the better the seal, the better the closure. To a degree that is true. We want to keep wine in the bottle, and we want to keep air out, hence "closure." The idea that the cork allows the wine to "breathe" is patently false. If oxygen were able to permeate the cork freely, the wine would oxidize rapidly. But if we are to extrapolate heroically the notion that the better the seal, the better the closure, we'd end up at the position that maintains that a perfect closure would be one which seals hermetically, with no gas transmission at all. In fact, advocates of screw caps frequently cite the writings of the late Émile Peynaud, who stated in *Knowing and Making Wine* that, "It is the opposite of oxidation, a process of reduction, or asphyxia by which wine develops in the bottle," or Pascal Ribéreau-Gayon, who in the *Handbook of Enology* asserts that, "Reactions that take place in bottled wine do not require oxygen." But another celebrated French wine scientist, the late Jules Chauvet, had this to say on the subject of closures:

"I believe that no one can ever replace natural cork, at least not currently. Cork is porous, enabling the realization of equilibrium of oxidation-reduction in the bottle. If you want to bottle a wine and drink it 15 days later, the closure has no importance at all. But for wines that are kept for a few months or a few years, you must use a cork, and a good cork ... We did an experiment in which we sealed some wine bottles with cork and some with ground-glass stoppers. We noticed that three months later the wine sealed by glass had a better appearance but it was already reduced. The cork-sealed wine was still ailing from its 'bottling sickness,' and was still a little oxidized. Later on we saw the cork-sealed wine improve and the glass-sealed one get worse. The latter became undrinkable. We are sure that a micro-exchange of oxygen is needed to induce an equilibrium that allows a light and pleasant aging."
Jules Chauvet, interviewed by Hans-Ulrich Kesselring in *Le Vin en Question*.

Most people would now agree that a degree of oxygen transmission is needed for the successful development of wines in bottle. "The point we've been making for a few years now is that it is possible to use different levels of oxygen—introduced into the wine either at bottling or post-bottling—in a creative way, to manage the development of the wine so it is at its optimum when it is consumed," says the Australian Wine Research Institute's Peter Godden. "I don't think zero permeation is ideal for many, if any, wines. I wouldn't use such a closure to seal my wines. But this is perhaps missing the point. Variable levels of oxygen ingress will create different wines. This is the key point from our closure trial. We took one wine and bottled it with 14 closures. Since then we have taken one wine and bottled it with various numbers of closures. You get different wines, and they look different after as little as three to six months. They are not all heading toward the same endpoint. They are going off in different tangents. We are past the question of whether or not the wine needs oxygen."

So why is some oxygen transmission by the closure necessary? There are two possible scenarios to explain the process of wine aging and the influence of the closure on this development.

First, it may well be that oxygen transmission isn't needed for a wine to develop, and that a wine will evolve in a pleasing way sealed hermetically. In this scenario, oxygen transmission is needed solely to avoid problems with reduction (caused by volatile sulfur compounds, see Chapter 15).

In the second scenario, wines will age in the total absence of oxygen, but in a way that we don't really like. In this scenario, for *successful* aging, some oxygen transmission is needed, not just to avoid reduction, but also to facilitate the complex chemical transitions needed to result in a wine aging to an optimal outcome.

But, as Godden points out, it is clear that the level of oxygen transmission will adjust the rate of aging, and it likely will also adjust the trajectory of aging. Therefore, a wine sealed with a very low oxygen-transmission closure might end up in a different place—never reaching the same destination—as one aged under cork in such a way that we know and appreciate. In discussions of desired closure performance, we need to distinguish between two types of wine. On the one hand there are wines destined for early consumption, which make up the vast majority of wines produced these days. They need to drink well on release, and hold their quality for as long as it takes them to get through the supply chain. On the other hand are fine wines that are destined for aging. In some cases, the expectation is that these wines will improve in bottle for many years, and then hold this quality for perhaps as long again. In other cases, the wines aren't expected to improve dramatically, but they are supposed to retain drinkability for several years as they loiter on restaurant wine lists or in consumers' cellars. The closure requirements will differ for these different types of wines.

Thus we arrive at a slightly more nuanced pair of questions. What do we want from an alternative closure to ensure successful wine aging for fine-wine styles, and what is the ideal closure for a short-rotation wine? Then, if we dig a bit deeper, it emerges that red and white wines respond differently to oxygen. With their high phenolic content, red wines can absorb a lot more oxygen than white wines without showing an oxidized character. Indeed, red wines appear to need some limited oxygen exposure during the winemaking process in order to develop optimally. And white wines that are differently treated in their early stages will show different levels of resistance to oxidation. Those that are protected from oxygen exposure all the way from the crusher to the bottle will be more fragile than those that have been exposed to more oxygen during the winemaking process, for example, through deliberate pre-fermentation juice oxidation or barrel fermentation. Add to this the effects of different levels of free sulfur dioxide at bottling and wine pH, and the picture that emerges is that different wines are likely to have different requirements from closures. It seems remarkable that cork has done so well as a one-size-fits-all closure.

To date, the most important closure study has been that of the AWRI whose important closure trial began in 1999. This trial uses a battery of measurements to test oxygen transmission, including detailed sensory analysis and measurement of free and total sulfur-dioxide levels. The study included 14 different closures, including a number of synthetic corks, two different screw caps, two natural corks, and a range of technological corks (such as Amorim's Twin Top® and Sabaté's Altec). Of these, the synthetic corks allowed the most oxygen ingress, with the results that all the wines had oxidized within the first few years. The two corks used in the study showed varying degrees of oxygen transmission, but on average performed better than the synthetics. The Twin Tops® and Altecs provided fairly consistent tight seals, but there were problems with all the cork-based closures in that many of them showed TCA taint. The screw caps provided the tightest seals, but there were reduction problems.

Perhaps the most significant observation at the 63-month time point is that wines sealed with different closures look quite different, and these differences track fairly closely the degree of oxygen ingress, as deduced from the remaining levels of free sulfur dioxide.

Other trials have looked at closure performance. Paulo Lopes and colleagues have studied a range of closures using nondestructive colorimetry, involving an indigo-carmine dye. This dye changes color in contact with oxygen. Currently, synthetic closure manufacturer Nomacorc are partnering with several academic institutions and other industrial partners to undertake significant oxygen in wine study, based on the use of a luminescence probe. At conferences, Jim Peck of G3 Enterprises has reported other, as yet unpublished data on closure-oxygen transmission based on the use of MOCON measurements. The hope is that soon winemakers will have a better idea of the requirements of various wines in terms of closure-oxygen transmission, and also the actual physical properties of the various alternative closure types.

There is an extra level of complexity here, in that closure OTR is in many cases quite dynamic. When a wine is bottled with a cork there are a number of different phases of oxygen influence. First of all, there is oxygen that is present in the headspace, introduced during the bottling process. Then there is the diffusion of oxygen present actually dissolved in the closure. And then there is the steady state OTR level through the closure. This dynamic exposure to oxygen might be quite important, but it makes discussion of closure OTR very complicated. (This issue of post-bottling oxygen exposure is discussed in detail in Chapter 10.)

Recent (2012) estimates are that screw caps now have more than 10% of the bottled wine market. This total market is estimated to be some 20 billion annually worldwide. Of this, screw caps are now around 2.5 billion, with synthetic corks around 2 billion and natural cork accounting for most of the remainder. This is a dramatic rise from the situation five years ago, when screw caps were estimated to be around 200 million worldwide.

Technical corks are proving to be increasingly important closures for wine bottles, but compared with synthetics, screw caps, and natural cork, they get relatively little coverage, to the point that many people in the trade aren't even familiar with the term "technical cork." It refers to a natural cork-based closure made by combining disks or granules of natural cork to produce an inexpensive closure solution. Their significance is shown by the fact that they now represent around half of the business of Amorim, which is by some distance the world's largest cork company.

The most significant technical cork is the Twin Top®, developed by Amorim (whose products are distributed in the US by its wholly owned subsidiary Portocork) in the mid-1990s. It's a closure that adapts the technology used in Champagne cork production. "Champagne corks are not only an important product for cork companies," says Carlos de Jesus of Amorim, "but also led to the development of the Twin Top®." Cork companies realized that Champagne corks were performing well, with low levels of taint and consistent physical performance. "We asked ourselves how we could transfer this technology to still wine closures and get good results?" reports de Jesus.

The Twin Top® has an agglomerate core sandwiched between two disks of good-quality natural cork. The advantage of having natural cork at both ends of the closure is dual. First, no orientation machine is needed before the cork is applied on the bottling line, and second, the consumer sees nice-looking natural cork when the capsule is removed, and not agglomerate. Launched just in the mid-1990s, the Twin Top® is now the best-selling technical stopper in the world, with sales of 650 million units per year (2012 data). It is usually employed for large volume, price-sensitive wines where the cost of good natural cork would be prohibitive.

While the one-plus-one is the most popular technical cork, it is rapidly being caught up by a new generation of closures known as micro-agglomerates. While the agglomerate portion of the one-plus-one is made up of relatively large particles of cork, giving quite an ugly appearance, micro-agglomerates are made up of much smaller granules and have an altogether more attractive, uniform look. Because the agglomerate portion of the micro-agglos (as they are commonly referred to) is much more attractive, it is not sandwiched between cork disks.

The first micro-agglomerate was the Altec, which, though revolutionary at the time, turned out to be fatally flawed. Introduced in 1995 by French cork company Sabaté, Altec was made of finely ground cork flour glued together with synthetic microspheres to produce a cork-based closure that looked pretty classy and had uniform properties. The synthetic microspheres were needed to provide a degree of elasticity to the Altec. Without them, the closures would have been too rigid, because of the very small size of the cork granules employed.

The market, dissatisfied at the time with the quality of cheap natural cork and not convinced by the first generation of synthetic corks, endorsed Altec by buying two billion units over the following years. (This is before the screw-cap revolution, which started in Australia in 2000.) But grumbling about the organoleptic impact of Altec soon started. Some people complained of a glue taint. In fact, the problems were caused by a consistent low level of cork taint, caused by the presence of 2,4,6-trichloroanisole at very low levels. What had happened is that the manufacturing process using small granules of cork had averaged out the TCA naturally present in the cork, so that instead of having a few contaminated closures in a batch, every closure was contaminated to a low degree. And unfortunately for Sabaté, this level of TCA was above detection threshold for some tasters. It was a major disaster, and by 2002 sales plummeted.

Sabaté, to its credit, responded well. It looked at ways of cleaning the cork component from any contamination and with the help of the French atomic energy commission it came across a process that actually worked. It involved the use of carbon dioxide in a state known as its critical point. At a particular combination of pressure and temperature, the liquid/gas interface disappears, and you then have a substance that can penetrate like a gas, but clean like a liquid. For carbon dioxide, the critical point is not too hard to achieve: it's 88°F and 73 bars of pressure—a conveniently low temperature, even if the pressure is on the high side. This technique has been used to remove caffeine from coffee and by the perfume industry to extract fragrances.

After trials, the new version of Altec, called Diam, was released commercially in 2005, and Sabaté has since changed its name to Oeneo-Bouchage, after some restructuring. The Diam range has expanded and currently consists of Diam versions 2, 3, 5, 10, and Grand Cru in ascending order of cost and impermeability to oxygen, as well as the Mytik sparkling-wine closure. Oeneo has improved the appearance of the closure by adding a grain effect, to make it look more like natural cork. Diam's great benefit is that it is taint-free, because of the effectiveness of the production process. However, because of the added cost of the supercritical carbon dioxide treatment, Diam costs a little more than some of the other technical corks on the market, and is really aiming to compete not with synthetic corks and screw caps (which is the target market for most technical corks), but with natural cork.

"The price of Diam is €50–300 [$65–400] per 1,000," says Bruno de Saizieu, marketing director with Diam (2012 figures). "Clearly, where we play is against natural cork. Our market is more and more for high-quality wines. We now have the Diam Grand Cru, for wines of €15 [$20] plus. In Burgundy we already have some *grand cru* wines, as well as 50% of the Chablis market." Diam is doing well. "Our increase in sales last year was 20%," says de Saizieu. "I expect the same increase this year." In the U.S., Diam made an agreement for G3 Enterprises to take over all distribution in April 2010.

Following the initial market success of Altec, other cork producers began work on micro-agglomerates, and most now offer them as part of their portfolio. For leading cork manufacturer Amorim, its micro-agglo, called Neutrocork®, has been experiencing rapid growth in sales. "It is now the fastest-growing technical stopper," reports Amorim's de Jesus. He says that sales have grown over the last four years at 20% per year, with a current figure of 410 million units. For the micro-agglo category as a whole, corresponding growth has been 12%. "It's a workhorse in getting wines back to cork from plastic closures," says de Jesus. "It can undercut the synthetic corks by as much as 50%, depending on the market and the quantity." MA Silva's micro-agglomerate is Pearl®, while Corksupply offers the Vapex microextra®.

In conclusion, there now exist several alternative ways of sealing a bottle of wine. Oxygen transmission by these closures seems to be very important in determing the way the wine develops in bottle, but there is still few solid data on what is taking place here, and what sort of levels of oxygen transmission are appropriate for different wine styles. This makes it tricky for winemakers to choose the right closure, from a purely scientific point of view.

Section 3
Our Interaction with Wine

19 Flavor and its perception: taste and smell in wine tasting

OUR INTERACTION WITH WINE

There's an unspoken assumption in the world of wine: one that needs challenging. Take a critic and a bottle of wine. The wine possesses certain characteristics. If you take an array of scientific measuring devices you can make a description of the wine in terms of its physical parameters, which will be correct within the margins of error of the devices used. You could prepare a document with this description, sure in the knowledge that if someone else were to measure these physical properties, or you were to come back later and reassess the wine, you would end up with a pretty similar description. Now it's time to let the critic assess the wine. A critic assesses a wine very differently from a set of scientific measuring devices. Indeed, what critics are actually doing in describing or rating a wine is telling us about their interaction with what is in the glass. The outcome tells us about them as well as the wine, and it is very hard to separate the two. Critics bring to wine not only their own cultural and contextual differences, but also their perception of the wine itself depends on complex physiology and neurobiology. Just as with scientific instruments, we need to know the error margins in their performance. How consistent is their perception of the same liquid sampled at different times, in different contexts, on different days? You'd expect the best critics to be fairly consistent, but has anyone actually measured this? More importantly, the assumption is that our perceptions of the same wine are broadly similar, and that we each share the same physiological and neurobiological "equipment" for assessing wine. But what if the difference between critics' opinions is not simply a matter of competence or preference, but instead reflects a more fundamental biological difference? This would change the way we look at wine tasting.

Along similar lines, we are familiar with the idea of continuous variation in the population for a variety of biological traits. Take height, for example. A group of 100 people will range from tall to short, with most people somewhere in the middle. We're comfortable with the idea that variation in taste and smell is similar, in that sensitivity will vary across a group such that there'll be a spread, with most people in the middle and progressively smaller numbers as we move toward the ends of the distribution. But what if variation in taste is a discrete, discontinuous distribution, a little like eye color? That is, rather than a spectrum of small differences in sensitivity, we are looking at separate groups, where the individuals in each group have similar sensitivity but the groups differ markedly from each other? We need to bear these kinds of issues in mind as we consider the biological basis of wine tasting.

Most attempts to understand our interaction with wine focus on the sense of taste and smell, and on the detection of flavor molecules by the receptors in the mouth and nasal cavity. But this limited level of understanding just won't do if we want a proper explanation of how we actually perceive wine. Take a sip of wine. Your experience of this wine is a unified conscious event—a representation in the brain. There's a lot that happens to the signals coming from the tongue and nose before they are assembled into this mental construct. Taking things one step further, the way we share our experience of wine is through words. The transition from conscious flavor experience to verbal description is another complex one, but we're going to have to grapple with it if we are to gain a useful insight into the human response to wine tasting. Indeed, it seems that the verbal descriptors we use are instrumental in shaping our conscious representations. The subject of flavor and its perception is the focus of the next four chapters.

We begin by looking at the senses of taste and smell, and how they interact. From this we move to look at the mouth environment, and the role of saliva in wine tasting—a neglected subject but, I feel, and important one. We explore how the brain makes sense of flavor, and then we finish with a discussion of the way that we construct a representation of "wine" in our brains, and the role that language plays in this. It's a complicated story and one that is still in the process of being unraveled—but also a thoroughly interesting one.

TASTE AND SMELL

Linda Bartoshuk, Yale university professor and respected authority on the science of taste, is giving a lecture. Midway through, she interrupts her presentation and begins to hand out strips of blotting paper that have been soaked in a solution of propylthiouracil (a thyroid medication, known more simply as PROP). The audience is surprised because science lectures aren't usually this interactive. Each person is told to place the paper on his or her tongue, and the result is surprising. One-quarter can taste nothing at all. Of the others, most find the paper to taste quite bitter, and a sizeable minority experiences an intense bitterness that is extremely unpleasant. What Bartoshuk is illustrating is the now thoroughly documented individual variation in the ability to taste bitter compounds. Her research, building on an accidental discovery that took place in the 1930s, has shown that people can be separated into three different groups according to their ability to taste PROP. Of the total population, one-quarter are PROP nontasters, half are medium tasters, and the remaining one-quarter is supertasters, or hypertasters. The latter group is exquisitely sensitive to PROP and certain other bitter compounds. This hard-wired difference is thought to be genetic. Anatomically, Bartoshuk has shown that supertasters have an extremely high density of taste papillae—the structures that house the taste buds—on their tongues, with nontasters having relatively few.

This means that individuals in each group are living in different "taste worlds" than the others. Although the main difference between these populations relates to bitter-tasting ability, the taste differences also extend (albeit less dramatically) to other flavor sensations. According to Bartoshuk, "Supertasters perceive all tastes as more intense than do medium tasters and nontasters." As you'd imagine, these research findings could have significant implications for the way we approach wine. Bartoshuk certainly thinks so: "It is important for winemakers to test their wines on all three groups." She adds, "It would be interesting to see if we could find systematic differences in preferences for specific wine types across the three populations."

But this work on hypertasters and nontasters is not without criticism. Is this simply a case of a specific aguesia (an inability to detect a specific chemical by the tongue) rather than being more generally applicable? And what is the connection between the mechanism that allows an individual to taste these bitter compounds and the density of taste buds on the tongue?

Then there is the phenomenon known as "thermal taste," which is the ability to taste sweetness from thermal stimulation alone, and it has been associated with higher responsiveness to four "prototypical" taste stimuli but, interestingly, not to PROP. Could it be that there is a level of hyper- or supertasting that is unrelated to PROP-taster status, and that PROP is just part of a bigger story? A study by Juyun Lim and colleagues at Oregon State University examined this, comparing individual differences in perception of four prototypical taste stimuli (sucrose, NaCl, citric acid, and quinine), as well as PROP. They also looked at creaminess ratings of three milk products that had different fat contents, in order to look at the relationship between taste and oral touch. They found that intensity ratings across two trials were significantly correlated for all five of the taste stimuli, and that averaging across replicates led to significant correlations among the four prototypical stimuli. In contrast with these findings, the bitterness of PROP was correlated only with the bitterness of quinine. Interestingly, none of the four taste stimuli, nor PROP, were significantly correlated with ratings of creaminess. The authors postulate that rather than just using PROP-taster status, a combination of intensity ratings of as few as two of these prototypical taste stimuli could reveal individual differences in overall taste perception.

In agreement with these observations on individual differences are the experiences of companies that use large consumer sensory panels. Jane Robichaud, who at the time I interviewed her worked for U.S. company Tragon, used these panels to help producers tune their wines to consumer preferences, by uncovering the drivers of these preferences. As such, Robichaud has lots of experience assessing people's sensory perceptions of wine. Robichaud, a trained winemaker as well as a sensory scientist, was one of the authors of the celebrated wine aroma wheel, and worked for large Californian wine producer Beringer before moving to Tragon. Her work there involved using a process known as product optimization, which consists of quantitative consumer sensory research that then provides winemakers with practical information that can help them make "consumer-defined" wines. They recruit members of the public and put them through a two-day sensory-analysis boot camp. "They don't need to be connoisseurs, but they need to be regular consumers who are good at group dynamics," says Robichaud. Typically, 70% of these people will pass and 30% will have to drop out. The job of this trained panel is to come up with the descriptive language that will be used to describe the wines being examined.

The second stage is known as the optimization phase, and for this typically, 150 to 200 or more of the "target" consumers are recruited. "Interestingly, we find that people are wired quite differently such that they like different things," says Robichaud. "Taking coffee as an example, some people like it strong, dark, and rich, some like it medium and coffee-ish, and some like brown water." Robichaud explains that it is possible to find discrete "preference segments" (defined as groups of consumers exhibiting similar and distinct likes for specific combinations of attributes).

"About 30% of the population are not good measuring devices," concludes Robichaud. This seems to tally well with the work on PROP sensitivities, but Robichaud doesn't think that PROP status is all that useful a measure. "We did a bit of PROP testing at Beringer and it didn't work very well. It bore no relation to who was a good bitterness taster." The problem here seems to be that there are so many chemical compounds that elicit bitter tastes, with quite different structures, and PROP is just one of them. However, Robichaud does think that about one-third of people don't seem to get bitterness in wines at all.

There are important implications of observations such as these for wine tasting. It seems that we may be living in rather different worlds as wine tasters, and that however much we try to iron out differences in preference or variation in critics' appraisals, some of this could have its basis in biology. All this is, of course, to do with different perceptions relying on the data-gathering apparatus of the taste buds and olfactory receptors.

U.S. master of wine Tim Hanni has taken work on individual differences in taste perception and fine-tuned it for work with retailers and wineries. One of the challenges for large wine retailers selling wine to normal people (as opposed to geeks, who know what they want) is that the wine offering is usually a confusing wall of wines. How do retailers guide customers to wines they will like? Most people are afraid of making a mistake with their wine purchases, and so stick to the safe ground of the familiar. Many simply pick up what's on special offer. Hanni's idea is to use peoples' taste preferences—their biological taste worlds—to guide them to wines that match. Working together with Cornell University researcher Virginia Utermohlen, Hanni surveyed some 1,600 consumers, and was able to cluster consumers into one of four categories: hypersensitive, sensitive, tolerant, and sweet. Hanni has also developed a self-assessment protocol, which allows people to test themselves to see which group they fall into (see www.myvinotype.com).

In early 2010 Bibendum, an UK wine agent and distributor, started working with Tim Hanni to see how it could use this research to help real consumers discover new wines. Bibendum was interested in finding ways of communicating wine that focused on flavor rather than grape varieties, countries, and soil types. Bibendum has since developed its own taste test (www.tastetest.co.uk). This has also been adapted for use by UK supermarket Morrisons. (I just took it and found out my taste profile is "6 Fresh.")

If we are going to interpret how these results relate to wine tasting, we'll need a grasp of the basic science involved. What we commonly think of as "taste" or "flavor" is actually a complex mix of four different sensory inputs: taste, smell, touch, and vision (and possibly also hearing, bizarre as it sounds). Strictly speaking, the sense of taste involves just the inputs from specialized taste buds on the tongue. We can perceive just five different tastes: sweet, salty, bitter, sour, and a fifth taste known as "umami" (a Japanese word that loosely translates as "meaty" or "savory" and refers to the taste of amino acids, the chemical building blocks of proteins such as glutamate). The receptors for each of these different tastes are spread more or less evenly across the tongue. This may come as a surprise to those familiar with the tongue map beloved of school biology textbooks, which shows sweet, salty, bitter, and sour flavors to be located in different regions. But taste provides us with relatively little information compared with the sense of smell, known as "olfaction" in the trade. While there are just five basic tastes, we can discriminate among many thousands of volatile compounds ("odorants"). Indeed, much of the character and interest in wine stems from the complex odors detected by the olfactory system because our taste buds alone give limited detail. So, how does olfaction work? Our olfactory epithelium, located in the top of our nasal cavity, contains olfactory receptor cells, each of which express just one type of olfactory receptor. Each of these receptors—and there are hundreds of them in humans—is tuned to recognize the particular molecular structure of different odorants.

In fact, the precise details of how we "smell" are, at a molecular level, still mysterious. The two competing theories are proposing rather different mechanisms. The orthodox view is that olfactory receptors recognize shapes of molecules. This seems logical, and fits with many results, but there are anomalies that have led others to propose a very different mechanism. For example, two molecules of almost exactly the same molecular shape can smell rather different. A researcher called Luca Turin, who is the thorn in the olfaction research community's side, devised the alternative theory of smell. His theory is that the olfactory receptors detect not the shape but the vibrational properties of odorant molecules.

So where does the sense of touch come in? As well as detecting the texture of foods (very important to their palatability) touch is used by the brain to localize flavors perceptually. When you put a piece of steak in your mouth, the input from both the taste buds and the olfactory epithelium is combined in the brain such that you think this information is coming from where you can feel the steak to be in your mouth. Likewise, take a swig of wine and the taste sensation appears to come from the whole mouth, not just where the taste buds are found.

Bartoshuk emphasizes that it is important to distinguish between "retronasal" and "orthonasal" olfaction. Orthonasal olfaction refers to what we typically call smell. When we sniff an odor, it moves through the nostrils into the nasal cavity, where it is detected by the olfactory receptors. In contrast, retronasal olfaction occurs when we chew and swallow food, or slurp a wine. Odors are forced behind the palate and into the nasal cavity by a back-door route. "We think that the two forms of input are even analyzed in different parts of the brain," says Bartoshuk. "We have evidence that taste plays an important role in telling the brain that the odor is coming from the mouth and should be treated as a flavor." She has found that in patients with taste damage, flavors are often diminished. Work in her lab involving the anesthesia of taste shows that the intensity of taste also plays a role in the intensity of retronasal olfaction. The implications? "If this is so, then supertasters with their more intense taste sensations may also experience more intense retronasal olfaction." Taste and smell therefore overlap.

Because olfaction is a key element of wine tasting, do individuals differ in their ability to smell? We've seen that there is good evidence that people can be separated into different groups according to their ability to taste; do we also we live in different "smell worlds"? This question is harder to answer because of the increased complexity of olfaction, but the answer seems to be a qualified yes, although to a much-reduced degree than is true for taste.

This is where we need to appreciate the significant role of the brain in processing the information detected by our senses. So far, we have just been looking at the way the tongue and olfactory epithelium detect the chemical environment they are exposed to. For us to use this information, the brain has to interpret it and pull out the useful parts from the middle of all the noise, a function known as "higher-order" processing. It's a complex field of psychology, and one where experiments that provide firm answers are rare. For now, it's sufficient to note that the brain does rather a lot to the information that it receives from the tongue and nose. The role of learning is key here, and we take particular notice of what we have learned to be relevant information, and ignore what we think is unimportant. This is discussed in greater detail in the next chapter, on wine and the brain.

An example of the latter is habituation. Repeated or constant exposure to an odor reduces peoples' ability to detect it. Dr. Charles Wysocki, an expert in olfaction from the Monell Chemical Senses Center in Philadelphia, thinks that this could relate to the performance of professional wine tasters. "If individuals are constantly exposed over a lengthy session, they become less sensitive to odorants that repeat themselves, such as oak." This overlaps with a phenomenon known as sensory-specific satiety, discussed in the next chapter, which is the brain's way of telling us that we have had enough of something, and causes us to find a particular flavor or smell less appealing in a very specific manner.

It is clear that people differ in their sensitivity to different odors. Wysocki comments, "If a large enough sample of people is tested—say 20, the range in sensitivity to a single odorant can be 10,000-fold on a single day." Others put this figure a little lower. Dr. David Laing, from the University of Western Sydney, suggests that in a sample of 100 people, "You could expect a variation of about 100 times between the most and least sensitive persons." Still, it is a significant difference, and enough to explain the reported differences in perception of the cork-taint compound trichloroanisole. But Laing adds that, as with all things biological, "The natural distribution of sensitivities means that many of us differ in sensitivity by only a few times, for example, fewer than ten." Another olfaction researcher, Professor

Tim Jacob from Cardiff University, supports this idea. "It is possible that we do each have different smell universes, but it is remarkable that we agree about smells to the degree we do."

A more extreme variation is where individuals are completely unable to detect certain odors, a condition known as specific anosmia. An example of this familiar to diabetes doctors is the ability to smell ketones, found in the breath of patients with poorly controlled diabetes. This ability is an all-or-none phenomenon, with about one-quarter of doctors failing to detect this smell. Anosmias such as this (and it is not clear how many there are, or whether any relate to odors commonly found in wine) are usually genetic in origin. Fascinatingly, though, Wysocki points out that environmental exposures to certain odors can influence gene expression, turning on receptors in the olfactory epithelium. "Some people who cannot smell androstenone [a pig pheromone found in pork meat] can be induced to perceive its odor by repeated, short exposures to the odorant over a few weeks." The implication for wine-tasting ability here would be that we could learn to detect new smells that we have been previously unaware of. Again, it is not clear how widespread this is.

Our noses are relatively temperamental performers. According to Jacob, women have a heightened sense of smell at ovulation. Appetite will also stimulate the ability to smell, making us more perceptive when we're hungry. "There are centrifugal neuronal pathways leading from the brain to the olfactory bulb which modulate odor perception," says Jacob. "These act as a sort of gate, allowing more or less information through." Jacob also suspects that humidity affects the perception of smell, and in his research has noticed as yet unquantified seasonal and weather-associated differences. Intriguingly some odors can also counteract others, such that small quantities can cancel the smell experience of another, unrelated odor. Age also modulates the senses of taste and smell, although in different ways. There is a clear loss of ability to smell with age, and while there is a much smaller loss in taste ability over a lifetime, it affects men and women differently. Males show a steady decline in the ability to taste bitter substances, whereas women show a sharp decline in this ability during menopause.

So the picture emerging is a complex one. We see that there are significant individual differences in tasting ability, with three distinct populations each living in different "taste worlds." We also see that there are complex and less clear-cut individual differences in the sense of smell, with the two senses of taste and smell overlapping to a certain degree. But how do these rather surprising results relate to wine? Would supertasters make the best wine tasters? "No," says Bartoshuk. "There is too much learning involved. Much of the skill of a wine expert comes from learning the odor complexes produced in wine." She maintains that, "We know that learning plays a very important role in the naming of odors." Jacob agrees that learning is crucial, stating that, "The inexperienced person does not have a smell vocabulary. This hugely restricts the ability to describe and define odors." Even for wine experts, a common problem is the impoverished language we have for describing tastes and smells. In Jacob's opinion, "A large part of the wine taster's skill comes from being able to develop some sort of classification system and then to associate words/categories with smells."

One researcher has addressed this directly. Gary Pickering, a professor of oenology at Brock University in Ontario, Canada, has begun studying whether PROP-tasting status has any effect on wine perception and appreciation. The initial findings are illuminating. "We've just shown in our lab for

Above Wines ready to be judged at the Royal Melbourne Wine Show. Competitions and shows are a common feature of the wine landscape, and rely on the ability of skilled judges to taste and rate numerous wines without knowing their identity.

the first time that PROP supertasters and tasters perceive the acidity, bitterness, and astringency of red wines more intensely than nontasters," he explains. "Also, these differences appear to be moderated by the red-wine style being evaluated."

So, the key question. If supertasters live in a world of enhanced flavor sensations, are they better wine tasters? And are nontasters disadvantaged? Perhaps a significant observation here is that women are more likely than men to be supertasters. In the U.S., where this has been studied, about 35% of females are supertasters, but just 15% of men. Does this mean that women are at an advantage when it comes to wine tasting?

Are you a supertaster? Test yourself

The way scientists assess taster status is by using a piece of blotting paper soaked in 6-propylthiouracil (PROP). But PROP is a prescription-only drug, so it's unlikely you'll be able to get hold of any. Unfortunately, scientists with access to PROP won't be able to supply PROP papers to interested parties because they'd need ethical clearance from their institution's review board. The good news is that there is a simple method you can use at home. It's a little messier, and not as dramatic, but it kind of works.

What you'll need
blue food coloring
piece of paper with a hole punched in it, about ¼ inch in diameter
hand lens or magnifying glass

Method
Swab some blood food coloring onto the tip of your tongue. Your tongue will take up the dye, but the fungiform papillae, which are small round structures, will stay pink. Place the piece of paper on the front portion of your tongue, lift it off, and then count how many pink dots there are inside the circle with the aid of a magnifying glass. I've tried it, and it works because the result correlates with my PROP taster status.

Results
fewer than 15 papillae: nontaster
15–35 papillae: taster
more than 35 papillae: supertaster

Not according to Pickering. It might even be the opposite. "I would speculate that supertasters probably enjoy wine less than the rest of us. They experience astringency, acidity, bitterness, and heat (from alcohol) more intensely, and this combination may make wine—or some wine styles—relatively unappealing," he explains.

Finally, a humbling thought for those of us who evaluate wine professionally. Judged by our mammalian peers, we humans have a pretty poor sense of smell. Our olfactory epithelium covers just one-fifth of the area of that in cats. And dogs can distinguish between the smells of clothing worn by nonidentical twins. In fact, for most mammals the smell world is just as vivid and important to them as the visual world. Think again of the example of taking the dog for a walk and observing how keen that animal's sense of smell is. And as for discriminating among a number of aromas in a complex mixture, such as wine … well, we're just not very good at it. According to Laing, "Humans can only identify up to a maximum of four odors in a mixture, regardless of whether the odors are single molecules (e.g. ethanol) or a complex one (e.g. smoke)." Worth bearing in mind next time you are tempted to write a flowery tasting note?

Specific anosmias

Rotundone is the name given to a rather interesting wine flavor compound. Discovered by the scientists at the Australian Wine Research Institute in 2007, it is the molecule responsible for the "black pepper" aroma in some wines made with the Shiraz/Syrah grape variety. Technically speaking, rotundone is a bicyclic sequiterpene and, as well as making wine taste peppery, it is also responsible for similar aromas and flavors in herbs and spices, including peppercorns. It's also incredibly powerful, being detectable at tiny concentrations.

The reason I mention it here is that one of its most remarkable properties is that one-fifth of people can't smell it at all. Think of the implications of this. The AWRI study showed that while most people detect rotundone at the miniscule concentration of 8 ng/liter in water, 20% of the panelists they used for sensory analysis failed to detect it at 4,000 ng/liter. This sort of smell blindness is known as a specific anosmia.

While there are many such anosmias known, it wasn't until 2007 that the first study was published linking an anosmia with a specific mutation in an olfactory receptor gene in humans . This study began with the precept that people's perceptual variations of certain odors, as well as their assessments of the pleasantness or otherwise of these odors, differs markedly. As an example, different individuals variously perceive a steroid called androstenone as offensive ("sweaty, urinous"), pleasant ("sweet, floral"), or simply as odorless. The authors combined a psychophysical study of a large group of people (i.e. one in which they look at the perception of a range of odors) with a genetic study in which they looked at how these individuals differed in the expression of the genes encoding 335 putative human olfactory receptors. They demonstrated that a human odorant receptor, OR7D4, is activated in test-tube experiments by androstenone, and then showed there were genetic differences among humans in this receptor, which could account for their different responses to this odorant. It is likely that similar genetic differences also occur for other odorant receptors.

It is not surprising that anosmias should exist. While we each have a suite of a few thousand different olfactory receptors (which are transmembrane proteins; each olfactory epithelial cell in the nasal cavity has just one type of olfactory receptor), we can discriminate many more different aroma molecules (known as odorants).

20 Wine and the brain

The central focus of any philosophical study of wine is the perceptual event that occurs when we "sense" the wine that is in our glass by sniffing it, or putting it in our mouths, or a combination of these two processes. The goal of this chapter is to explore the nature of this perceptual event from a biological perspective. That is not to say that (largely reductionist) science is the only legitimate way of framing questions about taste. There are limits to the types of questions that the scientific method can address. But what biology has to say about the perception of wine is of great use, because its insights can usefully constrain our thinking. I've entitled this chapter "Wine and the brain" because here I am making the assumption that the perceptual event of wine "tasting" is an event that is occurring in the brain, and that it is one and the same as the electrical communication between neurons occurring here as we process the signals generated by our sensory apparatus when we encounter wine. To biologists, this suggestion—that conscious events are explicable in terms of neuronal activity in the brain—is uncontroversial. If they are aware of the mind-body problem at all, biologists frequently assume it has been solved. But I'm not a scientific fundamentalist, and I recognize that the reductionistic language of neurobiology is just one way of approaching the complex subject of conscious awareness, and that it doesn't necessarily exclude other descriptive approaches.

I need also to add a word on terminology. Although the term "wine tasting" is technically inaccurate (we are referring to the combination of several different sensory inputs here, chiefly smell, taste, touch, and vision), I'm going to continue using it simply because there isn't a better description for the practice of assessing and comparing wines. And while "smell" stands largely as a sense on its own (you can smell a wine without tasting it, although there will be some visual input), taste doesn't because you can't taste a wine without smelling it. Much of the sensory information we receive when wine is in our mouths comes from the senses of olfaction and touch, which can't therefore be dissociated from the sensations coming from the taste buds. Thus I prefer to use the term "flavor" to describe this multiple sensing of wine in the mouth that results in a seamless, unified perception of wine.

A NEW THEORETICAL FRAMEWORK FOR WINE TASTING?

While many of these issues are as yet unanswered, what we already know of how the brain processes flavor suggests that we should revise the theoretical framework we use to approach wine tasting. I argue that the way rating or scoring wine is currently practiced is based on a false premise: that when critics rate a wine, they are assessing the wine, and that any score thus produced is a property of this wine. This is incorrect. We need to make the subtle yet important paradigm shift of seeing a critic's assessment as a rating of that critic's perception of the wine. The critics are actually describing a conscious representation of their interaction with the wine, and therefore the score or rating is a property of that interaction, and not of the wine itself.

TASTING WINE

Uncork a good bottle, pour yourself a glass, and then take a sip. We call this process wine "tasting," but this is actually a fairly misleading term. In this interaction between you and the wine, the impression you form is a conscious experience that involves the fusion of inputs from at least four different senses, coupled with some sophisticated brain processing. It's a unified "representation" that can't easily be dissected into its component sensory inputs, which we commonly try to do with taste and smell.

INTRODUCING THE CHEMICAL SENSES

In a famous article, philosopher Thomas Nagel posed the question of what it might be like to be a bat. The world would seem very different to an entity relying on a sophisticated sense of touch, which operates like sonar, for spatial navigation. Along similar lines, any readers who have ever walked a dog will be aware that for canines, the world of smell is as important and dynamic as the visual world we are so familiar with. In contrast, we humans live in a vastly diminished smell world, having traded our olfactory acuity for enhanced color vision—an evolutionary change that took place some time during primate evolution. Many of the genes encoding olfactory receptors in our genome are now redundant, that is, the DNA coding for these receptors is still there, but it no longer makes functional proteins. Rather than being important for orientation, social organization, and mate choice, as it is for many mammals, our use of the chemical senses, taste and smell, is largely restricted to food choice, a reduced but still vital role.

As discussed in the previous chapter, taste and smell begin with receptors in the mouth and nasal cavity that turn chemical information into electrical signals, which can then be processed by the brain. But it's wrong to think of our sensory systems as complicated measuring instruments that give a read-out of the taste and aroma molecules they encounter. Instead, what the brain does is to model the world around us. Our sensory systems are bombarded constantly by a mass of information, which, if attended to uniformly, would swamp our perceptive and decision-making processes. So the brain is able to extract from this sea of data just those features that are most relevant. This is done by a procedure known as higher-order processing.

We often think that our sensory system is revealing to us the world around us in an accurate and complete way. But in reality, what we experience is an edited version of reality based on the information most relevant to our survival and functioning. For almost all purposes it does no harm for us to think of the world around as revealed to us to be "reality." In fact, life would become extremely complicated if we operated any other way. But for this discussion it's useful to realize that the version of reality we experience is an edited and partial one. (And quite a personal one, too.)

This can be illustrated in a number of ways. Think again about your household pets, if you have any. As I mentioned earlier, dogs live in a smell world that is almost completely closed to us, and which is just as vivid to them as the visual world is to us. Rats and mice, like many small mammals, get almost all the information they need about their environment from a combination of sniffing and using their whiskers. They are nocturnal, so vision isn't all that useful at night. Now switch on your radio or television, or take a call on your cellphone. It's clear that the air is full of information that we can't access unless we have a device to decode it. Third, take a look at the visual illusion in the figure, known as the café wall illusion. It's just one of many tricks that "fool" the visual system. These sorts of illusions give us clues about the sort of higher-order processing that is taking place, and demonstrate that what we "see" is not always what is there.

The higher-order processing in the human visual system is the best understood of any of the senses. Scientists have worked out how visual processing extracts features of the environment most likely to be relevant. For instance, our peripheral vision is sensitive to motion. Moving objects immediately stand out, because neurons are tuned to respond to them. This ability to detect motion is much stronger in the periphery than it is in the central visual field. Faces are also likely to be significant cues, so our visual systems have special brain mechanisms for face processing. This is the reason why so many advertisements and magazine covers rely on human faces, even where the face isn't particularly relevant to the publication.

Although it is less well-studied, this sort of higher-order processing is also important in flavor

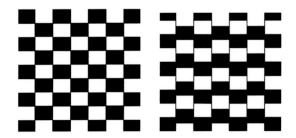

Above The café wall illusion. The lines are parallel, but that is not what we "see."

detection. We are bombarded with chemical stimuli all the time, and the brain has to filter this information so that only the important parts get through. It seems that much of the brain is dedicated to producing a suitably edited view of reality, just as the staff in a newsroom work hard all day sifting through the output of their journalists to produce a 15-minute news bulletin for broadcast that evening.

GATHERING DATA

So we come to the key question. How is it that electrical currents from nerve cells are translated into a unified conscious experience in the brain? Science is a long way from being able to address this complex question directly. But a relatively new technique, functional magnetic resonance imaging (fMRI), has transformed brain research in recent years by allowing researchers to visualize the brain in action. During an MRI scan, a subject is placed inside a large cylindrical magnet and exposed to a massive magnetic field. A sophisticated detection device then creates three-dimensional images of tissues and organs from the signals produced. fMRI is a variation on this theme, where the technique is used specifically to measure changes of blood flow in the brain. When a group of brain cells becomes more active, they need more blood, and this generates a signal in the scan. Although there was some initial controversy about whether a direct correlation exists between the blood flow detected in an fMRI scan and actual brain activity, the consensus in the field is that this is indeed the case. (This raises a more complex question, alluded to above, and beyond the scope of this discussion: can our conscious thoughts be reduced to an electrical signal in the brain? Is the brain activity viewed by MRI the same as the conscious thought we are having? Most biologists would assume this; few would have ever questioned the implications of this assumption, or realized that it was an assumption.) The power of fMRI is that it can show how we use our brains when, for example, we think of chocolate or move our middle finger; the limitation is that to be able to detect these signals reliably, a subject must lie inside a large metal cylinder and keep his or her head completely still.

Because of the practical and experimental difficulty of these types of studies, it's an area where there's still a lot of uncertainty. However, even the limited data obtained so far is highly relevant for wine tasting and is important if we want to provide a robust theoretical basis for the human interaction with wine.

Let's pick up on the theme of flavor-processing in the brain. The senses of taste and smell work together to perform two important tasks: identifying nutritious foods and drinks, and to protect us from eating things that are bad for us. The brain achieves this by linking food that we need with a reward stimulus—it smells or tastes "good"—and making bad or unneeded foods aversive. To do this, flavor perception needs to be connected with the processing of memory (we remember which foods are good and those which have made us sick) and emotions (we have a strong desire for food when we are hungry that then motivates us to seek out a decent meal). Because seeking food is a potentially costly and bothersome process, we need a strong incentive to do it.

Hunger and appetite are thus powerful physical drives. They are also finely tuned. It is striking that we are able to eat what we need and not a lot more or less. Even a slight imbalance, over decades, would result in gross obesity or starvation. Taste begins on the tongue, where we have some 5,000 specialized structures called taste buds, each containing 50–100 sensory cells responding to one of five different primary tastes. These sensory cells convert this chemical information into electrical signals, which then pass through to the primary taste cortex in the brain. This is located in a region called the insula. Taste provides us with relatively little information compared with the sense of smell, more commonly referred to in scientific texts as "olfaction." Whereas there are just five basic tastes, we can discriminate among many thousands of volatile compounds ("odorants"). Our olfactory epithelium, located in the top of the nasal cavity, contains olfactory receptor cells, each of which express just one type of olfactory receptor. This information is also turned into electrical signals by these receptor cells, which is then conveyed to the olfactory cortex via a structure known as the olfactory bulb.

At this stage, with the information that exists at the level of the primary taste and smell areas of the brain, it is likely that all that is coded is the identity and intensity of the stimulus. But what the brain does next is this clever higher-order processing mentioned above—it extracts the useful information from this mass of data, and begins to make sense of it. This is where we turn to the work of Edmund Rolls, until recently a professor of experimental psychology at the University of Oxford, who has studied a region of the brain called the orbitofrontal cortex. fMRI is one of the tools used in his work.

The work of Rolls and others has shown that it is in the orbitofrontal cortex that taste and smell are brought together to form the sensation of flavor. Information from other senses, such as touch and vision, is also combined at this level to create a complex, unified sensation that is then localized to the mouth by the sense of touch. After all, this is where any response to the food or drink, such as swallowing it or spitting it out, will need to take place. Rolls has also demonstrated that the orbitofrontal cortex is where the reward value (the "niceness," known more grandly as "hedonic valence") of taste and smell is represented. That's another way of saying this is where the brain decides whether what we have in our mouths is delicious, dull, or disgusting. Another fMRI study has shown that the brain uses two dimensions to analyze smells: intensity and hedonic

valence. The amygdala responds to intensity while the orbitofrontal cortex is the region that decides whether the smell is good or bad.

CROSS-MODAL PROCESSING

Some nerve cells in this brain region respond to combinations of senses, such as taste and sight, or taste and touch, or smell and sight. This convergence of inputs, known as cross-modal processing, is acquired by learning, but it is a process that occurs slowly, typically requiring many pairings of the different sensations before it is fixed. This may be why we often need several experiences with a new food or wine to be able to appreciate it fully. It is also at this level where stimulus–reinforcement association learning takes place. This is the situation where, for example, you are faced with a new food (stimulus) that tastes good, but then it makes you vomit violently (association). Next time you put some of this in your mouth, you immediately spit it out in disgust. It saves you the bother of vomiting again, and is therefore a protective mechanism. However, this aversive mechanism is of somewhat limited power. For example, a student may drink 17 cans of beer on a Friday night, throw up in the street, feel rotten the next morning, but return to drinking beer the next evening.

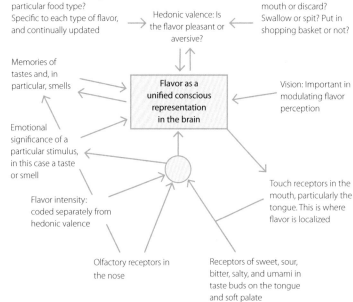

Satiety: Have I had enough of a particular food type? Specific to each type of flavor, and continually updated

Hedonic valence: Is the flavor pleasant or aversive?

Decision making: put in mouth or discard? Swallow or spit? Put in shopping basket or not?

Memories of tastes and, in particular, smells

Flavor as a unified conscious representation in the brain

Vision: Important in modulating flavor perception

Emotional significance of a particular stimulus, in this case a taste or smell

Flavor intensity: coded separately from hedonic valence

Touch receptors in the mouth, particularly the tongue. This is where flavor is localized

Olfactory receptors in the nose

Receptors of sweet, sour, bitter, salty, and umami in taste buds on the tongue and soft palate

Left Representing flavor in the brain. This diagram shows a rather approximate and idealized network of relationships among the various inputs and brain processes that contribute to the unified representation of flavor in the brain. The key point is that flavor is a consequence of some complex processing steps, and there is no simple linear pathway from the primary sensations of taste and smell to our conscious experience of these inputs. The gray circle illustrates the various higher-order processing steps taking place. The position of some of the arrows is open to debate: connections shown here are for the purposes of illustrating concepts and do not represent solid data on these interrelationships.

SENSORY-SPECIFIC SATIETY

One aspect of Rolls' research on the orbitofrontal cortex that has direct relevance to wine tasting is his work on sensory-specific satiety. This is the observation that when enough of a particular food is eaten, its reward value decreases. However, this decrease in pleasantness is greater than for other foods. Putting it more simply, if you like both bananas and chocolate, and then eat loads of bananas until you can't stomach the thought of another banana, you will still hanker after a piece of chocolate. This clever brain trick makes us desire the particular kinds of foods that we need at a given time, and helps us balance our nutritional intake. By fMRI, Rolls has shown that in humans the response in the orbitofrontal cortex to the odor of a food eaten to satiety decreases, but the response to another odor that has not been eaten doesn't change. The subjects' perception of the intensity of the smell of the consumed food doesn't change, but their perception of its pleasantness (hedonic valence) does. In another study, he showed that swallowing is not necessary for sensory-specific satiety to occur. At a large trade tasting it is quite common to taste as many as 100 wines in a session. If these results of sensory-specific satiety are extrapolated to this type of setting, then it's likely that the brain will be processing the last wine you taste differently to that of the first, assuming that there are some components to the taste or smell in common, for example, tannins, fruit, or oak.

TRAINED TASTERS EXPERIENCE WINE DIFFERENTLY

In 2002, researchers from the Functional Neuroimaging Laboratory of the Santa Lucia Foundation in Italy, headed by Dr. Alessandro Castriota Scanderberg, put together a simple yet elegant study addressing the key question of whether trained tasters experience wine differently from novices.

They took seven professional sommeliers and seven other people matched for age and gender but without specific wine-tasting abilities, and monitored their brain responses while they tasted wine. But getting people to taste wine while they are having their brain scanned is no trivial feat. "The experience was pretty uncomfortable," recalls Andrea Sturniolo, one of the sommeliers involved. "I was under a tunnel with four plastic tubes in my mouth, totally immobile". Through these tubes the researchers fed subjects with a series of four liquids, including three different wines, and a glucose solution as a control. Subjects were told to try to identify the wines and form some sort of critical judgment on them. They were also asked to judge when the perception of the wine was strongest—while it was in the mouth ("taste") or immediately after swallowing ("aftertaste"). "The experiment lasted a good 50 minutes," says Sturniolo, "which seemed endless." He added, "Certainly they were not the ideal conditions in which to carry out such a delicate experiment, but as these conditions were identical for all participants, I think the results are reliable."

So what did the scans show? Some brain regions—notably the primary and secondary taste areas, in the insula and orbitofrontal cortex—were activated in both sets of subjects during the "taste" phase. But during this initial period, another area was activated only in the sommeliers, and this was the front part of a region known as the amygdala-hippocampal area. In the "aftertaste" phase the untrained subjects also showed activation of this amygdala–hippocampal area, but only on the right side, while in the sommeliers this zone was activated on both sides. In addition, during the aftertaste the sommeliers exclusively showed further activation in the left dorsolateral prefrontal cortex.

Not surprisingly, given its importance in the processing of flavor, the orbitofrontal cortex is one of the regions activated in the brains of both trained and untrained wine tasters in this study. But what about the other areas that were highlighted specifically in the sommeliers?

First we have the amygdala-hippocampal area. This is a zone that plays a key role in processing motivation (the amygdala) and memory (the hippocampus). According to study leader Dr. Scanderberg, "The finding of an early and consistent activation of the amygdala–hippocampus complex in the sommelier group suggests a greater motivation for the recognition process." This may indicate that the sommeliers were expecting a reward and thus pleasure from the wine-tasting process. The other key area is the left dorsolateral prefrontal cortex, which is a zone involved in the planning and use of cognitive (thinking) strategies. The sommeliers' unique activation here is consistent with the idea that only experienced tasters follow specific analytical strategies when wine is in their mouths. The researchers speculate that these strategies might be of a linguistic kind, associating words with specific flavors. In parallel with fMRI studies on musicians, which have shown that music activates different areas of trained musicians' brains than those of casual listeners, it seems that the sommeliers are experiencing something different to the average person when they taste wine. "There is clear evidence that the neural connections of the brain change with training and experience," says Dr. Scanderberg. He explains that, "There are two apparently contradictory ways that the brain adjusts its structural network in parallel with the increasing expertise of the subject." The first, and most common, is to assign a specific function to a smaller cluster of cells higher up in the brain's hierarchy. For example, during the rehabilitation of stroke patients it is common to see a particular task activate a much smaller, but higher region in the brain at the end of rehabilitation than it did at the beginning. The second strategy is to recruit more brain areas to help with a complex task. Experienced wine tasters seem to follow this second strategy, pulling in new brain areas to help with the analysis of sensory stimuli.

The implications for wine tasting are clear. I'm assuming here that, as a reader of this book, you may well be someone who has consumed a fair bit of wine over a number of years. Do you remember one of the wines that first really appealed to you? If you were to go back in time now and taste that wine again, yet with your current wine drinking history, then you'd actually perceive something quite different as you sipped that wine the second time around. Drinking all that wine has changed your brain, and we aren't talking alcohol-induced neural degeneration. By paying attention while you've been drinking, just like the sommeliers in this study, your response to wine differs from that of untrained

subjects. This also underlines the importance of the learning component in wine appreciation. People versed in one culture of wine may need to relearn about wine when exploring another. Even if you have years of expertise in Australian reds, for example, you may have to start from scratch when trying to appreciate German Riesling.

As with the Italian sommeliers, another study reinforces the idea that knowledge changes perception. This study was carried out by a group of researchers from California working in the new field of neuroeconomics. They used fMRI to show that the information people are given about the wine can change their actual perception of the wine, and how pleasant they find it[1]. The authors discuss an economics term called experienced utility (EU), and describe how marketing efforts frequently attempt to change the EU of a particular good, without changing the nature of the good. The researchers chose to use wine as a test case of how price can modify EU. They fed a group of 20 subjects five different Cabernet Sauvignon wines while they were in an MRI machine. The subjects were told the retail prices of the wines they tasted and were told to focus on the flavor of the wine and say how much they liked them. However, there was a clever twist to this experiment: in reality, only three wines were being presented to the subjects, with two of the wines being presented as different wines at different price points. So, the five different wines the subjects tasted were: $5 wine (wine 1, its real price); $10 wine (wine 2, which was actually a $90 wine); $35 wine (wine 3); $45 wine (wine 1, at a fake price), and $90 wine (wine 2 at its real price).

Unsurprisingly, there was a correlation between price and liking. Significantly, subjects preferred wines 1 and 2 when they were told they were drinking the higher-priced wines. The brain scans, comparing the response of subjects when tasting the same wines but believing them to be differently priced, showed that the parts of the brain that experience pleasure are more active when subjects think the wine is a higher price. The price isn't just affecting perceived quality—it seems to be affecting the actual quality of the wine by changing the nature of the perceptive experience.

1 Plassmann H., O'Doherty J., Shiv B., Rangel A. *Marketing actions can modulate neural representations of experienced pleasantness.* Proc Natl Acad Sci USA 105:1050–1054 2008

The importance of these results are that they show that our expectation as we approach wine—perhaps caused by sight of the label—will actually change the nature of our wine-drinking experience.

WORDS AND WINE: HOW WE FORM REPRESENTATIONS OF THE TASTING EXPERIENCE

Moving fields slightly, Frédéric Brochet, a cognitive psychologist, has done some important work that is highly relevant here. He has studied the practice of wine tasting as carried out by professionals. His claim is that the practice and teaching of tasting rests on a fragile theoretical basis. "Tasting is representing," says Brochet, "and when the brain carries out a 'knowledge' or 'understanding' task, it manipulates representations." In this context a "representation" is a conscious experience constructed by the mind on the basis of a physical experience, in this case, the taste, smell, sight, and mouthfeel of a wine. Brochet uses three methodologies in his work: textual analysis (which looks at the sorts of words that tasters use to verbalize their representations); behavior analysis (inferring cognitive mechanisms from looking at how subjects act); and cerebral function analysis (looking at how the brain responds to wine directly through the use of fMRI).

TEXTUAL ANALYSIS: STUDYING THE WORDS THAT TASTERS USE

Textual analysis involves the statistical study of the words used in a text. Brochet used five data sets, consisting of tasting notes from *Guide Hachette*, Robert Parker, Jacques Dupont, Brochet himself, and notes on eight wines from 44 professionals collected at Vinexpo. Employing textual analysis software called ALCESTE, Brochet studied the way that the different tasters used words to describe their tasting experiences. He summarizes his six key results as follows. (1) The authors' descriptive representations are based on the types of wines and not on the different parts of the tasting. (2) The representations are "prototypical," that is, specific vocabularies are used to describe types of wines, and each vocabulary represents a type of wine. Put another way, when tasters experience a particular wine, the words they use to describe it are those that they link to this sort (or type) of wine. (3) The

range of words used (lexical fields) differs for each author. (4) Tasters possess a specific vocabulary for preferred and nonpreferred wines. No taster seems to be able to put aside their preferences when their representations are described. Brochet adds that this result, the dependence of representations on preferences, is well-known from the fragrance world. (5) Color is a major factor in organizing the classes of descriptive terms used by the tasters, and has a major influence on the sorts of descriptors used. (6) Cultural information is present in the sensorial descriptions. Interestingly, Brochet states that, "Certain descriptive terms referring to cognitive representation probably come from memory or information heard or read by the subject, but neither the tongue or the nose could be the object of the coding."

BEHAVIORAL ANALYSIS: PERCEPTIVE EXPECTATION

In the next set of experiments, Brochet invited 54 subjects to take part in a series of experiments in which they had to describe a real red wine and a real white wine. A few days later the same group had to describe the same white wine and this white wine again that had been colored red with a neutral-tasting food colorant. Interestingly, in both experiments they described the "red" wine using identical terms even though one of them was actually a white wine. Brochet's conclusion was that the perception of taste and smell conformed to color, and that vision is having more of an input in the wine-tasting process than most people would think. Brochet points out a practical application of this observation, which has been known for a long time in the food and fragrance industries. No one sells colorless syrups or perfumes any more.

In a second, equally mischievous experiment, Brochet served the same average-quality wine to people at a week's interval. The variation was that on the first occasion it was packaged and served to people as table wine, and on the second as a *grand cru* wine. So, the subjects thought they tasted a simple wine and then a very special wine, even though it was the same wine both times. He analyzed the terms used in the tasting notes, and it makes telling reading. For the *grand cru* wine versus the table wine, "a lot" replaces "a little"; "complex" replaces "simple";

and "balanced" replaces "unbalanced"—all because of the sight of the label.

Brochet explains the results through a phenomenon called "perceptive expectation," which means subjects perceive what they have pre-perceived, and then they find it difficult to back away from that. For us humans, visual information is much more important than chemosensory information, so we tend to trust vision more. Brochet uses these results to explain Dr. Émile Peynaud's observation that "Blind tasting of great wines is often disappointing."

VARIATION IN REPRESENTATIONS

A further study in this series examined how the qualitative ratings of a series of wines differed among a group of wine tasters. This group of eight tasters was asked to rank 18 wines they tasted blind in order of preference. The results differed widely. With a similar methodology to that employed by the Italian researchers, Brochet then used MRI to assess the brain response of four subjects to a series of wines. One of the most interesting results obtained with this technique was that the same stimulus produced different brain responses in different people. In terms of brain area activated, one was more verbal, another more visual. Also, when a subject tastes the same wine several times, the images of each tasting are somewhat different. Brochet concludes that this demonstrates the "expression of the variable character of the representation." The representation is a "global form, integrating, on equal terms, chemo-sensorial, visual, imaginary, and verbal imagination."

Dr. Charles Spence from the Department of Experimental Psychology at Oxford University has also gathered relevant data on this sort of cross-modal sensory processing. I quizzed him about how his studies might apply to wine. "Here in my lab we do indeed do a number of studies looking at how what people see influences their perception of the flavor identity and intensity," says Spence. "However, unfortunately we can't give people alcohol, because of danger of litigation, so we do most of our studies with colored soft drinks, looking into which colors are particularly effective in modulating flavor perception. It turns out that one of the reasons why red coloring turns out to be such a powerful driver of what we experience

(both in terms of smell and taste) is that redness typically equates with the ripening of fruits in nature." Spence thinks that both semantics and experience are also important. "Expectations or labels about what something might be can play a key role in how you interpret an ambiguous or ambivalent odor," he adds. "Surprisingly, expertise doesn't seem to help with the red-wine color effect. I have seen experts completely fooled, perhaps even more than novices. The thing is that many of these multisensory interactions occur pre-attentively. In other words, given the overload in the amount of sensory stimulation that is constantly bombarding each of our senses, our brains try to help out by binding what we see, hear, taste, etc., automatically, and only giving us awareness of the result of this integration. Hence, attention also fails to impact on many of these cross-modal illusions." Spence also has something to say about the importance of learning, which he thinks plays a very important role. "I have not seen this studied for the case of wine tasting but for other combinations of taste and smell, your previous experience (e.g. in terms of cultural differences in exposure to certain foods, or to certain combinations of tastants and odorants) critically determines how your brain will bind the different sensory cues. The brains of Westerners, for example, are especially geared up to binding sweet tastes in the mouth with almond-like odors, but not salty tastes with almond odor. Go to Japan and the reverse is true, as Japanese people never experience the combination of sugar and almond, but instead get a lot of exposure to almond and salt taste in pickled vegetables and condiments."

CONCLUDING REMARKS

While there's still a lot to be learned about how the brain constructs our experience of wine, it is already clear that this is a complex area that we often try to simplify. It is our attempts to simplify the concepts underlying wine tasting and iron out the very real inter- and intra-individual variation that lead to problems in the interpretations of results from tastings. There is a lot more to the wine experience than just smell and taste. The basic information from these chemical senses is supplemented in a very real way by other input, for example, from vision, touch, and memory.

Added to this, the higher-order integration of all this input is a flexible and complicated processing stage that then forms our unified perception (or "representation") of the tasting experience. The important results of Brochet and others show that factors such as whether or not we are tasting blind make a crucial difference to the nature of this representation, and that representations of the same wine differ quite markedly among tasters. Added to this, the past experiences of tasting will change the nature of our current experiences. Information of this nature should help us in our understanding of the scientific underpinnings of the wine-tasting process, and help in the design of tastings. It is likely that further studies using similar techniques to those described here will give us a greater understanding of the rather complex business of tasting and describing wine.

In closing, I'll offer some tentative ideas that I haven't been able to develop here because of space constraints. First, I think that we already have enough evidence to warrant a paradigm shift with regard to rating wines. What critics are scoring is not some intrinsic property of the liquid in the bottle, but a perceptual representation that is to some degree specific to them. Does this mean that we can't have a shared experience when we taste the same wine? While it's helpful to acknowledge the individual nature of these representations, we also need to bear in mind that one of the remarkable properties of the human mind is its ability to exploit shared space, thanks to language and the development of writing and other recording technologies. The laptop I am writing this manuscript on is effectively acting as an extension of my brain. It gives me the ability to take my thoughts, in word form, and then develop them over an extended period of time. Most importantly, I can then share these thoughts with others, and in turn access extensions of their mental landscape in a similar fashion. With wine tasting, our sharing of experience through a common culture of wine enables a degree of calibration of perceptual representations to occur. In particular, we develop a language for sensory terms—a way to encode and share our representations. The language we use for describing wine is intrinsic not only to sharing those ideas, but also to forming them in the first place. By possessing an extended vocabulary for taste, smell, and flavor sensations, we are able to approach wine tasting in a structured fashion, and in a way that generates a detailed verbal description of the wine being analyzed. It follows that the nature of this vocabulary will shape the description of the experience, and even the experience itself.

The Pepsi challenge and what it tells us about wine tasting

Read Montague, a neuroscientist at Baylor College of Medicine in Texas, devised a fascinating experiment that has implications for wine tasting. It stemmed from a series of TV commercials in the 1970s and 1980s where individuals were subjected to the "Pepsi challenge." In this test Pepsi was pitted against Coke blind, with subjects not knowing which was which. They invariably preferred the taste of Pepsi, but this wasn't reflected in their buying decisions. Montague wanted to know why.

He re-enacted the Pepsi challenge with volunteers, this time scanning their brain activity by MRI machine. On average, Pepsi produced a stronger response in the ventral putamen, a region thought to process reward. In people who preferred Pepsi, the putamen was five times as active when they consumed Pepsi than it was in Coke-preferring subjects drinking Coke.

In a clever twist, Montague repeated the experiments, this time telling subjects what they were drinking. Remarkably, most of them now preferred Coke. The brain activity also changed, with activity in the medial prefrontal cortex, a region that shapes high-level cognitive powers. The subjects were allowing what they knew about Coke—its brand image—to shape their preferences. Remarkable.

The implications for wine tasting are clear. When we don't taste blind, our preferences are liable to be shaped by pre-existing information we have about the wine. Try as hard as we might to be objective, it isn't possible. What we know about wine will mold how we perceive the wine, and will even shape how much we enjoy a particular bottle. This brings another fascinating level of complexity to wine tasting.

21 Saliva, tannin, and mouthfeel

The subject of the inside of our mouths—more formally the oral environment—is one that has been neglected in discussions about the taste of wine. But the mouth is where we interact with wine most closely, at least until it is swallowed.

Most wine drinkers with more than a passing interest in what is in their glass usually sniff it before they taste it, but then any preliminary judgment from the aroma is supplanted by a more concrete judgment made once the wine is in their mouth. The way we experience the wine once we have begun to interact with it in this way is multimodal, in that it involves the senses of taste, smell, and touch. For this reason, saliva is a vital component that mediates how we experience wine.

WHAT IS IN SALIVA?

Saliva is a watery secretion from three different glands: the parotid (in the cheek, under the ear); the submandibular (under the jaw); and the sublingual (under the tongue). Even without any stimulation, there is a low but steady flow of saliva. This unstimulated flow is important to keep the mouth moist, and for protecting the teeth and surfaces inside the oral cavity.

Most saliva, however, is secreted because of stimulation from a taste or smell, or from a mechanical stimulus, such as chewing. An association can even stimulate salivary flow. Think of Pavlov's dogs who, after a while, began slobbering at the ringing of a bell even in the absence of a food stimulus. The average person secretes between 1 and 3 pints of saliva a day, most of which is swallowed. Flow decreases at nighttime (during sleep) to virtually zero, and can be affected by medication or dehydration.

So what is special about the composition of saliva, in addition to its majority component, water? First, it contains proteins, called mucins, which have a high-carbohydrate content, are very slippery, and are capable of absorbing large amounts of water. Mucins aren't unique to the mouth. These mucins form a covering over the oral tissues, which helps lubricate the mouth (useful for talking and for masticating food) and also protects against irritation and unwanted microbes. The ability of mucins to absorb quite a lot of water into their structures helps keep this protective lubricating layer reasonably thick, and in concert with the liquid nature of saliva, clears unwanted microorganisms and food debris from the mouth.

Saliva also contains high concentrations of calcium and phosphate ions, which help protect the dental enamel, and allow for remineralization of the teeth. It also protects the teeth by washing away or diluting any potentially harmful chemicals and buffering against pH changes like those caused by acids in the diet. A covering layer of salivary proteins, called a pellicle, further protects teeth.

All this protection is necessary because the inside of the mouth is an incredibly vulnerable place. It's warm and wet, an ideal condition for harmful microbes to grow in. Teeth are sensitive to acidity, which can destroy them. Saliva is therefore playing an essential protective role, and one that we take for granted until salivary flow is diminished or stops altogether, as occurs in some diseases, with some medications, or after radiotherapy treatments for cancer. If saliva flow is severely diminished, people are forced to use artificial saliva, which must be sprayed into the mouth at regular intervals. However, such artificial saliva is relatively unsophisticated and doesn't carry out all the functions of normal human saliva.

PRPS AND HRPS

The saliva components of most interest to us in this context are two groups of proteins, known as the proline-rich proteins (PRPs) and histatins or histadine-rich proteins (HRPs). The PRPs make up about 70% of salivary proteins overall, and have a high proportion the amino acids proline, glycine, and glutamine. Acidic PRPs. Altogether, the three groups have around 20 different members. Acidic

PRPs bind calcium strongly and are important in both forming the tooth pellicle layer and also making sure there is enough calcium present for tooth remineralization. This is because they bind calcium when it is present at high levels, and then are able to release it gradually during conditions when there is less of it around. Glycosylated PRPs are lubricants and also interact with microbes. In contrast, basic PRPs have only one role, which is to bind to tannins, forming precipitates. HRPs are small proteins, rich in histidine, and only found in saliva. Twelve of them are known in humans, and they account for just 2.5% of saliva protein. They have antibacterial and antifungal properties, but they are also very good at binding tannins. From a wine-tasting perspective, the ability of PRPs and HRPs to bind with tannins is where the real interest comes in.

Plants form tannins as defense molecules, both for defending against microbial attack and, because plants are extremely vulnerable to being eaten, also acting as antifeedants (a chemical that causes a pest, such as an insect, to stop eating). Plants are literally rooted, and so they have to make themselves unpalatable, acting as chemical factories to produce a wide range of toxic defensive secondary metabolites, as well as developing physical defenses, such as thorns and stings. It is only a relatively restricted set of plant species that are suitable for human consumption. In many cases, the only part of the plant we are able to consume is the part that it wants us (or other animals) to, such as the fruits it produced, in order to assist in seed dispersal. In the case of grapes, they are camouflaged green and made unpalatable by high tannins, high acidity, and no sugar until the seeds are mature enough to be dispersed, at which point the ripening process has made them attractive to eat, sweet, and easy to find.

"Tannin" is a blanket term for a group of rather different polymers that historically were used to convert animal skins into leather in the process of tanning. Tannins in wine are found in a variety of states: they are intrinsically "sticky" molecules and join up with other wine components, such as anthocyanins (the colored pigments found in grape skin), to form pigmented polymers, or can combine with other chemicals. Tannin size is referred to by the term DP (degree of polymerization), and this indicates the number of tannin subunits that are joined together. Seed and wood tannins are typically smaller than skin tannins, and it is thought that these smaller tannins possess a bitter taste, rather than express astringency. A chief goal of the process of *élevage* is to manage tannins in such a way by means of limited oxygen exposure to produce a wine that is harmonious. It can be speculated that the ability of tannins to undergo modification in the presence of oxygen is because this is one of their roles in plant defense. For example, when you bite into an apple and leave it, after a while the "wound" area becomes brown, with the tannins in the tissue interacting with oxygen to make the damaged area less inviting for colonizing microbes.

One of the key roles of salivary PRPs and HRPs is to protect us from the harmful effects of tannins by binding to them and precipitating them before they reach the gut. If the salivary PRPs didn't cause this precipitation, the tannins would interact with digestive enzymes (which are also proteins) in our gut and render them ineffective. This would reduce the palatability of plant components by making them much less digestible. The aversive taste of unripe fruits is in part due to high tannin concentrations, with the plant using this as a way of keeping the fruit from being consumed before the seeds are ready for dispersal, along with color changes and high acid/low sugar. We find the bitter taste and astringent sensation of tannin aversive and, as with such unpleasant oral sensations, the aversion can protect us from harmful consumption. Thus the PRPs and HRPs are potentially filling two roles. They allow us to detect tannins in food and to reject the food if the concentrations might be dangerous, and also help neutralize any tannins present in food to be ingested.

THE SENSATION OF ASTRINGENCY

We sense tannins in wine largely as an astringent sensation, but there is also a contribution from taste in some circumstances. Astringency is not principally a taste, in the sense that it is not one of the primary taste modalities of sweet, sour, bitter, salty, and umami. Instead, it is chiefly detected by the sense of touch in our mouths (there is still some discussion in the scientific literature about whether or not astringency is tasted). The feeling of the wine in the mouth, referred to commonly as mouthfeel, is detected by mechanoreceptors in the mouth. Anyone who has had dental work that has modified the inside of their mouths in some way will know that we are very sensitive to changes in the mouth environment. Interestingly, one of the things the sense of touch does is to localize tastes and smells to where the food or drink is located within the mouth. This makes us ascribe the properties of taste and smell to the food/drink, even though they might be sensed elsewhere in the mouth. It is useful if we need to remove the food from our mouths when it tastes bad. Proteins present in saliva bind dietary tannins entering the mouth and form precipitates. These proteins include the PRPs and HRPs, whose role is to carry out this binding and protect us from the potentially harmful effects of tannins in inhibiting digestive enzymes. But it also includes one of the other important protein types in saliva, the mucins. As mentioned earlier, these are involved in forming a lubricated, slippery protective layer over the internal surface of the mouth. Tannins remove this lubrication, causing a sense of dryness, puckering, and loss of lubrication in the mouth. This is what we describe as "astringent."

Related to astringency is the taste of bitterness. The majority of tannins are chiefly sensed as astringent, but they can also be tasted as "bitter" when they are small enough to interact with bitter tongue receptors. Tannins seem to reach their most bitter taste at a DP of 4 (four subunits joined together, which is small), and then decrease in bitterness and increase in astringency, with this astringency peaking at a DP of 7 (according to some studies, at least), before becoming less astringent as they become larger. The astringent nature of tannins can be moderated by the presence of polysaccharides (sugars) or other wine components. It is also modified by the chemical adornments that tannins can grab, and there are many of these. In wine, tannins are continually changing their length (DP) and adding things to their structure. So, structurally, wine tannins can be incredibly complicated, and researchers are still trying to correlate mouthfeel properties with structure.

Interestingly, tannins are more astringent with lower pH (so, wines with higher acidity taste more astringent, even with the same tannin content) and less astringent with increasing alcohol. However, the bitterness of tannins rises with alcohol level, and is unchanged by pH changes.

It is worth noting that acid is known to stimulate salivary flow. If salivary flow is increased, then there is more protein present in the saliva to form precipitates with tannins. The implication of this is that two red wines with identical tannin composition but different pH will have a different mouthfeel. This could be part of the explanation for the observation that lowering pH increases the sense of astringency. But it could be that there is some sort of additive effect between sensing acidity and astringency.

As we drink wine, the wine itself will increase the flow of saliva, which in itself will change the perception of the wine. The binding of tannins may well reduce their ability to reach the bitter receptors, and thus their bitterness may decrease and their astringency increase at the same time.

IMPLICATIONS FOR WINE TASTING

This is all well and good, but what are the implications for tasting wine? Next time you taste and spit red wine, take a look at the spittoon. It's actually quite an unpleasant sight, with large strings of congealed, red/purple/black strings of saliva. This is the result of interactions between wine and saliva, and chiefly the binding of salivary proteins by tannins to form precipitates. The mucins in saliva play their part by helping create these visco-elastic strings of colored spit.

In the normal situation of drinking wine it seems likely that the production of saliva is able to keep pace with the rate the wine is consumed. With red wine, the challenge to the palate is the repeated exposure to tannins. With white wines, the tannic content is much lower, and the challenge will be the acidity, which is usually much higher than in red wines (that is, the pH of whites is lower). With Champagne and sparkling wines, the acidity is lower still. None of these should provide a major challenge to the perception of wine, unless exposure is repeated rapidly over a short space of time.

Just such an exposure takes place in many situations where wine is tasted professionally. Whether the situation is a trade tasting, or competition judging, or a critical assessment of a region's wines, it is common to find professionals tasting upward of 100 samples a day.

With red wines, tannins will be interacting with salivary proteins, precipitating and causing a sensation of astringency and mouth drying. The initial layer of mucins lubricating the mouth will be stripped. Then, with repeated exposure the deeper layer of mucins will be stripped; this is something that would normally not happen with a typical wine drinking as opposed to tasting scenario.

Normally, repeated exposure to the same taste or smell will result in a degree of adaptation. However, with astringency, repeated exposure results in the sensation of astringency increasing. Wine ingestion stimulates saliva production, but this is not sufficient to deal with repeated samples of red wine in close succession, in that it fails to replenish the lubricating layer of salivary mucins on the mouth surfaces. The result is that the sense of astringency increases with each fresh sample, to the point where it can become uncomfortable.

This is not meant to read as a counsel of despair for professional wine tasting. But it is an observation that should encourage us to approach tasting with a degree of humility. Tasting many wines in succession carries risks, not just of palate fatigue, but also of the effects of presentation order. The perception of any one wine can be influenced by the nature of the preceding wine. For this reason, it is good practice to have different tasters on a panel taste in different order, even if it is as simple as some tasting in reverse order. Given that saliva is unable to cope well with the sort of frequency of wine tasting typically carried out by professionals, what can we do to help? At the most basic level, we need to hydrate properly. Dehydration reduces saliva flow. If we spit, we not only eject the wine, but also the saliva our mouths have produced. If we are producing around a quart a day of saliva, and then spitting and not swallowing it, then this fluid shortfall needs to be made up. The effectiveness of palate cleansers in restoring the palate to baseline conditions in sensory analysis has been examined. One study compared astringency buildup using a number of different cleansers: deionized water, a 1 g/liter of pectin solution, a 1 g/liter CBMC

(carboxymethylcellulose) solution, and unsalted crackers. Subjects tried the same wine six times, with a cleansing process after the third wine. The unsalted cracker was found to be the most effective at reducing astringency buildup while water alone was the least effective, but astringency built up whatever cleanser was used. Another study showed that pectin rinse was the most effective cleanser, followed by unsalted crackers. Carboxymethylcellulose has been shown to be effective in some instances.

For fine wines, small differences in quality are significant. And for top red wines, mouthfeel is one of the key components of the wine. Elegance and harmony, much prized in older wines in particular, depend in large part on mouthfeel. For assessing these types of wines, the number of samples that can be reliably assessed is reduced.

There is one further point that must be made in relation to saliva and wine tasting, and this concerns inter- and intra-individual differences in saliva production. People differ in the composition and production of saliva, and each person's salivary flow rate will change with a number of factors including hydration state, time of day, emotional state, and the influence of medication. In addition, 10–15% of the population breathes largely through their mouth, and this will result in significant evaporation of saliva (estimated to be a loss of 12 fluid ounces per day). Both intra- and inter-individual differences in saliva are likely to affect the mouthfeel of red wines. This adds another level of inter-individual variation to wine tasting, in addition to factors such as taste-bud density, olfactory receptor repertoire, and knowledge, and experience. The extra level of intra-individual variation is something we must be aware of as tasters. The fact that our experience of wine can differ from day to day or even hour to hour should help us remain humble in the face of wine.

In conclusion, we encounter wine most profoundly in the mouth. And the internal environment of the mouth clearly has a significant impact on wine perception. Saliva is vital in mediating our experience of wine, and so attempts to understand wine appreciation need to consider saliva and the mouth environment as an intrinsic component of flavor perception.

22 Synesthesia, language, and wine

There is a lot of talk about wine. In the wine trade we spend a good deal of time putting the sensory experience of wine into words—the medium we use to share our perceptual experience. This verbal sharing of our own private worlds is of great interest to those neuroscientists and philosophers who have an interest in sensory perception, but to others outside the trade, the language of wine can be a source of amusement, bewilderment, and even derision.

In 2005, in an article in the London newspaper, *The Times*, Jonathan Meades took a shot at the way language is used in wine descriptions. He pointed out that when Hugh Johnson's iconic *World Atlas of Wine* was first published in 1971, Johnson offered a lexicon of fewer than 80 descriptors for tasting notes. Meades clearly disapproves with the way this list has since burgeoned. "The globalization of wine-making and the type of people now buying it have caused that lexicon to be vastly augmented. A new, qualitatively different language has evolved. The old one, founded in the certainties of St. James's and St. Estèphe, was a code. It was as hermetically precise and exclusive as the jargon of any other self-regarding profession," says Meades. "This has largely disappeared, drowned by a clamorous demotic which, far from being codified, attempts to express (rather than classify) a wine's qualities and, equally, to demonstrate the verbal and invention of the merchant, sommelier, writer, buff, casual drinker." Meades adds that such talk "is frequently characterized by the hedged bets of jest and self-parody."

Meades' unease with winespeak raises an important question. In describing wine, what are we trying to do? Are we using a learned code, where we associate particular standardized terms with physical attributes of the wine? Or, do we attempt to describe what is *actually* there in ways that others, unschooled in our "system" can relate to or understand? In practice, it is likely that a little of both takes place. Nonetheless, it is important to work out which system we are attempting to implement: the code, or the real-life description. This then leads to further questions. Which linguistic tools are appropriate or permissible in descriptions of wine? Some argue that we should go no further than simile. Others see metaphor as a crucial tool in our endeavor to share what we are experiencing. Should we be attempting to describe a wine as plainly and accurately as possible, breaking it down in a reductionistic manner into its constituent flavors and aromas, or do we use more figurative and creative language to build a more holistic description?

The goal of this chapter is to begin to explore the use of language in sharing our perceptual experience of wine, using the intriguing phenomenon of synesthesia as a springboard. On the way we'll revisit how the brain processes sensory stimuli, and look at how a descriptive pseudo-synesthesia is emerging as a useful tool to enable us to compare sensory experiences among individuals. Finally, we focus on how academics have begun to study the use of language in tasting notes, and report on some of their findings.

SYNESTHESIA

Synesthesia, the stimulation of one sense by another, is a bizarre phenomenon. Indeed, it's hard for the majority of us who do not experience this to imagine what it must be like to see green when we read the word "wheelbarrow," or taste sweetness when we hear the note C. Initially, we'd be inclined to suggest that synesthetes are making it all up or fooling themselves, but rigorous study by psychologists has shown that for true synesthetes, this mixing up of the senses is very real.

The most common sort of synesthesia is when the sensation of particular colors is associated with written or spoken letters, digits, or words. This is known as lexical synesthesia, where words (the inducing stimulus here) confer color experiences (the concurrent perception) on an individual in a reliable, reproducible fashion. This is quite different from the way the word "red" might cause us to "see" in our mind's eye objects that are red, or the color red itself. In this synesthesia the subject will actually see (as in experience) a particular color given a particular verbal cue. Around one in 2,000 individuals possess this ability, and it is remarkably robust. If people are tested at intervals, they score very highly matching the same colors to certain words.

Perhaps more relevant to this current chapter is a much rarer sort of synesthesia, involving taste sensations linked to cues from other senses. Researcher Lutz Jäncke and his colleagues from the University of Zurich published findings[1] on a synesthete called Elizabeth Sulston, a 27-year-old professional musician. As she began to learn music formally, she noted that particular tone intervals caused characteristic taste sensations on her tongue. Interestingly, besides this highly unusual interval-to-taste synesthesia, she also has a more common tone-to-color synesthesia. Could there be synesthetes out there for whom taste sensations stimulate other senses?

But there's also a much more common form of synesthesia that is largely ignored, probably because almost all of us experience it. How often have you described a smell as sweet? For example, consider the smell of strawberry. But sweet is actually a taste. We link certain smells with certain tastes in our brains without even realizing it, and so it seems perfectly normal to talk of a "sweet" smell.

Modern brain-imaging techniques have begun to shed light on potential mechanisms underlying synesthesia, demonstrating that it is the result of cross-activation of different brain areas. A paper[2] in scientific journal *Neuron* reports studies on subjects with what is known as grapheme-color synesthesia (where grapheme refers to letters and numbers). Using functional magnetic resonance imaging, the researchers showed that this is caused by cross-activation between different areas of a brain region called the fusiform gyrus that are involved in processing colors and graphemes.

Psychologists are fascinated by synesthesia, not just because it's interesting in its own right, but also because it is a tool for helping answer difficult questions about perception. One of these issues that synesthesia can help address is called the binding problem. Different aspects of sensory experience are represented in widely distributed brain areas. Even within a particular sense, different features may be represented separately. In vision, for example, shape, motion, color, and size are all registered initially in separate brain areas. Yet we perceive these rather widely spread pieces of sensory information as a seamless, unitary perception. How are they brought together?

In effect, synesthesia is a pathology; it's faulty wiring that yields clues about how the normal system works. In fact, it is a recurring theme in biology that disease or malfunction reveals useful information about how the system is assembled. Synesthesia may seem very strange, but it reinforces the fact that we experience a highly edited version of the world around us. In effect, our sensory systems "model" the reality out there, extracting the useful pieces of information and presenting them to us in the most efficient way. This is discussed in Chapter 20 on wine and the brain.

SYNESTHESIA IN WINE DESCRIPTIONS

So far what we have been discussing is what I'm going to call "true" synesthesia. There also exists a pseudo-synesthesia, which is common among artistic types and can be a bit of a literary affectation. In this case, people use a consciously controlled, imagined synesthesia as a way to express ideas in a fresh or impactful way. Wine writers often employ this pseudo-synesthesia in describing wine. Some examples? I find myself describing some red wines as dark, not just because they are dark in color, but also because the flavors have a "dark" quality to them. Or a white wine I recently tried which was so fresh and delicate, I described it as "transparent."

Vision shapes our sensory perception. If we come to a wine and see it has a deep red/black

1 Beeli G., Esslen M., Jäncke L. "When colored sounds taste sweet." *Nature* 434:38 2005
2 Hubbard E.M., Cyrus Arman A., Ramachandran V.S., Boynton G.M. "Individual differences among grapheme–color synesthetes: brain–behavior correlations." Neuron 45:975–985 2005

color, this influences the types of descriptors we will use. It could be argued from the above that the use of synesthetic descriptions of wine isn't all that far removed from the way our sensory processing operates. Our perception is a unity. As we have discussed, many separate sensations from different senses are brought together in a binding process that yields a seamless, unitary perception. Of course, it is possible in any situation to focus on particular, rather discrete, sensory aspects, such as smell in sniffing a flower, or touch in rubbing our hands across a piece of wood, or hearing in listening for an announcement. But there are other aspects of a momentary sensation that are more universal, such as the pleasantness of an experience. We usually don't have to think about whether something is pleasant or not. Plunging into the sea around the British Isles at any time of the year is an uncomfortably cold experience. Our bodies say to us in no uncertain terms that this is unpleasant. But dipping into a hot tub feels nice. We don't have to think about these contrasting experiences to have some immediate feedback in our inner worlds about their niceness. But in the same way that these thermal transitions can be pleasant or not, so can olfactory experience (for example, sniffing a rose versus finding yourself downwind of a landfill site) or touch (stroking the smooth fur of a cat versus brushing your hand against some rough brickwork) or hearing (birdsong versus the roar of traffic). Pleasantness is a sensory quality that encompasses all sensations. This is mirrored by the way we use language. A strawberry can be sweet, as can a well-hit seven iron, as can another person's action. Think also of the myriad of different modalities that can be described as "nice." Thus to use descriptors from one sensory modality to describe another isn't as implausible and affected as it sounds.

TASTE PSYCHOPHYSICS AND SYNESTHESIA

Synesthesia appears once more in a branch of psychology known as taste psychophysics. This field of study concentrates on how physical taste stimuli are perceived by the mind. Linda Bartoshuk's work on the psychophysics of taste was discussed in Chapter 19. In this work she has addressed the difficult question of how we can compare sensory experiences among different individuals.

Are two people drinking from the same bottle of wine are having a common experience? "In my view, this is one of the most interesting questions in sensory science," responds Bartoshuk. "It taps into an important philosophical issue: since we cannot share experiences directly, is there a way to make comparisons across individuals (or groups) indirectly?" One of Bartoshuk's contributions to this field is that she has devised a reliable scale for making intersubject comparisons that makes use of cross-sense comparison.

Part of the problem of comparing taste experiences among individuals stems from differences in taste sensitivity. Some people are much more sensitive to taste experiences, and they are known as hypertasters or supertasers. But it is not just taste that is affected by these genetic differences; it is also the sense of touch in the mouth. Supertasters also perceive more intense retronasal olfaction, presumably because they perceive more intense oral sensations.

Given the individual differences in taste perception, how does Bartoshuk make sensory comparisons among individuals and groups? Initially, she used responses to varying dilutions of salt solutions (NaCl) as a taste standard, but she found that this varied with PROP-tasting status. The answer was to take advantage of the surprising observation that experiences from different sensory modalities can be matched for perceived intensity: herein lies the connection with synesthesia. Putting this more simply, using appropriate standards from an unrelated sense that shows less individual variation than taste, such as the brightness of a light or the loudness of a sound, can make between-subject comparisons in taste intensity possible. For example, in one experiment, non-tasters matched the bitterness of black coffee to the brightness of low-beam headlights at night, while supertasters matched it slightly above high-beam headlights at night. A deliberate, voluntary synesthesia such as this enables us to break free of the noise and confusion brought about through genetic differences in taste, making recourse to descriptions from senses for which we live in much more closely similar worlds. Without the use of an appropriate standard from another sense, scales labeled for taste intensity produce invalid comparisons across groups and individuals.

THE LANGUAGE OF WINE

From here we move to the language of wine, and literary devices used to communicate sensory experience. Let's focus first on written language, where letters, which are visual sensations, are turned into words. We are so familiar with this that to have it pointed out to us seems absurd. As soon as we see words on a page, these visual sensations become loaded with meaning. Think of a love letter, or a tax demand: the visual sensations almost immediately stimulate an emotional response in us. As an aside, the written word has enabled the development of a complex society. It enables us to use pencil and paper, or laptop computers, as an extension of our mental space. We can share with others our thoughts, and use these devices as tools to store, and then add to, moment-to-moment thoughts, thus in time building an article or idea in a way that wouldn't be possible otherwise.

In wine writing, we do the opposite. We attempt to turn our conscious perception elicited by a flavor, but added to by our memory and learning, into letters on a page, which we hope will in some way convey our perceptions to

others who lack the same flavor stimulus. We are attempting to share, in as transparent a way as possible, our own private world of perception. What are the most effective and legitimate ways of doing this? Should we enlist figurative language in descriptions of wine? This is where we turn to the field of cognitive linguistics.

A fascinating academic project is underway at the Department of Modern Languages, University of Castilla-La Mancha in Spain, involving Dr. Ernesto Suarez-Toste, Dr. Rosario Caballero, and Dr. Raquel Segovia, entitled *Translating the Senses: Figurative Language in Wine Discourse*. The initial stage of the project involved collecting a data set consisting of 12,000 tasting notes from a range of British and American publications (*Wine Advocate*, *The Wine Spectator*, *Wine Enthusiast*, *Wine News*, *Decanter*, and *Wineanorak.com*). This text is cut, pasted, and cleared of all extra information. The types of metaphors used are tagged, and then a concordance is used to track each instance of any type of metaphor of interest.

"Wine folks use metaphor all the time," say Suarez-Toste. "Aroma wheels are OK for identifying aromas, but the structure and mouthfeel almost always demand the use of figurative language. It has nothing to do—at least not necessarily—with waxing poetical about something sublime; we use metaphor even for the most average grape juice around," he states. "For one thing, we personify wine most of the time. Not simply by saying it has a nose instead of a smell. It has character, it's endowed with human virtues and vices. It can be generous, sexy, voluptuous, whimsical, shy, demure, bold, or aggressive. We almost cannot conceive wine without personifying it."

We reach for metaphors because of the impoverished language we have for describing tastes and smells. "Because there is no single lexicon with the expressive potential to cover all the range of sensorial impressions, the intellectualization of sensorial experience is inextricably linked to the figurative uses of language," explains Suarez-Toste. "There is no problem with this as far as such areas of human life as poetry are concerned, but the inherent subjectivity of sensorial experience represents innumerable difficulties when technical discourse is under scrutiny."

Above A synaesthetic wine label. "I started describing wines this way for myself about 10 years ago," says Jason Lett, proprietor of the Eyrie Vineyards in Oregon. "My written descriptions of wine were making me miserable. Now, I often pair my notes with scribbled drawings rather than sentences. This label is one of my first decent painted tasting notes, rather than one that is simply drawn with a pencil. The introduction of color and brushwork has given me yet a way to describe wine that I continue to explore."

What about the good old tasting note? "This relies heavily on a combination of terms articulating the remembrance of the taster's repository of aromas and flavors, connotations, and, above all, figurative language which, although may be perceived by the layman as deliberate obscurity, is a valuable tool that allows the (only partially satisfactory) communication of the experience of tasting wine. The vocabulary used points to various figurative phenomena (synesthesia, metonymy, metaphor), all of which are indispensable tools for articulating what is an intrinsically sensorial experience."

Suarez-Toste and colleagues break these metaphorical wine descriptions into various categories. So we have wine as a living creature, wine as a piece of cloth, and wine as a building. It's easy to make fun of this sort of description, but such metaphors are borne of necessity. While we'd like to have a more exact way of sharing our experience of wine in words, such precision doesn't exist, and those who restrict themselves merely to naming aromas and flavors end up missing out on some of the more important aspects of the character of wines that can't be described in this way, such as texture, structure, balance, and elegance.

"Currently we're obsessed with structure and mouthfeel," Suarez-Toste explains. "These usually demand architecture and textile metaphors. One curiosity that our audiences enjoy is that a wine can be described in the same tasting note as silky and velvety. Of course the terms are for them mutually exclusive. The idea is that both are different (but almost synonymous for the critic's purpose) realizations of a textile metaphor. The connotations are smooth and expensive, fresher in silk (more used for whites) and warmer in velvet (more frequent in reds), but essentially the same. And that's just the beginning. Lists of materials are boring when compared to words that unconsciously betray the textile metaphor: this wine is seamless; is bursting at the seams; the fruit is cloaked by tannin; wears its alcohol well; a core of tannin is wrapped in layers of fruit; and so on."

Others have studied the cognitive linguistics of wine. French researcher Isabel Negro has written about the use of the French language in wine tasting. She notes the heavy use of synesthetic wine descriptions in French, and suggests that this is because in France, winespeak conveys a cultural view of wine tasting in which, uniquely, all the senses are used. Compared with English winespeak, French winespeak also has a unique feature, in that hearing is invoked. "Wine tasting is metaphorically represented as listening to a music composition, as evidenced by the metaphorical expressions notes, register, harmony, and finale, among others," says Negro. "This is a specific feature of the French discourse."

Negro tallied up the different sorts of wine metaphors used by type in French wine descriptions. The most popular was wines are people (476), with the second most common being synesthetic metaphors (147). Wines are food is next (70) followed by wines are clothes (45), objects (31) and buildings (28).

CONCLUDING REMARKS

So where does this leave us with descriptions of wine? When we try to share our perception of wine by means of words, we are attempting to do something very difficult. It's worth remembering that as we come to taste a wine, we are bringing a lot of our own "stuff" to the table, such as our culture of wine, the context of our experience of wine, and our expectations about the tasting experience. Then, as we taste, our experience of flavor is a multimodal sensory experience consisting of taste, smell, touch, and vision inputs, which are frequently bound together at different pre-attentive levels, and which then can even feed back to modify the unimodal sensations themselves.

But taste is imprecise, and shows a lot more individual-to-individual variation than other senses. Thus as we taste wine we use the "stuff" we bring to the tasting experience to help frame our thoughts about the wine. As Frédéric Brochet has shown, our mental representations of wine are prototypical. As we taste, we decide, from our experience, what sort of wine we are tasting, and this then helps us pick our descriptors. Or we might know that a particular wine we are tasting is a Pauillac, and this will lead our thoughts in certain directions as we taste it. It is an intriguing thought that the language we have developed (and each taster will have evolved his or her own lexicon) actually shapes our perception, in part through attentional effects, in part because our language of wine will give us a framework to build our descriptions on. How do we find something if we aren't looking for it? Having a vocabulary for wine will likely direct and shape what we "get" from each wine.

But pointing out the complexity of the tasting process is not meant to be a counsel of despair—writing tasting notes and communicating about wine is still a useful endeavor. Rather, as we understand more about the biology of perception, it helps us make sense of the results of this process and causes us to be more realistic about the degree of precision or consensus that is possible in assessing wine. While there is such a thing as expertise in wine tasting, we should taste humbly and not seek to champion a uniform, one-size-fits-all model of wine assessment. Incompetence and individual differences in perception are quite separate entities that are often mixed in together. Tasting well is a difficult skill, and some of the anomalies between different tasters' views on the same wine are undoubtedly because some wine tasters aren't very good at it. Others are likely due to cultural differences or degrees of experience, and still others may be because of real differences in perception that have a biological basis. There's still an important place for critics, but there is no room in this new synthesis of wine assessment for the über critic. However, this new understanding of wine perception emphasizes the importance of a learned culture of fine wine, where what comes next builds on the foundations of what went before. Many of the current conflicts in the world of wine are caused by a failure to recognize that there is a culture of aesthetic appreciation of fine wine that is to a large degree learned by the process of comparison and benchmarking. It is a system of aesthetics. But that's another story.

Is figurative language appropriate in technical tasting notes, or should we aim for more technical and analytical descriptors? I would argue that if we are seeking a communication of our perceptions that is as accurate as possible, then the use of figurative language is essential, simply because we lack any other effective way of communicating vital aspects of wine, such as structure and texture. In addition, the move to technical-sounding language is usually accompanied by a reductionist dissection of wine into its component parts. The use of metaphor brings us back to a more holistic description of wine, which is more appropriate because our perception is, after all, a multimodal, unitary one. There is also an important place in wine description for the use of synesthetic descriptions. While these run the risk of sounding somewhat contrived, they can bring a fresh, altered perspective that keeps us from getting into a rut with over-reliance on stale tasting-note terms.

23 Wine flavor chemistry

Wine is a complicated chemical soup, the nature of which scientists are only just beginning to understand. It is estimated that wine contains around 1,000 volatile flavor compounds, of which yeasts produce more than 400. Despite decades of research, scientists are only now beginning to get a fuller understanding of the true nature of the chemical composition of wine. Part of the difficulty is that the picture is a dynamic one. The volatility of various flavor compounds can be altered by other components of the wine. Added to this, human perception of various flavor chemicals is altered by their context—the suite of other chemicals present in the wine. Thus chemical "A" might be below detection level in one wine, and above it in another, even though its concentration is the same in both.

One further complicating factor is that many of the most important chemicals that shape a wine's specific character are present at very low concentrations. The most prevalent constituents of wine are often relatively unimportant in terms of the sensory qualities of the wine. And the actual compounds we currently know most about reflect not necessarily their importance in determining wine flavor, but rather our ability to sample them with the techniques available.

What is the goal of wine flavor chemistry? Do we need to be able to put a chemical name to all the nuances of a fine wine, in order to be able to appreciate it? No, clearly not. But if we understand the precise mechanisms by which certain components of the grape must are transformed into beneficial flavor molecules, for example, by the metabolic action of yeast, barrel- aging, or bottle maturation, then winemakers can adapt their techniques to maximize positive flavor development. In a similar vein, viticulturists can adapt their techniques to encourage the formation of precursors of positive flavor molecules and avoid the development of grape constituents that have a negative impact on wine characteristics.

THE TASTE OF WINE: FACTORS INFLUENCING WINE FLAVOR

The factors contributing to the flavor of wine are still far from being fully understood. Here, I'm using the term "flavor" to refer to our perception of wine. People also talk about the "taste" of wine, but this is ambiguous, because they could be referring to either the specific information that comes from the sense of taste itself, or they could be referring to the global "taste" of wine where this term is used to describe all of what we sense when we drink wine.

The emerging understanding of flavor is that it involves the combination of a number of senses, and draws on input from the senses of taste (technically known as gustation, referring to the information that comes from taste buds in the tongue and mouth), smell (technically olfaction, with the information coming from the olfactory receptors in the nasal cavity), touch (the feel of the food or liquid in the mouth), and even vision (visual cues have been shown to shape the perception of flavor). In addition, there is input in the form of our context or previous experience— what we know about what we are about to taste can alter our perception. Thus flavor is a "multimodal" sensation, one in which all these different sensory inputs are combined together in the brain to give a unified perception of flavor, with perhaps the strongest influence being that of olfaction.

Any study of wine flavor has to consider both the physical properties of the wine and also the way that humans sense these. A wine has physical properties, which can be measured. These are "real," in that the wine has a chemical composition. If several researchers examine the wine using analytic tools, such as spectrophotometry or gas chromatography-mass spectrometry (GC-MS) we can expect them to get the same results. (Any differences in the measurements will be because of the calibration and accuracy of the tools they are using.) It follows that wine has a large number of aromatic compounds that can be detected by the human olfactory system. It also has chemicals that elicit a taste response. But while this chemical composition is a property of the wine, the "taste" of a wine is not. It is a result of an interaction between the taster and the wine, with the taster bringing something to the encounter that significantly alters the wine's perception.

THE CHEMICAL COMPOSITION OF WINE

Wine is a complicated mix of many hundreds of flavor-active compounds. The exact number of volatile molecules found in wine is unknown, but estimates generally fall in the range of 800–1,000. There are clearly many of them, although in each wine only a limited number of these are found above the level at which most humans would detect them, known as the perception threshold. But the complexity of wine flavor and aroma goes beyond just the large number of volatile compounds it contains. In particular, there are two factors that increase this complexity. First, people differ in their sensitivity to various taste and smell molecules. Some people lack the ability to sense specific taste or aroma compounds (the term for this deficit is "aguesia" with regard to taste and "anosmia" with regard to olfaction). As well as this, people have a range of different thresholds for compounds, meaning that some may be a little more sensitive to a specific smell than others. Second, wine aroma and flavor isn't additive,

in that it is simply a sum of the different smells and tastes of the various chemicals it contains. Instead, there are many interactions between the different components, including masking interactions (where one compound interferes with the perception of another) and synergistic interactions (where a perception is created by a combination of two or more different compounds).

Thus there are two elements here that anyone studying the flavor of wine must grapple with. First, what is chemically present in the wine—and because many of the active compounds are potent and thus their presence at even tiny levels can be significant, this presents a challenge to analytical chemists. Second, we need to understand how this mixture of chemicals is actually perceived. Our final perception of a wine is the result of complicated processing in which all the different tastes and aromas of wine are integrated into a single perception, with us bringing quite a lot to the wine-tasting process. Vicente Ferreira of the University of Zaragoza in Spain is one of the leading experts working on wine composition and aroma. Ferreira sums this up by saying that wines don't have a single characteristic aroma, but "Rather they have a palette of different aromas, which are difficult to define and which surely are perceived differently by different people."

Ferreira separates the various flavor compounds of wines into three different groups, which is a structured way of thinking that helps us grapple with this subject. In addition, there is

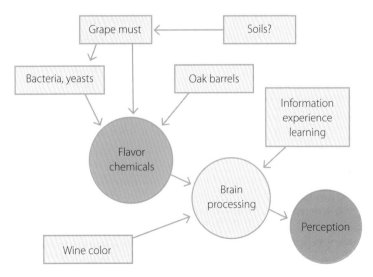

Left Flavor chemicals come directly from grape must, from bacteria and yeasts directly, and also from bacteria and yeasts modifying precursors in the must. There's also the possibility that some elements of flavor could come from the soils, although this is debated, and any soil effect is likely to be indirect. Oak barrels also impart flavor, as would grape juice concentrate if this were added, or tartaric acid.

the important concept of the wine matrix. While some wines contain what he terms "impact" compounds, many lack these, and instead contain a large number of active odorants, each adding nuances to the wine. One of the challenges for sensory scientists is that in many cases it simply isn't possible to establish a clear link between a sensory descriptor and a single aroma molecule. Instead, what tasters refer to by specific descriptors is often the result of the interaction of two or more odor-active chemicals. "When I started my PhD, wine aroma was about finding a molecule to explain everything," says Laura Nicolau, one of the researchers involved in the New Zealand Sauvignon Blanc Program. "By the end, people started to think it is not only one, it should be a combination. Vicente Ferreira was the first, to my knowledge, to talk about this."

WINE ODOR

Ferreira describes the basal composition of what he refers to as "wine aroma," which is the result of 20 different aromatic chemicals that are present in all wines to make a global wine odor. Of these 20 aromas, just one is present in grapes (β-damascenone); the rest are produced by the metabolism of yeasts, in many cases working on precursors present in the grape juice. These are:

- Higher alcohols (e.g. butyric, isoamylic, hexylic, phenylethylic)
- Acids (acetic, butyric, hexanoic, octanoic, isovaerianic)
- Ethyl esters from fatty acids
- Acetates and compounds, such as diacetyl
- Ethanol

The influence of alcohol (ethanol) is quite strong. Ethanol has been shown to modify the solubility of many of the aroma compounds, and makes them more reticent to leave the solution, thus making the wine less aromatic. A study by Whiton and Zoecklein from 2000 showed that as alcohol rose from 11% to 14%, there was reduced recovery of typical wine volatile compounds in an analytical chemistry experiment. In 2007, Ferreira's group identified a range of esters responsible for the fruity berry flavors in a series of red wines. But when they added more of these

to the wine, it didn't increase the fruity impact, because of the suppressing effect of other wine components, including alcohol. They showed this in another experiment in which they added increasing levels of ethanol to a solution of nine esters at the same concentration they are found in wine, and discovered that the fruity scent quickly falls as alcohol rises, to the point that when alcohol reached 14.5%, the fruity aroma had been totally masked by the alcohol.

CONTRIBUTORY COMPOUNDS

There are also another 16 compounds present in most wines, but at relatively low levels, which here I've labeled as "contributory compounds." Their odor activity value (OAV; the ratio of the concentration of the compound to its perception threshold) is usually below 1, but they have odor activity that is synergistic, contributing to characteristic scents despite being at lower concentrations that would normally lead to them being smelled. These include:

- Volatile phenols (guiaicol, eugenol, isoeugenol, 2,6-dimethoxyphenol, allyl-2,6-dimethoxyphenol)
- Ethyl esters
- Fatty acids
- Acetates of higher alcohols
- Ethyl esters of branched fatty acids
- Aliphatic aldehydes with 8, 9, or 10 carbon atoms
- Branched aldehydes, such as 2-methylpropanol, 2-methylbutanol, 3-methylbutanol, ketones, aliphatic γ-lactones
- Vanillin and its derivatives

IMPACT COMPOUNDS

Impact compounds are a group of chemicals that are responsible for giving characteristic aromas to certain wines, even when they are present at extremely low concentrations. These are of great interest because they often contribute to distinctive varietal aromas. However, many wines lack distinct impact compounds.

For example, Sauvignon Blanc is very interesting as a grape variety because much of its characteristic aroma is believed to come from a small number of impact compounds, chiefly methoxypyrazines (of which the most significant is 2-methoxy-3-isobutylpyrazine) and three thiols (4-mercapto-4-methylpentan-2-one [4MMP], 3-mercaptohexan-1-ol [3MH], and 3-mercaptohexyl acetate [3MHA]). These have therefore been the focus of intensive research

EXAMPLES OF IMPACT AROMAS

Methoxypyrazine: the most important one is 2-methoxy-3-isobutylpyrazine (MIBP; known widely as isobutyl methoxypyrazine), which has a detection threshold of 2 ng/liter in water and white wine (slightly higher in reds), and is responsible for green, grassy, and green-pepper aromas. 2-isopropyl-3-methoxypyrazine (isopropyl methoxypyrazine) is also important, but likely secondary to MIBP. The methoxypyrazines are one of the few classes of impact compounds formed in the grapes, and are highly stable through fermentation and aging.

Monoterpenes, such as linalool, which is important in many white wines, such as Muscat, and has floral, citric aromas.

Rose-*cis* oxide: characteristic of Gewürztraminer, this has a sweet, flowery, rose-petal aroma.

Rotundone: a sesquiterpene that gives pepperiness to Syrah, at incredibly tiny concentrations. Remarkably, one-fifth of people cannot smell this.

Polyfunctional thiols (mercaptans): These include 4MMP, which has a box-tree aroma (4.2 ng/liter detection threshold), 3MHA, which has a tropical-fruit scent (60 ng/liter), and 3MH. These three are important in the aroma of Sauvignon Blanc. A number of other thiols are also important in wine aroma, and are discussed in Chapter 15.

THE NONVOLATILE WINE MATRIX

But in addition to the actual aroma molecules, some of Ferreira's most interesting recent work has been on what is called the nonvolatile matrix of wine. The idea here is that wine constituents that don't have any aromatic characteristic of their own nevertheless exert a strong influence on the way that the various aromatic molecules present in wine

are perceived. In effect, the nonvolatile matrix influences how we interpret the smell of wine.

Ferreira and his colleagues recently reported an interesting experiment in which they showed that the nonvolatile matrix is critical in determining the aromatic character of wine, even to the point that when the aromatics from a white wine are put into a red-wine matrix, the wine smells like a red wine.

"Knowledge of volatile and nonvolatile composition alone is not enough to understand completely the overall wine aroma and in general its flavor," state the authors in the introduction. "Interactions among odorants, perceptual interactions between sense modalities, and interactions between the odorant and different elements of the wine nonvolatile matrix can all affect the odorant volatility, flavor release, and overall perceived flavor or aroma intensity and quality."

For this study, they selected six different Spanish wines, all of which are available commercially (from Somontano producer Viñas del Vero).

1 Aromatically intense unoaked Chardonnay (fruity white)
2 Barrel-fermented Chardonnay (protein-rich white)
3 A young, light Tempranillo (neutral red)
4 A four-year-old barrel-aged 90% Tempranillo/10% Cabernet Sauvignon blend (highly structured polyphenol-rich red)
5 A three-year-old Tempranillo with marked astringency (very astringent wine)
6 A three-year-old Tempranillo with marked woodiness (typical woody aroma)

The aromatics from samples of each of these wines were removed by the use of a process called lyophilization (freeze-drying) and then any remaining aromatics were removed by using a chemical called dichloromethane. The dichloromethane was itself removed by passing nitrogen through the sample until it was all gone. The extract was then dissolved in mineral water to produce the wine matrix.

In a separate series of manipulations, aromatic extracts were collected from each of the wines, producing six aroma extracts. Then combining different wine matrices with different aroma extracts made a series of reconstituted wines. 20 ml (.7 fl oz) of aroma extract and 120 ml (4 fl oz) of nonvolatile extract were combined, both equivalent to 600 ml (20 fl oz) of wine, along with 52 ml (1.75 fl oz) of ethanol to bring the alcohol level up to 12%, and enough mineral water to bring the sample to 600 ml (20 fl oz) in all. In total, 18 reconstituted wines were made, which a trained sensory panel then analyzed.

The results showed that the nonvolatile extract (the matrix) had a surprisingly large impact on aroma perception of the wine. As an example, when the aroma extract from wine 1 (fruity white) was reconstituted with the nonvolatile extract from the other white, there was relatively little effect. But when it was combined with the red wine nonvolatile matrix, there was a large difference, and the sensory panel started using terms relating to red fruits, rather than terms typically used to describe white wines. Other red wine terms that started appearing were "spicy" and "woody".

The effect was most marked when the nonvolatile matrix came from the astringent red wine, number 5. But a similar effect occurred when red-wine volatiles were added to a white-wine matrix, and white, yellow, and tropical fruits all start appearing in judges' tasting notes.

These results surprised the authors. Previous studies have shown that nonvolatile components of wine can affect wine aroma, but this is largely through binding to them and making them less releasable. The remarkable thing about this study is that it demonstrates that the nonvolatile matrix is having an important effect in actually modifying the perception of the volatile components of wine. It is well-known that cross-modal sensory effects can modify perception, especially when vision is involved (experts describe white wines colored red using red-wine terms), but this was avoided in this study by using black glasses and asking participants to describe the aromas before they tasted the wine. The authors also demonstrated that the non-volatile matrices were free of traces of aromatic chemicals. They found just tiny quantities of a few of the most polar odorants, but at quantities well below their perception thresholds.

What is emerging from these types of studies is a more holistic view of wine. While the reductionist approach attempts to study wine flavor by breaking it down into its constituent chemical compounds and then studying these in isolation, the field of wine flavor chemistry is maturing with the realization of the limits of reductionism. The reductionist approach has been incredibly useful, but a view of the flavor of wine that treats the wine as a whole, and takes the human side of perception into account, is likely to lead to a more complete and satisfying understanding of wine flavor.

Maurizio Ugliano, previously a researcher with the Australian Wine Research Institute, but now with closures company Nomacorc, agrees with this sentiment. "I did a lot of work with fermentation and nutrients when I was in Australia," he said. "People always had a tendency to simplify these stories. For example, they'd say, these esters have been increasing with fermentation, and the threshold of these esters if we don't add nitrogen was below the threshold, and when we add nitrogen it is above threshold. The approach has been to isolate each compound from its context and talk about it individually, saying the effect of this variable is important on this compound because it is becoming above threshold and before it was below threshold. The reality is that when you have some manipulation that affects many things at the same time you can't study one compound in isolation and talk about an individual threshold." He thinks this is a problem with the current view of wine flavor chemistry. "We currently don't look at these changes in a holistic matrix way. When we have a wine in a bottle with a high dose of oxygen, we are not forming reductive sulfur compounds; we are probably killing some of the existing compounds, including the good thiols, and we are forming some aldehydes and we might be forming some lactones, such as sotolon.

"If we want to predict from a threshold point of view what is happening to hydrogen sulfide, to reduction, we can't just say that before hydrogen sulfide was 3 micrograms a liter, now it is 1, before it was above threshold and now it is below and there is no reduction any more. This might not be the case. One of the compounds that has formed might still support the expression of reduction from hydrogen sulfide, even when the hydrogen sulfide is below threshold. We are not studying them in a system way in this example. There are lots of people talking about systems biology and metabolomics—these '-omic' types of approach. The funny thing is that aroma chemistry, from an analytical point of view, has been very 'omic,' since the beginning. People always took the approach that they needed to analyze as many compounds as possible in one single analysis. There is no point in analyzing the way that one ester changes over time. But from a sensory point of view, we struggle to introduce the concept of systems in the way we approach the changes of aroma compounds over time. It is difficult when you have all these combinations of things that you need to test in sensory work."

MORE ON THE DIFFERENT WINE FLAVOR COMPOUNDS

This isn't the right context for a detailed discussion of the complex chemistry of wine flavor, so I will spare you the chemical structures and most of the long names—these can be found elsewhere. Wine chemistry has already been discussed to some extent in other chapters of this book, particularly with respect to reduction (Chapter 15), *Brettanomyces* (Chapter 17), yeasts (Chapter 16), and barrels (Chapter 12). For the purposes of this chapter, it will be useful to take an overview of the key classes of wine flavor compounds, highlighting a few that are of particular interest. These compounds can be divided neatly into five groups: acids, alcohols, sugars, polyphenols, and volatile compounds, although there is some overlap.

ACIDS

Acid is a vital component of wine, helping to make it taste fresh, but also helping to preserve it. White wines with higher acidity usually age better than those with low. Red wines can get by with a little less acidity because they contain phenolic compounds that help preserve them.

The main organic acids found in grapes are tartaric, malic, and citric. Tartaric acid is the key grape acid, and can reach levels of 15 g per liter in unripe grapes. It's quite a strong acid and is specific to grapes. In musts it is found in the range of roughly 3–6 g per liter. Malic acid is abundant

in green apples and, unlike tartaric acid, is widely found in nature. Before *veraison* it can hit levels of 20 g per liter in grapes. In warm climates, it is found in musts in the range of 1–2 g per liter, and in cooler climates it occurs at 2–6 g per liter. Citric acid is also widespread in nature, and is found in grapes at 0.5–1 g per liter. Other organic acids present in grapes include D-gluconic acid, mucic acid, coumaric acid, and coumaryl tartaric acid. Further acids are produced during fermentation, such as succinic, lactic, and acetic acids. In addition, ascorbic acid may be added during winemaking as an antioxidant.

This is the part where it gets quite confusing. There's no single measurement for acidity in wine. There are two measures, both with the initials "TA," but which are different. And there's pH. And also volatile acidity (VA, largely acetic acid), but we are not going to consider this here, because it is smelled rather than tasted.

Let's begin with pH. It refers to the concentration of hydrogen ions (known as protons) in a solution. It's expressed as a negative logarithmic value, which means the lower the number, the higher the acidity. And it also means that a solution at pH 3 has 10 times more acidity (defined as protons) than one at pH 4 (corresponding to roughly the range of pH values found in wine, although it can sometimes drop a bit lower than 3). This is where we need to get a

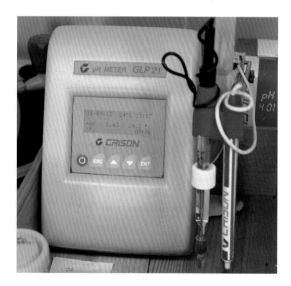

Above A pH meter, an important piece of lab equipment.

bit technical. The "acidity" of an acid depends on something known as its dissociation constant, or pK_a. The lower the pK_a, the more dissociated the acid is, which means it releases more protons into solution. Sulfuric acid has a pK_a of around 1, so it is almost completely dissociated, making it a very strong acid (in terms of protons in solution). Of the organic acids, tartaric has a pK_a of 3.01, which means it is pretty strong. Malic is 3.46, lactic is 3.81, and carbonic acid is 6.52 (which means it has very little dissociation, and is thus a weak acid).

If malolactic fermentation takes place, then the malic acid will be largely converted to lactic acid by the action of lactic acid bacteria. Lactic acid tastes less acidic than malic acid, contributing just one proton per molecule while malic contributes two. As a result, the pH of the wine shifts upward through malolactic fermentation by 0.1–0.3 units.

Musts and wines are known as acidobasic buffer solutions. This means you have to work quite hard to change their pH levels. If you add acid to water, you can shift its pH very quickly, because there is none of this buffering effect. But the presence of other compounds in musts and wines makes it less easy to shift the pH, and it's a little easier to shift pH in wine than must. It's actually tricky to predict the pH of the final wine by looking at the pH of the must, because several things occur during the winemaking process that can change pH. Where acidification is needed, it is usually done with tartaric acid, and as a rule of thumb, 0.5–1 g per liter of tartaric acid is needed to shift pH by 0.1 units. Legally, you could change pH with malic or citric acid, but because these are weaker acids, it would require quite a bit more. And adding citric acid isn't a great idea where malolactic fermentation is going to take place, because the bacteria turn citric acid into diacetyl, which has a buttery taste and can be rather repellent. However, I know of some winemakers who use malic acid to make small changes in pH because it doesn't fall out of solution in the same way that tartaric acid tends to, especially when there is potassium in the must or wine. Some winemakers in warmer climates have illegally used sulfuric acid to change pH, because it is very effective at doing this.

Typical pH levels for a white wine would be 3–3.3, while for reds they would be 3.3–3.6. However, I recently had a New Zealand

Riesling with a pH of 2.65, and a while back a South African red that was delicious (and had aged well) despite a pH of 4.0.

High pH isn't necessarily a bad thing. It can confer on a wine a deliciously smooth mouthfeel (think of some Provençal rosés or northern Rhône whites, for example). Generally, though, winemaking at lower pH levels is safer because of the reduced risk of oxidation and microbial spoilage. pH affects the amount of sulfur dioxide (SO_2) that is present in the active molecular form. At pH 3.0, 6% of SO_2 is in the molecular form, whereas at pH 3.5 only 2% is. If the wine gets up to pH 4, then 0.6% of SO_2 is in the molecular form, and so a lot would have to be added for it to have any significant influence in protecting the wine. One famous New Zealand boutique winery is known for its rather interventionist red winemaking, acidifying to low pH and then, before bottling, de-acidifying to get the desired pH. This reduces *Brettanomyces* risk considerably, and helps in other ways, such as fixing color.

So what about TA? This stands for both total and titratable acidity. Total acidity is the total amount of organic acids in the wine. Titratable acidity looks at the ability of the acid in the wine to neutralize a base (an alkaline substance), which is usually sodium hydroxide. The endpoint is typically pH 8.2, and is indicated by the change of color of a reagent such as bromophenol blue or phenolphtalein. Total acidity is the best measure to use, but it is hard to measure in practice, so titratable acidity is used as an approximation of this, but it is by definition always going to be a lower figure than the total acidity. So when you see the "TA" of a wine given, you can assume it is the titratable acidity. The unit it is expressed in is grams per liter, but here's another potential source of confusion. Most countries use "tartaric acid equivalent," but in some European countries it is given in "sulfuric acid equivalent," which will be two-thirds of the value of tartaric acid equivalent.

When it comes to the taste of acidity, what is more important, pH or TA? Most of the literature on this suggests that it is the TA that gives the taste of acidity, and so the figure that's important to look for is not pH but TA. The confounding factor here is that pH and TA are usually correlated so they are hard to separate, in that low pH wines usually have high TA. But you can get higher pH wines with high TA, and here the acid would taste quite sour. The different organic acids do seem to have different flavors. Tartaric is hard, malic is green, and lactic is softer with some sourness. I find that often where warm-climate wines have their pH adjusted by tartaric acid, the levels of tartaric acid necessary can mean that the acid sticks out as very hard and angular, even where the pH isn't especially low. Another issue is that added tartaric acid reduces potassium concentrations in the wine (they bind to form potassium bitartarate), and potassium is thought to play an important part in contributing to the weight or body of the wine.

SUGARS AND SWEETNESS

Sweetness in wine is a combination of three factors. First of all, there is sugar itself. This is sensed by sweet taste receptors on the tongue. Second, there is a sweetness that comes from fruitiness. While "sweet" is tasted, some wines can also smell sweet, even though sweetness is a taste modality. Most commercial red wines are dry in terms of sugar content, but many have sweet aromas from their fruitiness. Very ripe, fruity flavors taste and smell sweet even in the absence of sugar. The third source of sweetness is alcohol itself, which tastes sweet. It's really instructive to try the same red wine at different alcohol levels, where reverse osmosis or the spinning cone has removed the alcohol. As the alcohol level drops, with all other components remaining the same, the wine tastes drier and less rounded and full. Where alcohol has been reduced substantially, such as in the new breed of 5.5% alcohol wines, which are lighter, it's necessary to add back some sweetness, usually in the form of residual sugar. It helps if the starting wine had a very sweet fruit profile to begin with, too. For lower alcohol whites, blending in some Muscat or Gewürztraminer, which have sweet aromas, helps substantially.

There are a number of ways of making a wine with some residual sugar levels. For some white wines, fermentation stops naturally, or slows to a point where simply simply chilling and/or adding a little sulfur dioxide very easily stops it. It can, of course, be deliberately stopped in this way at any stage, but if fermentation is still ticking along nicely then more of both (chilling and sulfur dioxide addition) will be needed. Blending in must or grape juice concentrate to a dry wine can also make a sweet wine. For commercial wine styles where just a few grams per liter are needed to round off the wine, this is most easily done on the blending bench than by attempting to stop the fermentation at an exact point.

In sweeter white wines and also Champagnes, sugar and acid balance are vital. The two play against each other. Sweetness is countered by acidity, such that a sweet wine with low acid seems much sweeter (and often flabbier) than the same wine with high acidity. In Champagne, a typical dosage for brut (dry) Champagne is 8–10 g per liter, which helps offset the acidity but doesn't make the Champagne taste sweet. Botrytized sweet wines are prized because, as well as concentrating sweetness and flavor, the shriveling process of noble rot concentrates the acid levels, and the great sweet wines of the world have very high sugar levels as well as high acidity.

ALCOHOLS

Ethyl alcohol is the most important component of wine, and is produced by fermentation of sugars by yeasts. On its own, it doesn't taste like much, but the concentration of alcohol in the final wine has a marked effect on its sensory qualities. This is evidenced by the "sweet-spot" tastings carried out during alcohol reduction trials. If a wine with a high natural alcohol level is subjected to alcohol reduction via reverse osmosis, a series of samples of the same wine can be prepared differing only in alcohol levels—say at half-degree intervals from 12–18% alcohol. Panels of tasters show marked preferences for some of these wines over others, and different descriptors are commonly used to describe the sensory properties of the different samples. Excessive alcohol can lead to bitterness and astringency in a wine. It may also taste "hot."

Meillon and colleagues conducted an interesting study in 2010. They took an Australian Syrah and reduced it in alcohol from its original strength (13.4%) to 8%, with three wines at intermediate levels in between these extremes. They then showed these wines to 71 French consumers who drank red wines at least once a month, measuring their liking and perceived complexity. The Syrah at 8% was liked significantly less, but for the other wines there was no significant difference. Adding sugar to the 8% Syrah increased its likeability considerably. They found they could segment this population by their reaction to the wines. Group 1 (18 individuals) liked the 11.5% wine the best, and liked the lowest two alcohol wines much less than the other groups. Group 2 significantly disliked the Syrah at 8% and 11.5%, but they liked the 13.5 and 8% wines. Group 3 preferred the two lowest-alcohol wines. The more consumers dislike the alcohol-reduced wine, the more bottles they have in their cellar—an interesting finding.

POLYPHENOLS

These are probably the most important flavor chemicals in red wines, but are of much less importance in whites. Polyphenols are a large group of compounds that use phenol as the basic building block. An important property of phenolic compounds is that they associate spontaneously with a wide range of compounds, such as proteins and other phenolics, by means of a range of noncovalent forces (for example, hydrogen bonding and hydrophobic effects). Phenolic compounds are widely thought to have health-enhancing effects, but their propensity to bind with proteins such as salivary proline-rich proteins (PRPs) (see Chapter 21) will conspire against them reaching active sites in the body where they might be active.

NONFLAVONOID POLYPHENOLS

There are two types of these smaller nonflavonoid polyphenolic compounds, the benzoic acids (such as gallic acid) and cinnamic acids. They are often present in grapes in a conjugated form (e.g. as esters or glycosides).

FLAVAN-3-OLS

These are important in wine, and include catechin and epi-catechin. They are particularly important in their polymeric forms, where they are called procyanidins (often referred to as condensed tannins).

FLAVONOIDS

Comprising the flavonols and flavanonols, these are yellow-colored pigments found in red and white grapes.

ANTHOCYANINS

These are the red/blue/black pigments in grapes, which are almost always found in the skins. Five different anthocyanin compounds are found in red wines, the dominant one being malvidin. They are not stable in young wines, but react with tannins to form complex pigments, which gradually become larger as wine ages, to the point where they become insoluble and precipitate out. Oxygen has an important role in facilitating the process of phenolic polymerization. The color of pigments depends on the acidity of the grape must and the concentration of sulfur dioxide. They tend to be redder at lower pH (more acid) and purpler at higher pH.

TANNINS

The term "tannin" is chemically rather imprecise, but it is one used by almost all wine tasters. It is used to describe a group of complex plant chemicals found principally in bark, leaves, and immature fruit, which form complexes with proteins and other plant polymers such as polysaccharides. As discussed in Chapter 21, it is thought that the role of tannins is one of plant defense. They have an astringent, aversive taste that is repulsive to would-be herbivores. In wine, tannins come from grape skins, stems, and seeds, and their extraction is heavily dependent on the particular winemaking process involved. Some tannins also come from new barrels when these are used to age wine. Tannin management is a crucial step in red winemaking. Tannins are thought to taste astringent because they bind with salivary proline-rich proteins and precipitate them out, and they may also react directly with tissues in the mouth.

VOLATILE COMPOUNDS

This is where things get really complex, but it is also where much of the action is. It's the volatile compounds that give wine its smell, referred to more respectfully as bouquet or aroma. Volatile compounds come directly from the grapes themselves, but more commonly are secondary aromas arising from fermentation processes, or even tertiary aromas developing during maturation and aging of wine. Most are present in extremely low concentrations, which, before the advent of highly sensitive analytical techniques, made their study a difficult business.

Rather than list the 400 or so thought to be important in wine, here follows a description of the main classes, with one or two specific examples.

Esters are especially important to wine flavor. They are formed by the reaction of organic acids with alcohols, and are formed during both fermentation and aging. Ethyl acetate (also known as ethyl ethanoate) is the most common ester in wine, formed by the combination of acetic acid and ethanol. Most esters have a distinctly fruity aroma, with some also possessing oily, herbaceous, buttery, and nutty nuances.

Although aldehydes are present in grape must, they are of relatively minor importance in wine flavor, with the exception of acetaldehyde (ethanal), which is a component of some sherries. Vanillin (4-hydroxy-methoxy-benzaldehyde) can be an important aroma molecule in wines aged or fermented in oak barrels.

Ketones include diacetyl (butane-2,3-dione), which gives a buttery odor at higher levels; this can be negative at higher levels. Acetoin (3-hydroxybutan-2-one) has a slightly milky odor. β-damascenone and β- and α-ionone are known as the complex ketones, or isoprenoids. The former has a rose-like aroma and is most commonly found in Chardonnay. The latter occur in Riesling grapes and are said to smell like violets. They are also present in other wine types. Benzoic aldehydes are taint compounds with a bitter, almond flavor that are sometimes produced as a result of the incorrect application of epoxy resin vat linings.

Some 40 higher alcohols, also known as fusel oils, have been described in wine. The most important of these are the amyl alcohols. With their pungent odors they are negative at higher levels, but kept in check they can be positive. For example, hexanol has a grassy flavor.

Lactones (furanones) have been identified both in grapes and oak barrels during wine aging. The oak lactones (cis- and trans-β-methyl-γ-octalactone) are particularly important in barrel-aged wines, imparting sweet and spicy coconut aromas, together with woody characteristics. Sotolon (3-hydroxy-4,5-dimethyl-5(H)-furan-2-one) is a lactone associated with botrytized wines, and has sweet, spicy, toasty, nutty aromas.

The most significant volatile acid in wine is acetic acid, produced during fermentation but more significantly a result of *Acetobacter* activity. It tastes sour and smells like vinegar.

Volatile phenols are important in wine aroma. 4-ethylphenol and 4-ethylguaiacol, found predominantly in red wines, are formed by the action of the spoilage yeast *Brettanomyces*, and have distinctive gamey, spicy, animally aromas (see Chapter 17). 4-vinylphenol and 4-vinylguaiacol are rare in red wines and more common in whites, and also have largely negative aromatic properties. These are formed by the enzymatic decarboxylation of cinnamic acids, a process inhibited by some grape phenols in red wines.

Terpenes are a large family of compounds that are widespread in plants. Grapes contain varying amounts, and these survive vinification to contribute to wine odor. More than 40 have been identified in grapes, but only half a dozen are thought to contribute to wine aroma. They are highest in Muscat wines, and the distinctive floral, grapey character is down to the likes of linalool and geraniol. Other varieties, such as Gewürztraminer and Pinot Gris, also have a terpene component to their aromas.

Methoxypyrazines are important in the aroma of certain wines. They are heterocyclic compounds that contain nitrogen and are formed by the metabolism of amino acids. 2-methoxy-3-isobutyl-pyrazine is a distinctive element of the aroma of varieties such as Cabernet Sauvignon, Cabernet Franc, and Sauvignon Blanc. At higher concentrations this can be excessively herbaceous, and it is generally seen as a problem in red wines, but an asset in certain styles of white when it contributes less of the green pepper and more of the fresh, grassy type of aromas. Methoxypyrazines have extremely low detection thresholds.

Volatile sulfur compounds are important in wine aroma (see Chapter 15). Mercaptans (thiols) are negative at higher amounts, but in controlled quantities they are important in the aroma of wines made from Sauvignon Blanc and some other white varieties. Some sulfur compounds have positive effects on wine aroma at extremely low levels.

24 Wine and health

The health claims made for wine are truly amazing. If you believed everything you heard about wine's benefits, you'd wonder why people bother with other drugs. A quick trawl through the scientific literature comes up with reports suggesting that wine—and red wine in particular—protects to a certain degree against heart disease, strokes, various cancers, AIDS, dementia, diabetes, benign prostatic hypertrophy, and osteoporosis. There have even been a couple of reports showing that teetotal women take longer to conceive, and that drinking two glasses of red wine counteracts the damage to the cardiovascular system caused by smoking one cigarette. But can this all be true?

MODERATE DRINKING SEEMS TO BE GOOD FOR YOU

It's well-established that in Western populations, moderate drinkers live longer than nondrinkers, who in turn live longer than heavy drinkers. It's a consistent finding in what is known as "epidemiological" studies—those that look at the incidence and distribution of diseases, and their causal factors.

This phenomenon is known as the "J-shaped curve," which is the shape of the line you get if plot mortality (the risk of dying) against alcohol consumption on a graph. Studies have shown that moderate drinking increases life expectancy, mainly through its protective effects on the cardiovascular system—your heart and blood vessels. Heavy drinkers also enjoy this benefit, but their risk of death starts to increase because they are more likely to suffer from the various conditions related to heavy drinking, such as cirrhosis of the liver, stroke, certain cancers, and increased risk of accidental or violent death. It is a pretty robust finding that has been replicated in countless studies to the degree that it is no longer controversial. It's also quite a significant effect: one large study looking at research spanning back 25 years on the subject

indicates that moderate drinkers cut their risk of heart attack by as much as one-quarter.

This message was reinforced by two papers published in the *British Medical Journal* in 2011, both from William Ghali and colleagues. These papers represented what is known as a meta-analysis, which is a study that attempts to bring together all published evidence on a particular subject from the medical literature in order to draw a more robust conclusion. In the first paper, Ghali carried out a review of the literature looking at studies that had examined the effect of alcohol consumption on biomarkers of coronary disease. They screened almost 5,000 articles, and included the results from 44, which were the relevant studies that met their criteria for suitable data. Overall, 13 biomarkers were included in the analysis. Alcohol was shown to significantly increase high-density lipoprotein (HDL) cholesterol, with a dose-response relationship, and it decreased fibrinogen levels. It didn't change triglyceride levels but it increased adiponecting and apoplipoprotein A1. All of these changes are reported to be cardioprotective. The authors noted that these changes are "well within a pharmacologically relevant magnitude," meaning that alcohol is acting as a prescribed medicine might. They point out that the degree of HDL cholesterol increase is better than can be achieved with any single therapy. Alcohol, consumed moderately, seems to be acting as a good drug. The second paper looked at selected cardiovascular disease outcomes. It examined 4,235 studies, and 84 turned out to be suitable for inclusion in the meta-analysis. The results examined the relative risk of dying for drinkers versus nondrinkers, and once again came up with some significant results. A moderate drinker has 0.75 risk of dying of cardiovascular disease compared with a nondrinker, and 0.71 risk of incident coronary heart disease. An alcohol consumption of 2.5–14.9 g/day (roughly one or two drinks) results in a 14–25% reduction of risk of cardiovascular disease compared with abstainers. Both studies together suggest that alcohol may be having a causal role here: there is a dose-response relationship, and the association is specific, in that alcohol is not uniformly protective for other diseases, such as cancer.

While the overall message is clear, though—that moderate drinking is healthier than heavy drinking or being a teetotaler—there has been a lot of debate about the details, and also the significance of this finding. Should doctors be advising nondrinking patients to take up moderate drinking? Is it just red wine that has this beneficial effect, or do all alcoholic drinks share it? And what are the mechanisms that could explain alcohol's (or wine's) health-promoting properties?

A QUESTION OF MECHANISMS

Let's deal with the last issue first—the mechanisms by which moderate drinking might protect against certain diseases. This has proved particularly difficult to unravel, and there is little consensus. There is currently a range of plausible theories, each supported by varying degrees of evidence.

ANTIOXIDANT EFFECTS

One of the most strongly advocated claims for red wine's health benefits stems from the antioxidant compounds it contains. These could prevent the oxidation of molecules in the circulation such as LDL (low-density lipoprotein, the "bad" cholesterol, which in its oxidized state is a major contributor to atherosclerosis). In the laboratory, phenolic compounds from red wine, such as resveratrol, have been shown to have potent antioxidant effects, protecting against the oxidation of LDL. However, large clinical trials of dietary antioxidants have failed to demonstrate any health benefits. It's a large step from showing activity of a compound in the laboratory to demonstrating that they are taken up by the gut, find their way to target tissues in the body, and have the required activity in real people. Added to this, some of the largest epidemiological studies have suggested that moderate consumption of all alcoholic drinks, and not just red wine, confers health benefits.

Above Intact clusters of red grapes. It is commonly thought that many of the polyphenolic compounds present in red grape skins account for the health benefits of modest, regular wine consumption.

DIRECT EFFECTS OF ALCOHOL

A more plausible mechanism concerns the effects of alcohol itself on the lipids in the bloodstream. Several studies have shown that alcohol consumption has the favorable effects of increasing the concentration of HDL (high-density lipoprotein, the "good" cholesterol), which is anti-inflammatory and decreases the risk of atherosclerosis. The degree of increase is similar to that achieved with other interventions, such as exercise programs. While the evidence for this mechanism is quite good, critics suggest that the benefit is fairly small and can be negated by other factors such as being overweight or smoking. It's also a hard idea for people to accept: the other components of wine sound much healthier.

THE PLATELET STORY

The third mechanism concerns the effects of wine on reducing the "stickiness" of platelets, specialized cells in the blood that are responsible for clot formation. This is known as an antithrombotic effect. This may be important in preventing events, such as heart attacks (caused by clots in the blood vessels supplying the heart) and strokes (clots in vessels supplying the brain). The antithrombotic benefit of moderate drinking is similar to the beneficial effects of taking aspirin.

PROTECTING THE ENDOTHELIUM

A fourth mechanism has been suggested by the work of London-based scientist Roger Corder, who is professor of experimental therapeutics at the William Harvey Research Institute. His research group has shown that polyphenols from red wine inhibit the formation of a compound called endothelin 1 (ET1) in cultured cells. The effect is a potent one, requiring only relatively small amounts of these compounds. ET1 is an important molecule because it causes the constriction of blood vessels, encouraging the development of atherosclerosis. This fits in with other research showing that red wine modifies the function of the endothelium, a layer of platelike cells lining the inner surfaces of blood vessels. The first stage of atherosclerosis in vessels is damage to the endothelium. If red wine can prevent these early events, then it is likely that it can inhibit the development of heart disease. While this ET1 effect seems to be a very promising explanation for red wine's beneficial properties on cardiovascular disease, it needs to be shown that this is what is happening in the human body.

THE SIRTUIN STORY

Perhaps even more speculatively, a further potential mechanism is suggested by a recent study looking at the effect of red-wine extracts on yeast cells. Resveratrol, a polyphenolic compound in red wine, has been found to boost the levels of an enzyme called Sir2, which is thought to stabilize DNA. The net effect is that the lifetime of the yeast is extended by up to 70%. Other studies apparently show the same effect in fruit flies. The important background to this story is that severe calorific restriction (typically one-third fewer calories than normal) is known to extend lifespan in mammals, and Sir2 elevation mimics this process. An even more recent study has uncovered the mechanism by which Sir2 has its life-prolonging effects. Of course, it's a long way from yeast or flies to humans, but if you wanted to extend your lifespan by one-third, most people would choose to drink wine as opposed to cutting their diet to near-starvation levels!

A COMPLEX PICTURE

It's likely that the mechanisms underlying the health-giving effects of moderate alcohol consumption, and specifically red wine, will be complex, involving more than one of the above systems. This will certainly need some unraveling by researchers, although they seem to be on the right track. Part of the difficulty is that it's very difficult—and expensive—to do the kinds of trials that need to be done on human populations to settle this issue once and for all.

CONFOUNDING

But before we get too carried away, it should be pointed out that there exist naysayers who suspect the health benefits of moderate drinking are actually an artifact of what is known as "confounding." Two reports, one from Sweden and the other from Denmark, have studied the characteristics of the different drinking groups in various studies. They've found that moderate wine drinkers tend to be wealthier, better educated, and possess a more favorable psychological profile, all of which are factors that tend to be associated with better health. Although most serious studies try to control for confounding, for instance, by trying to balance the different consumption groups in terms of socioeconomic status, it's possible that some of the associations between wine drinking and various diseases are a little muddied by this factor.

PRESCRIBING WINE?

So should doctors be prescribing wine? On the one hand, there's strong evidence that a couple of glasses of red wine a day could dramatically cut the risk of heart disease, particularly in susceptible groups such as middle-aged men. On the other hand, alcohol has a dark side, and most doctors are reluctant to start encouraging nondrinkers to start imbibing because of the risk that their patients will end up with a drink problem.

25 Concluding remarks

It is my hope that anyone who has read at least a few chapters of this book will come away with some answers, but even more questions. There is a pressing need for more good-quality research on many of the issues covered in these pages. The wine world is still burdened by too much received wisdom, folklore, and practice based on tradition and anecdotal observation. While many traditional practices have a sound scientific basis, it is likely that others don't, and it would be helpful to know the difference. At the same time, an important part of wine's appeal, at least at some levels of the market, is tradition—its historical context and cultural roots. Application of new technology needs to be sensitive to this. To finish with, I'd like to touch on two subjects that have formed common threads running through many of the subjects in this book: the common practice of blaming tools, and the limits of reductionist science.

DON'T BLAME THE TOOLS

Much of this book has been focusing on the science behind various tools, techniques, and manipulations open to the winemaker and viticulturalist. But more important than the specific techniques used is the will or intent of the producer. Good producers who make interesting wine will be able to utilize scientific and technological tools, should they choose to, to assist them in their goals. Indeed, the choice of which technologies to adopt and which to leave alone is an important one, and is aided by an understanding of the likely benefits and any drawbacks of these technologies. Producers aiming to make wine as cheaply as possible will no doubt find technologies that will help them in their goals, too. Just because technology puts tools into the hands of producers who we might view as misguided—for example, those wanting to enhance the flavor of cheap wines to make them taste expensive, or who want to make soupy, overextracted, concentrated, point-chasing wines,

or even those who want to cheat blatantly—doesn't mean that the technologies themselves are bad. They are neutral. Of course, some technologies might be provoking a near-irresistible temptation for winemakers to cheat, in which case perhaps they should be regulated. But for most tools, there are legitimate as well as dubious uses.

Let's use an analogy to illustrate this concept. Henry Cartier-Bresson took some of the twentieth century's most compelling photographs with a 35 mm Leica rangefinder, which was a new style of camera back in the 1930s. This camera was a tool, albeit one that with its compact dimensions, facilitated his sort of photojournalism. The tool was important to Cartier-Bresson, but much more significant was his genius as a photographer. He could have taken bad pictures with the Leica. That he didn't is down to his skill and intent more than just the characteristics of the camera he used. Perhaps, some might argue, in wine we focus a little too much on the tools and not enough on the intent and ability of the winegrower.

ARE REDUCTIONIST APPROACHES USEFUL FOR UNDERSTANDING WINE?

Reductionism, the splitting down of a system into its component parts, and then studying these parts in isolation, has been a tremendously useful way of doing science. Most science is done this way, but researchers are now beginning to realize that what the philosophers of science have been saying for a while—that there are limits to reductionism—is actually true. Reductionist science has allowed biologists to unravel the human genome. But making sense of this genetic code is another matter altogether, and will be a process that requires more than reductionist approaches. And while neurobiologists have uncovered in minute details the working of the nerve cells in the brain, how much does this tell us about consciousness?

In a similar vein, how useful are reductionist approaches in yielding understanding about

Glossary

Entries in **bold** are cross references.

Abscisic acid Important plant hormone (also known as plant growth regulator), involved particularly in signaling during episodes of stress, such as cold or drought. See Chapter 7.

Acetaldehyde The most common **aldehyde** in wine, formed when oxygen reacts with **ethanol**. Present in small amounts in all wines. Not nice: smells bad, and one of the reasons that oxidized wines aren't very pleasant. It's also the initial breakdown product of alcohol in the body, and responsible in large part for the unpleasantness of hangovers. Acetaldehyde is important because it is involved in the copolymerization of phenolic compounds. Combines readily with **sulfur dioxide**.

Acetic acid Volatile acid that is the main signature compound of vinegar. Formed by the action of *Acetobacter* bacteria on alcohol in the presence of oxygen. All wines have a little of it because it is a natural product of fermentation. But you don't want too much.

Acetobacter Bacteria that spoil wine by turning it into vinegar in the presence of oxygen.

Acids Important flavor constituents of wine, which provide tart, sour flavors that balance the other components. Wine contains a range of acids, most notably tartaric and malic acid (present in grapes), and lactic acid and succinic acid (produced during fermentation). Acidity is important in wine; as well as the flavor, it affects the color, and also the effectiveness of **sulfur dioxide** additions. It is commonly measured as total acidity, which is the sum of fixed and volatile acids. Acidity is also measured as **pH**, though this doesn't correlate exactly with total acidity. The relationship between the acid composition of wine and its actual acidic taste is a complex one: some acids are naturally more acidic than others for chemical reasons. The perception of acidity is also strongly influenced by other flavor components of the wine, notably sweetness, which counters the perception of acidity quite markedly.

Alcohol Common name for **ethanol**.

Aldehydes Also known, along with **ketones**, as carbonylated compounds. Rapidly combines with **sulfur dioxide** in wine. These are formed whenever wine is exposed to oxygen. The most important aldehyde in wine is **acetaldehyde**. Other aldehydes present in wine can be important in terms of flavor development: some higher aldehydes contribute to wine aroma, such as vanillin, a complex aromatic aldehyde that can be present in wine because of fermentation or aging in oak barrels.

Alleles A number of different forms of the same gene.

Amino acids Building blocks of proteins, present in wine at appreciable levels, and responsible for the taste of **umami**. There are just 20 of them, responsible for the many thousands of different proteins produced by living creatures.

Ampelography Science of vine identification.

Anthocyanins Phenolic compounds responsible for the color of red and black grapes. In wine, they interact with other components to form pigmented polymers and are responsible for wine color. See Chapter 23.

Antioxidant Chemical compound that prevents oxidation by reacting with oxidation: it takes the hit to protect other compounds. See Chapters 7, 23, and 25.

Apiculate yeast Group of yeasts involved in wild or indigenous yeast fermentations.

Ascorbic acid Known also as vitamin C, ascorbic acid is sometimes used in wine as an antioxidant. Acts synergistically with sulfur dioxide, but controversially the two together have been implicated in some incidences of **random oxidation**.

Astringency Perceived in the mouth by the sense of touch, astringency in wine is contributed by **tannins**—a drying, mouth-puckering sensation.

Autolysis In wine, autolysis is the self-destruction of yeast cells, releasing flavor components into the wine.

Balling another term for *Brix*.

Bâtonnage Stirring the lees, the yeast-cell deposit at the bottom of a fermentation vessel.

Baumé A system for measuring the sugar content of juice by means of density. Each degree represents about 1.75% sugar. It is measured using a hydrometer.

Bentonite Type of clay used to "fine" (remove suspended solids from) wine.

Biodynamie Controversial souped-up form of organic viticulture with a cosmic slant. See Chapter 7.

Bitterness Taste sensation, not all that common in wine, and commonly confused with **astringency** and sourness.

Botrytis Genus of fungus, but also a common term to describe infection of grapes by *Botrytis cinarea*. If it affects grapes that are already ripe it can be beneficial, and is responsible for many of the world's great sweet wines. But it also has a malevolent influence, causing gray rot.

Brettanomyces Yeast genus that is, at sufficient concentrations, a spoilage organism in wine. Controversial. See Chapter 17.

Brix A density scale of measuring sugar content, also known as **Balling** (the two are used interchangeably). Each degree represents 1% of sugar in the grape juice. As an example, 20 degrees Brix corresponds to about 12% alcohol in a finished wine (this will depend on the conversion of sugar to alcohol by the yeast).

Cane One-year-old stem of a grapevine, used as the basis of cane (also known as **rod and spur**) pruning. See Chapter 9.

Carbon dioxide Well-known gas, vital to plant growth as the carbon source of photosynthesis, but that also contributes to global warming (see Chapter 4). Used in winemaking to protect grapes, must, and wine from oxygen. Naturally produced in fermentation and, while in most wines this dissipates, in some styles appreciable levels remain where it helps preserve freshness.

Catechin Flavan-3-ol (the other significant one in wine is epi-catechin), a group of **phenolic compounds** that are the building blocks of **tannins**. In their polymeric forms, they are called procyanidins (often referred to as condensed tannins). See Chapter 23.

Chaptalization The addition of sugar to must to boost alcoholic strength.

Chloroanisoles Group of chlorine-containing compounds largely responsible for musty taint caused by rogue corks. Best-known example is **2,4,6-trichloroanisole** (TCA).

Clonal selection Taking cuttings from a superior vine in the vineyard for further propagation. While vineyards are often planted by using genetically identical material (a single **clone** of a variety), after some years some vines are seen to perform better than others, over and above site-specific influences. This might be because of spontaneous mutations, but is commonly because of disease pressure. It's worth noting that the most vigorous, actively growing vines in a vineyard are usually not the ones producing the best-quality fruit, and shouldn't automatically be chosen to take cuttings from.

Clone In viticulture, a group of vines all derived from the same parent plant by vegetative propagation (cuttings), and thus genetically identical.

Colloids Very tiny particles, smaller than a micrometer in diameter, usually made up of large organic molecules, which are important for the body of a wine. These can be removed by filtration, which, if done clumsily, can have the effect of stripping flavor from a wine.

Congeners Imprecise term for impurities in a spirit, thought to be responsible in part for hangovers.

Co-pigmentation Trendy but controversial term used to describe the fixing of color in red wines by the presence of noncolored phenolic compounds. For example, one of the reasons Shiraz is sometimes fermented with a dash of Viognier is because co-pigment phenolics from the white grapes facilitate the production of a darker-colored wine.

Copper Element often present in the vineyard because of the use of Bordeaux mixture to combat fungal disease (which contains copper sulfate). Can also be used to remove volatile sulfur compounds from wine to prevent reductive taints (See Chapter 15).

Cordon Name for the woody arm or branch of a vine, growing horizontally from the main trunk, and bearing spurs (when spur pruning is adopted).

Cover crop Plants grown between the vine rows during the dormant season. They are plowed into the soil before the vine growth kicks in. See Chapter 6.

Crossflow filtration Known also as tangential filtration, this is the technique behind reverse osmosis. See Chapter 16.

Cryoextraction Controversial freezing of fresh grapes before crushing to extract only the sweetest, richest juice. Used by producers of sweet wines to enhance them, but some consider this to be cheating.

Cytokinins Plant hormones particularly involved in regulating cell division, thus affecting growth stages.

Dekkera Spore-forming form of the yeast **Brettanomyces**.

Diacetyl Common name for buta-2,3-dione, this is a ketone produced during fermentation or by the action of lactic acid bacteria. Smells buttery and slightly sweet.

Downy mildew Significant fungal disease of vines caused by *Plasmopara viticola*. Introduced to Europe from the USA in the 1880s, downy mildew caused significant damage, until spraying Bordeaux mixture contained it, but it is still a major problem.

Enzymes Proteins that catalyze chemical reactions by making them happen faster, or reducing the temperatures needed for them to occur. Commercial preparations exist that can be used in winemaking for various reasons, some more justified than others.

Epigenetics A way in which the genome of an organism can be modified other than by changes in the coding sequence of the DNA. These changes can alter the way that genes are expressed (are turned into proteins), and are very important.

Esters Important to wine flavor, esters are formed by the reaction of organic acids with alcohols, and are formed during both fermentation and aging. Ethyl acetate (also known as ethyl ethanoate) is the most common ester in wine, formed by the combination of acetic acid and ethanol. Most esters have a distinctly fruity aroma, with some also possessing oily, herbaceous, buttery, and nutty nuances.

Ethanal Another term for **acetaldehyde**.

Ethanol Common name for ethyl alcohol, even more commonly referred to as just alcohol.

Ethyl acetate A common **ester** in wine formed by the combination of acetic acid and ethanol.

Ethyl alcohol In volume terms, ethyl alcohol is the most important component of wine, and is produced by fermentation of sugars by yeasts. On its own, it doesn't taste of much, but the concentration of alcohol in the final wine has a marked effect on the wine's sensory qualities.

4-ethylguaiacol *See* **volatile phenols**.

4-ethylphenol *See* **volatile phenols**.

Extraction Removal of **phenolic compounds** from the grape skin during the winemaking process.

Fanleaf virus One of the most common vine diseases, a range of viruses causes fanleaf virus. It is a big problem, and a good reason for planting with specially treated virus-free rootstock.

Fixed acids Term that describes the nonvolatile acids, tartaric and malic. Some acids are intermediate between volatile and fixed, though.

Flavonoids Large group of plant **phenolic compounds**, including pigments, such as anthocyanins.

Flor Film of yeast cells that can form on the surface of a wine, important in the production of some styles of sherry and vin jaune.

Fructose One of the main sugars in grapes, along with glucose.

GC-MS Stands for gas chromatography–mass spectrometry, which is a sensitive analytical technique for separating and identifying volatile compounds or gases.

Gibberellins Group of **plant hormones** important for influencing growth and development of vines. Sometimes applied artificially. Important in shoot elongation and release from dormancy.

GIS Stands for geographic information system. In viticulture, GIS is used to gather data about physical characteristics of a vineyard for **precision viticulture**.

Glucose Produced by photosynthesis, this is the most important sugar of the grape.

Glycerol Produced during fermentation, this is a polyol that can make a wine taste slightly sweeter, but which, contrary to popular opinion, doesn't affect viscosity.

Higher alcohols Known also as fusel oils, these are products of fermentation, and can contribute some aromatic character to wines.

Histamine Chemical involved in allergic reactions in humans. Present in some wines, but not thought to be responsible for adverse reactions, sometimes termed wine "allergies."

Homoclimes Trendy New World viticultural term for areas with similar climates.

Hybrid vines Vines that are produced by crossing two different species. Also known as interspecific crosses. Typically, disease-resistant American species are crossed with *Vitis vinifera* vines to produce resistant vines without the foxy flavors characteristic of the American species. These are also commonly called American hybrids or French hybrids.

Hydrogen sulfide Smells like rotten eggs, and a potential spoilage element formed during fermentation. It is caused by a nitrogen deficiency in the must. See Chapter 15.

Integrated pest management Known simply as IPM, this is an agricultural system that aims to reduce inputs of herbicides, fungicides, and pesticides through intelligent use. See Chapter 6.

Internode The part of a stem between two nodes (where the buds occur).

Ketones Usually produced during fermentation. β-damascenone and α- and β-ionone are called complex ketones and are thought to exist in grapes. β-damascenone smells rose-like. α- and β-ionone occur most notably in Riesling grapes.

Laccase Another name for **polyphenol oxidase**.

Lactic acid A softer-tasting acid produced by the bacterial metabolism of the harsher malic acid during malolactic fermentation.

Lactones Compounds that can be present in grapes, but which more commonly come from oak (the oak lactones, see Chapter 12). Sotolon is a lactone characteristic of botrytized wines.

Leafroll virus Problematic viral disease that can only be eradicated through using virus-free planting material.

Maceration Important in red wine, this is the process of soaking the grape skins to remove **phenolic compounds**. There are many ways of doing this, including modern innovations, such as leaving time for an extended maceration period at cool temperatures before alcoholic fermentation is allowed to start.

Maderization The process of a wine becoming oxidized—to become maderized, usually by heat.

Malic acid Along with tartaric acid, this is one of the two main organic acids in grapes. Transformed to the softer lactic acid by the action of lactic acid bacteria during **malolactic fermentation**.

Malolactic fermentation Conversion of malic acid to lactic acid effected by lactic acid bacteria. See Chapter 16.

Mercaptans Sulfur compounds sometimes found in wine. See Chapter 15.

Methoxypyrazines Nitrogen-containing heterocyclic compounds produced in grapes by the metabolism of amino acids. 2-methoxy-3-isobutylpyrazine is the compound responsible for the bell

pepper, grassy character common in Sauvignon Blanc and Cabernet Sauvignon wines. Concentrations of methoxypyrazine decrease with ripening; sunlight on grapes also reduces the concentration.

Microoxygenation Technique of slowly adding oxygen to fermenting or maturing wine. See Chapter 10.

Monoterpenes Chemical group contributing to the aroma and flavor of wine varieties, such as Muscat and also Riesling.

Mouthfeel Fashionable tasting term used to describe textural characters, most specifically structure, of a wine.

Nodes Parts of the grapevine stem that contains the bud structures, separated by internodes.

Oeschle Measure of sugar concentration in grapes.

Oenological tannins Commercially produced tannins, usually of nongrape origin, added to wines. More commonly used than you would think.

Oidium Fungal disease commonly known as **powdery mildew**.

Oxidation Substances are oxidized when they incorporate oxygen and lose electrons or hydrogen. Oxidation is always accompanied by the opposite reaction, reduction, such that when one compound is oxidized, another is reduced. In wine, oxidation occurs on exposure to air and is almost always deleterious.

Pectins Carbohydrates that glue plant cell walls together.

pH Technically, the negative logarithm of the hydrogen ion concentration in a solution. A scale used to measure how acid or alkaline a solution is (more acid = lower pH). pH is very important in winemaking.

Phenolic compounds Known also as **polyphenols**, this is a large group of reactive polymers with the phenol group as the basic building block. They are very important in wine. See Chapter 23.

Physiological ripeness Known also as phenolic ripeness, this trendy term is used to distinguish the stage of maturity of the vine and grapes, as opposed to just the sugar levels. In warm climates, picking by sugar levels alone can result in unripe, herbaceous characters in the wine.

Phytoalexins Important antimicrobial compounds produced by plants in response to attack.

Pigmented polymers Known also as pigmented tannins, this complex group of chemicals is now implicated in red-wine color production. They are formed by the combination of **anthocyanins** with **catechins** during fermentation. See Chapter 23.

Plant hormones Group of signaling molecules that influence plant growth and development, and also stress responses. Also known in the trade as plant-growth regulators. This group includes auxins, cytokinins, gibberellins, abscisic acid, and ethylene. Some researchers also include the brassinosteroids and jasmonic acid in this club.

Polysaccharides Carbohydrate **polymers** with sugars (monosaccharides) as the main subunits.

Polymers Molecules that form as the result of **polymerization** of smaller subunits.

Polymerization Process of making larger molecules by joining together smaller subunits.

Polyphenols These are probably the most important flavor chemicals in red wines, but are of much less importance in whites. Polyphenols are a large group of compounds that have phenol as their basic building block. An important property of phenolic compounds is that they associate spontaneously with a wide range of compounds, such as proteins and other phenolics, by means of a range of noncovalent forces (for example, hydrogen bonding and hydrophobic effects). See Chapter 23

Polyphenol oxidase (PPO) Enzyme that causes browning by reacting with **phenolic compounds** and oxidizing them. Also known as laccase. Grapes that have been infected with fungi have high levels of PPO and this is can cause oxidation of the wine, both directly and by combining with free **sulfur dioxide**.

Powdery mildew Known also as oidium, this is a nasty fungal disease that devastated European vineyards when it was introduced from the USA in the late 1840s. It was eventually countered by dusting vines with sulfur, a treatment still used today by some. The fungus responsible is *Uncinula necator*.

Precision viticulture Selectively applying vineyard inputs according to relevant data. See Chapter 4.

Proteins Polymeric molecules made from combinations of the 20 naturally occurring **amino acids**. Encoded by genes.

PVP Shorthand for poly *N*-vinylpyrrolidone, sometimes used to remove bitter phenolic compounds from white wine.

Quercetin Common **flavonol** in grapes.

Quercus The genus *Quercus* comprises the various species of oak.

Random oxidation Known also as sporadic post-bottling oxidation, this term describes the premature browning that occurs to some white wines some months after bottling. It is a common enough problem that some industry figures have referred to this as the "new cork taint." The main explanation for this phenomenon is variable oxygen transfer through the cork. Wines are protected against oxidation through the addition of sulfur dioxide (SO_2) at bottling. If the free SO_2 falls to very low levels then the wine is unprotected, and browning can occur. Corks have been shown to vary dramatically in their oxygen-transfer properties, and random oxidation is thought to be caused by the subset of corks that let in significantly more oxygen than others. However, some scientists suspect that random oxidation may be caused by as yet poorly understood chemical reactions independent of the closure. It has been suggested that the addition of the antioxidant ascorbic acid to keep white wines fresh, a common practice, may have the paradoxical effect of rendering the added SO_2 less effective, and making some wines susceptible to oxidation. Another proposed cause of random oxidation is poor procedure or intermittent failure on the bottling line, allowing some wines to have much higher levels of dissolved oxygen from the outset. Random oxidation is mainly a problem with white wines: while oxygen ingress through the closure will certainly damage red wines, they are more resistant to oxidation because of their high phenolic content. Oxidation is also more likely to be detected in white wines because of the dramatic color change that accompanies it.

Redox potential Stands for reduction-oxidation potential. Can be measured. See Chapter 15.

Reduction Shorthand term used to refer to sulfur compound flavors in wine. Technically, it is the opposite of **oxidation**. See Chapter 15.

Resveratrol **Phenolic compound** that is also a **phytoalexin**, present in grapes and red wines. May have some health-enhancing properties. See Chapter 24.

Reverse osmosis Controversial filtration technique used to concentrate wines, and also for alcohol removal. See Chapter 13.

Rod and spur pruning Alternative name for cane pruning. See Chapter 9.

Saccharomyces cerevisiae The yeast species responsible for alcoholic fermentation, known colloquially as brewer's yeast or baker's yeast. Comes in many different strains. See Chapter 16.

Saignée Also known as vat bleeding, a winemaking technique for taking juice of skins to increase the solids-to-juice ratio in red-wine making, thus boosting the phenolic content of the resulting wine.

SO_2 Chemical formula for **sulfur dioxide**.

Sorbitol Type of alcohol present in low levels in wine with a hint of sweetness to it, occasionally used by fraudsters as an illegal addition.

Sotolon Lactone present in botrytized wines.

Spinning cone Technically a gas–liquid counter-current device. A way of stripping alcohol from wines without removing important volatiles, used commonly to concentrate fruit juices without taking out interesting parts. Widely used but, because of the expense of the equipment, this is just a service industry. See Chapter 13.

Spur Stubby grapevine shoot pruned back to just a few nodes.

Spur pruning Method of pruning vines that leaves just short spurs on a permanent cordon. See Chapter 9.

Succinic acid Acid present at low concentrations in grapes and wine.

Sulfides Reduced sulfur compounds that occur during winemaking. Usually negative but can add complexity at the right levels. See Chapter 15.

Sulfur dioxide SO_2. Hugely important molecule added in winemaking to protect wine from oxygen and microbes. See Chapter 14.

Tangential filtration See **crossflow filtration**.

Tannins Tannins are found principally in the bark, leaves, and immature fruit of a wide range of plants. They form complexes with **proteins** and other plant **polymers,** such as **polysaccharides**. It is thought that the role of tannins in nature is one of plant defense. Chemically, tannins are large polymeric molecules made up of linked subunits. The monomers here are **phenolic compounds** that are joined together in a bewildering array of combinations, and can be further modified chemically in a myriad of different permutations. There are two major classes of tannins: condensed and hydroly sable. Hydroly sable tannins aren't as important in wine. The condensed tannins, also known as proanthocyanidins, are the main grape-derived tannins. They are formed by the polymerization of the polyphenolic flavan-3-ol monomers **catechin** and epi-catechin.

Tartaric acid Most important grape-derived acid in wine, tartaric acid often precipitates out as harmless tartarate crystals.

TCA Abbreviation for **2,4,6-trichloroanisole**.

Terpenes Large family of compounds that are widespread in plants. Grapes contain varying amounts, which survive vinification to contribute to wine odor. More than 40 have been identified in grapes, but only half a dozen are thought to contribute to wine aroma. They are highest in Muscat wines: the distinctive floral, grapey character is down to the likes of linalool and geraniol. Other varieties, such as Gewürztraminer and Pinot Gris, also have a terpene component to their aromas.

Total acidity Important measurement for winemakers, made by titration. Given in terms of grams per liter of tartaric or sulfuric acid. Total acidity includes the measurement of both fixed and volatile acids.

2,4,6-Trichloroanisole Potent, musty-smelling compound largely responsible for cork taint. See Chapter 18.

Umami Fifth basic taste, only recently recognized. It is the savory taste that results from the detection of **amino acids**. See Chapter 19.

Veraison The process in grape ripening where the skins begin to soften and the berries change color (most obvious in dark-skinned grapes). It is the transition from berry growth to berry ripening. Post-*veraison*, acidity decreases and sugar begins to accumulate.

Vitis vinifera Species name for the Eurasian grapevine, to which the varieties we know and love all belong.

Volatile acidity Acidity contributed by the various volatile acids, the most significant of which is **acetic acid**. A little is acceptable, but too much is very bad and makes the wine smell like vinegar.

Volatile phenols Important in wine aroma. 4-ethylphenol and 4-ethylguaiacol, found predominantly in red wines, are formed by the action of the spoilage yeast **Brettanomyces**, and have distinctive gamey, spicy, animally aromas (see Chapter 15). 4-vinylphenol and 4-vinylguaiacol are rare in red wines and more common in whites, and also have largely negative aromatic properties

Yeasts Unicellular fungi important for fermenting grape juice to wine. See Chapter 16.

Index